Behavioural Neuroscience

Brain and behaviour are intrinsically linked. Animals demonstrate a huge and complex repertoire of behaviours, so how can specific behaviours be mapped onto the complicated neural circuits of the brain? Highlighting the extraordinary advances that have been made in the field of behavioural neuroscience over recent decades, this book examines how behaviours can be understood in terms of their neural mechanisms. Each chapter outlines the components of a particular behaviour, discussing laboratory techniques, the key brain structures involved, and the underpinning cellular and molecular mechanisms. Commins covers a range of topics including learning in a simple invertebrate, fear conditioning, taste aversion, sound localization, and echolocation in bats, as well as more complex behaviours, such as language development, spatial navigation and circadian rhythms. Demonstrating key processes through clear, step-by-step explanations and numerous illustrations, this will be valuable reading for students of zoology, animal behaviour, psychology, and neuroscience.

Seán Commins is a Senior Lecturer in the Department of Psychology at Maynooth University, Ireland. His research focuses on the biological basis of learning and memory, spatial navigation and the impact of stroke on cognition. Dr Commins has published extensively and has received numerous awards in recognition of his research.

Behavioural Neuroscience

SEÁN COMMINS
Maynooth University, Ireland

CAMBRIDGE
UNIVERSITY PRESS

University Printing House, Cambridge CB2 8BS, United Kingdom

One Liberty Plaza, 20th Floor, New York, NY 10006, USA

477 Williamstown Road, Port Melbourne, VIC 3207, Australia

314–321, 3rd Floor, Plot 3, Splendor Forum, Jasola District Centre, New Delhi – 110025, India

79 Anson Road, #06-04/06, Singapore 079906

Cambridge University Press is part of the University of Cambridge.

It furthers the University's mission by disseminating knowledge in the pursuit of education, learning, and research at the highest international levels of excellence.

www.cambridge.org
Information on this title: www.cambridge.org/9781107104501
DOI: 10.1017/9781316221655

First published 2018
Reprinted 2019

Printed in the United Kingdom by TJ International Ltd. Padstow Cornwall

A catalogue record for this publication is available from the British Library.

Library of Congress Cataloging-in-Publication Data
Names: Commins, Seán, author.
Title: Behavioural neuroscience / Seán Commins.
Description: Cambridge, United Kingdom; New York, NY: Cambridge University Press, 2018. | Includes bibliographical references and index.
Identifiers: LCCN 2017051827 | ISBN 9781107104501 (hardback) | ISBN 9781107506992 (paperback)
Subjects: | MESH: Behavior, Animal – physiology | Behavior – physiology | Neurophysiology | Psychophysiology
Classification: LCC QL751 | NLM QL 751 | DDC 591.5/1–dc23
LC record available at https://lccn.loc.gov/2017051827

ISBN 978-1-107-10450-1 Hardback
ISBN 978-1-107-50699-2 Paperback

For Sinéad, Sorcha and Dara

CONTENTS

The colour plates are found between pages 198 and 199.

PREFACE

Over the last 70 years, especially since the discovery of the genetic code, scientists have become very reductionist in their approach. Many look deep into the cell, examining specific genes, molecules and their underlying machinery. Using amazing techniques, genes can be turned on or turned off; molecular pathways can be silenced or enhanced. Such approaches have brought about amazing discoveries. However, at times it feels that the whole animal has been forgotten. An animal's behaviour is not controlled by a single cell, a single gene or a particular molecule. We must take a step back, and make sure we understand the animal, its behaviour in a particular environment, its underlying physiological machinery, and, importantly, how these are all interlinked. It is only then that we will gain a more complete understanding.

Brain and behaviour are intrinsically linked. Disruptions to various brain regions and their neural circuits (e.g. by head injury) can lead to changes in behaviour. Likewise, modifications of behaviour (e.g. exercise, learning) can change neural structure. While many neuroscientists try to understand the chemical and molecular make-up of neurons and their electrophysiological properties, as well as how they communicate with each other, one job of the behavioural neuroscientist is to link specific brain regions, and even specific neurons within a brain structure, to overall function and behaviour. This is not new. Patient studies from the late nineteenth century pointed to the idea of localisation of function, that is, the concept that specific brain regions are involved in a particular behaviour. All brain regions do not contribute equally to a given function. Through decades of research we now know that some brain areas are heavily involved in language processing (e.g. Broca's area), while other areas deal mainly with memory (e.g. hippocampus), emotion (e.g. amygdala) or other functions. Patient and lesion studies provide great insight into brain and behaviour, and the emergence of functional imaging technology over recent decades probes even further, allowing multiple structures and circuits to be visualised, mapped and aligned to behaviour.

Although one of the major goals of behavioural neuroscience is to try to understand how various behaviours are mapped onto the brain, this is not as straightforward as it may seem. Animals have a huge repertoire of behaviours. They show emotions, they show different forms of memory, they hear and respond to different sound frequencies and they see different wavelengths of light. Even if we consider only mammals, some can talk, some swim, some walk and some even fly. In short, behaviours are complicated. More than this, as the mammalian brain contains millions of neurons (more than the number of stars in the Milky Way galaxy) and trillions of connections between these neurons, and each neuron has different biochemical and molecular make-up, brains are complicated.

How do scientists marry the two and map a specific behaviour onto a neural circuit? One method is to study and use a model organism that has a very simple biological make-up. So, rather than trying to study a brain containing millions of neurons, many scientists have successfully looked at organisms that contain a simple nervous structure, with at most tens of thousands of neurons. For example, the action potential and its properties were originally elucidated by the Nobel Prize–winning physiologists Alan Hodgkin and Andrew Huxley in the giant squid. This invertebrate has very large axons, which can be observed with the naked eye. Similarly, in 2000 a Nobel Prize was awarded to Eric Kandel for his work elucidating learning and memory processes in another invertebrate with a simple neural circuitry, the *Aplysia* (sea slug). Some of this work is described in Chapters 5 and 6.

An alternative method of investigation is to look at the behaviours and break these down into simpler components. Psychologists over the last 100 years have very successfully worked out many of the rules and conditions that govern behaviour. For example, Ivan Pavlov in 1920s showed how a neutral stimulus (a stimulus that doesn't normally trigger a response) can evoke a strong response in an animal simply by associating this neutral stimulus with another that does trigger the response. Ringing a bell, for example, would not normally evoke a salivation response in dogs unless it was associated with another stimulus (e.g. food) that does trigger this response. So, despite the seemingly daunting task of mapping behaviour onto neural circuits, it can be achieved.

One of the aims of this book is to highlight the extraordinary work that has been achieved to date in the field of behavioural neuroscience. The book demonstrates how scientists have attempted, through the use of multiple techniques and the examination of various species, to understand our behaviour in terms of their neural mechanisms. More than this, the book aims to be student-centred; it acts as a resource to bring together experimental findings from many neuroscientists working in a particular area. Through the use of multiple illustrations and step-by-step explanations, the current understanding of the biological processes underlying a particular behaviour is clearly laid out.

Each chapter follows the same general format. Initially, there is a description of a specific behaviour; this is then followed by a detailed analysis of the brain regions and neural circuitry involved in the behaviour. Finally, cellular and molecular changes that occur within the circuitry are described, to offer explanations of how such a behaviour occurs. As well as mechanistically explaining a particular behaviour, the book aims to be thought-provoking. Students will, at the end of each chapter, realise that the explanation is not complete. They will hopefully get a sense that scientists are still working to understand more and more, and that this work is ongoing and always changing. This will provide students with an insight into the work of the scientist and allow them to think about how future experiments should be designed and what still needs to be addressed in the field. To promote this further, each chapter ends with a series of questions and discusses some of the unresolved issues in the field. A final aim of the book is to promote a sense of enthusiasm in a new generation of behavioural neuroscientists, who will hopefully be inspired to understand more.

The book contains 16 chapters. As the student moves through the book, the behaviours become more complex. Some behaviours are currently well mapped at the neural and cellular level, and these are described in detail. As the book introduces more complex behaviours, the current knowledge of the exact neural circuitry and molecular machinery underlying such behaviours is less complete. However, the key brain structures and the underlying properties of the neurons contained in these regions are described, as well as an analysis of how they contribute to a particular behaviour. In addition, the book does not limit itself to one particular species but includes work on many. The various species are selected on the basis that they represent the best model of a particular type of behaviour. Further, by using different species, students will get to know the anatomy and brain structure of a range of animals, including zebra finches, bats, owls, *Aplysia* (sea slugs) and rats. Each chapter is also associated with a major subdiscipline within behavioural neuroscience, including learning, plasticity, fear conditioning, motor response, taste aversion, navigation and circadian rhythms.

Specifically, Chapters 1 through 4 provide an introduction to neurons and the brain and how neurons communicate with each other, and discuss in general the various brain regions and their functions. In addition, a brief introduction to the various techniques used in behavioural neuroscience is provided, followed by a description of the principles of behaviour and how different behaviours can be measured in the laboratory. These chapters may provide a useful starting point and an introduction to the area of behavioural neuroscience for undergraduate students.

In Chapters 5 and 6 students are introduced to the *Aplysia*, a small marine invertebrate, and consider why this animal makes for a perfect model in understanding the mechanism of learning and memory, despite not having a brain! Chapter 5 describes how habituation and sensitisation, considered to be among the simplest forms of learning, play an important role in the survival of this marine animal. The chapter describes the work of neuroscientists, particularly the Nobel Prize–winning laureate Eric Kandel, who have shown how habituation and sensitisation can be produced in the lab, and how the neural circuitry and molecular mechanisms underlying these behaviours have been elucidated. Chapter 6 follows directly from the previous one and describes a more advanced form of learning: classical conditioning. While we accept that the *Aplysia* may have a very simple learning repertoire, is it possible that the *Aplysia* can also be classically conditioned, as Pavlov's dogs were? The chapter describes how the *Aplysia* can be conditioned, and the site of change within the neural network and the molecular events associated with learning are also examined.

Before examining various mechanisms of learning in the vertebrate brain, students are introduced to the concept of synaptic plasticity (Chapters 7 and 8). If learning and memories are to be stored in the brain, there must be a mechanism to allow this to happen. Long-term potentiation (LTP) is one very plausible mechanism (Chapter 7). The student is introduced to the basic concepts of LTP, how LTP is elicited and the molecular underpinnings of such a phenomenon. The latter part of the chapter looks at the evidence that LTP is a realistic model of how mammals learn. Chapter 8 describes the idea that not only can the strength of synapses be enhanced (potentiated), but synaptic strength can also be reduced. This synaptic depression or LTD is described in two areas of the mammalian brain (the cerebellum and hippocampus), along with a description and discussion of the cellular and molecular mechanisms.

Although much progress has been made in understanding learning and memory mechanisms in simple invertebrate systems, the task in localising learning and memory circuits in the mammalian brain is much more difficult, especially given the sheer numbers of neurons involved and the complexity of the neural system. The trick is to use a simple learning paradigm, like classical conditioning, but one in which the stimuli involved are well defined and provoke an automatic response. Our eyes are very sensitive; if someone blows into them we immediately close the lids to protect them (which parallels the defence reaction of *Aplysia*). This reaction is automatic but, importantly, it can paired with other stimuli to evoke a response. Chapter 9 introduces the student to eye-blink conditioning. It outlines the neural circuit in the cerebellum involved with this simple learned motor response and the cellular and molecular changes that occur with conditioning.

An eye-blink response is an automatic motor reaction and therefore does not require too much cognitive processing. What about higher-order brain processing such as emotions? Can we, for example, learn fear? Fear is fundamental to an animal's survival. Some animals have an innate fear of other animals, but fear can also be learnt. If we place a sharp object in an electrical socket, the chances are that we would never, ever do this again. We have learned to be afraid. In Chapter 10, the student is introduced to the amygdala, a brain region heavily involved in emotions, especially fear. The chapter describes how fear conditioning can arise through a network of circuits involving the amygdala. Then some of the known biochemical and molecular mechanisms underlying fear conditioning are described.

Animals have another very important defence mechanism. If we eat something in a restaurant and a couple of hours later we take ill, the likelihood of eating that food again or indeed even visiting that restaurant again is very low. Chapter 11 describes taste-aversion conditioning, a behaviour that also involves the amygdala, among other brain structures. While many of the previously described learning responses require multiple training trials, this form of learning is special; it typically involves only a single trial. We do not have to eat food and fall ill multiple times before we avoid it; once is sufficient. Like in previous chapters, in Chapter 11 the student is introduced to the neural circuitry involved in this behaviour and the molecular and cellular underpinnings.

The ability to locate the source of a sound is essential for most animals. In humans, the skill is clearly important in a range of scenarios from identifying the source of a speaker during a conversation to identifying the location of a car when crossing a busy road. In some animals this ability takes on an even greater importance. For example, in the case of the barn owl, identification of the origin of a sound as a potential food source is needed for its very survival. Chapter 12 shows how the physical features of an owl's head and how its brain deal with sound information coming at slightly different frequencies and timing combine to provide an aural map of space. This, in turn, allows the animal to identify exactly from which direction the sound has originated.

Having the ability to locate a sound source and react to it is certainly an impressive skill, but how do you use sound to navigate your environment? The impressive skill of echolocation is performed by many animals, but bats are most adept at this ability. Chapter 13 introduces the student to different bats and how they can identify various elements of complex environments using sound echoes. Specifically, the chapter describes the various neural pathways of the bat brain, with a particular emphasis on the auditory cortex. In addition, how neurons respond to different sound frequencies, allowing the animal to judge distances and velocity, is described. This knowledge is critical for a bat to identify moving prey as well as to avoid obstacles when flying.

Many animals navigate their environment without using echoes. The complex behaviour of knowing where you are, where you want to go and how to get there is explored in Chapter 14. Although complete knowledge of a circuit involving all aspects of spatial navigation is currently unknown, the role of a particular brain structure, the hippocampus, in navigation is described in this chapter. In addition, the student is introduced to specialised cells found in hippocampus and surrounding areas that aid this behaviour. The work described in this chapter was recognised in 2014 by the awarding of the Nobel Prize to John O'Keefe, Edvard Moser and May-Britt Moser.

Language is considered to be a distinctively human form of communication. It is so complex that it might be considered nearly impossible to understand its basic components and neural underpinnings. However, scientists do have a model that resembles language acquisition in humans: birdsong learning. When a baby hears sounds emanating from its parent's mouth, these sounds are repeated and remembered. With time, the sounds the baby makes, in the form of words, match exactly with those of the parents. In addition to hearing sounds from others, children also depend on hearing the sounds that they themselves emit. Can each of these components be matched in the songbird? Chapter 15 examines how some birds acquire their distinctive song by learning from their parents and by using their own aural feedback. The neural circuits of such behaviours are also examined.

Finally, many of our behaviours, implicit and explicit, are governed by biological rhythms. We are active during the day and fall asleep at night. Our sleep cycles between REM (rapid eye movement) and non-REM states. Body temperature, secretion of hormones and other physiological changes follow a set rhythm. Over a longer period, many animals hibernate during the winter months before waking up in spring. How are these rhythms governed? Chapter 16 introduces the student to the suprachiasmatic nucleus, the brain's master circadian clock, and describes some of the mechanisms that control our biological rhythms.

Because of the nature of science and the volume of work currently being done, there are many behaviours that I have not included in this book; furthermore, while I have done my best to capture the state of current knowledge in each chapter, there may be some relevant information that I have missed or have not fully captured. I apologise in advance for this.

ACKNOWLEDGEMENTS

There are a number of people that I would like to thank, particularly my friends and colleagues of the Department of Psychology, Maynooth University.

I would like to acknowledge my current and former postgraduate students whom I have had the privilege of working with including Drs Sarah Craig, Anne-Marie McGauran, Deirdre Harvey, Jonathan Murphy, Paraic Scanlon, Mairead Diviney, Daniel Barry, Sean Anderson, Francesca Farina, Liz Walshe, Kirby Jeter, Joe Duffin and Michelle Caffrey. I would especially like to thank Dr John Kealy for reading through the manuscript and helping me edit it. I would also like to thank Professor Shane O'Mara, who has been my academic mentor over many years and who introduced me to the world of behavioural neuroscience.

On a personal note, I would like to thank my parents, Jim and Rosemarie, and my family, who have always supported me in so many ways. To my wife, Sinéad, who has been a constant source of encouragement, inspiration and happiness, and my two children, Sorcha and Dara. Sorcha spent many weeks producing the beautiful drawings for the start of each chapter, and Dara drew a fine set of neurons for Chapter 3. Thank you for the joy you bring.

1 Neurons and Neural Communication

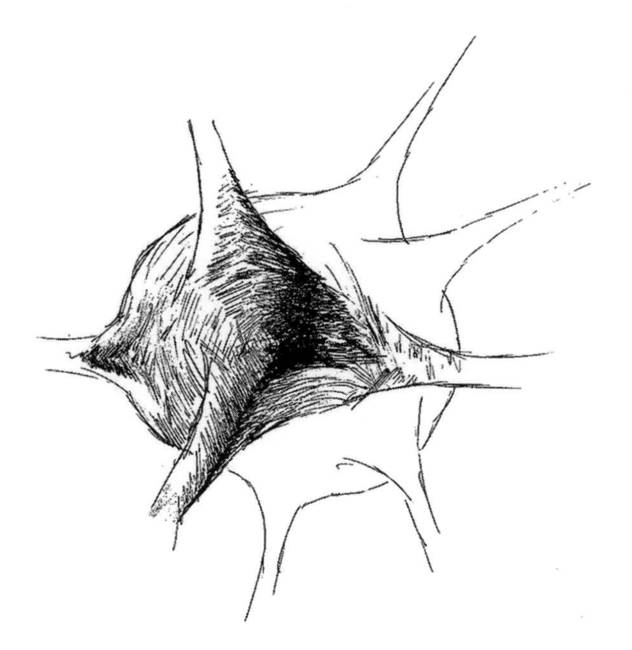

Brain Cells and Basic Structure

From the tallest tree to the simplest bacteria, the basic unit of any living organism is the cell. All structures within our bodies are composed of cells, which include heart cells, liver cells, skin cells, hair cells and brain cells. Brain cells can be divided into two main types: neurons, which allow us to feel, taste, see, move, feel emotions, remember and communicate; and glial cells, which help to support and protect neurons. Within the human brain there are billions of neurons and trillions of connections among them. When you look up at the night sky and consider that our Milky Way galaxy is thought to contain 200 billion stars, you can begin to appreciate the sheer number of cells located within your brain, a structure that is slightly larger than a closed fist.

Neurons come in all shapes and sizes, but there is a basic structure that is common to all (Figure 1.1). The cell body or *soma* contains the nucleus and other structures critical to keeping the cell alive and functioning. The *axon* is a long fibre that emanates from the cell body and is involved in sending information from the cell body to the end of the neuron. This information is passed as a small electrical pulse called an *action potential*, which will be discussed in more detail later. The axon is covered by a myelin sheath, which both insulates the axon so the information is kept within the one neuron and ensures that the action potential is passed quickly from one end of the neuron to the other. Some axons are small, but others are very long; for example, one axon runs from the base of your spine to your big toe. The axon can branch many times, but each branch ends with a small bulb called the *synaptic terminal* or *synaptic bouton*. It is at this junction, the synapse, that one neuron communicates to the next one. However, information must pass from one neuron to the next across a small gap called the synaptic cleft (see Figure 1.1); for this to occur, the electrical signal is converted into a chemical signal that readily crosses the gap and triggers a signal in the dendrites of the second neuron. *Dendrites* are branchlike fibres that serve as receivers of signals from multiple neurons. The message continues in the second neuron and keeps going until it is terminated. Therefore, the main function of the neuron is to process signals and pass the signals along.

Glial cells make up the majority of cells in the brain. In fact there are about 10 glial cells to every neuron. There are four main types of glial cells: oligodendrocytes, Schwann cells, microglia, and astrocytes.

Oligodendrocytes are cells whose extensions, composed of a fatty substance called myelin, wrap around the axons of neurons. It is these cells that allow for the fast transmission of an action potential down a neuron. Oligodendrocytes are found mainly in the central nervous system (CNS; brain and spinal cord).

Schwann cells perform a similar function by wrapping around neurons, but they do so in the peripheral nervous system (PNS, nerves extending from the spinal cord to other parts of the body – heart, lungs, toes, fingers, etc.) as opposed to the CNS. Schwann cells can also help with the regeneration of neurons, a task that oligodendrocytes are unable to do. This explains why regeneration is easier following damage to peripheral nerves, compared to neurons in the brain. Another major difference between these two classes of glial cells is that each Schwann cell wraps around a segment of the neuron, whereas the extensions of an oligodendrocyte can wrap around multiple neurons (Figure 1.2).

Microglial cells, or *microglia*, are smaller than the other cells types and are involved in the inflammation response. They protect the brain from invading microorganisms. In addition, if neurons become damaged and die, the job of the microglial cells is to break down these cells even further, and then clean them up by engulfing the remaining debris by a process termed *phagocytosis*.

Astrocytes, as their name suggests, are cells that are star-shaped. These cells, similar to microglia,

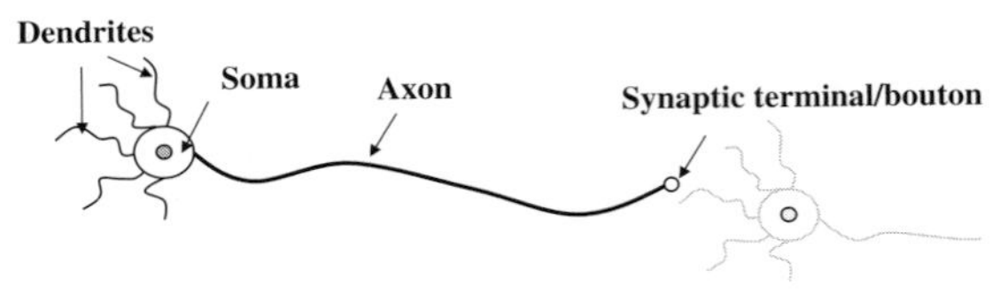

Figure 1.1 Diagram of a neuron and its various components.

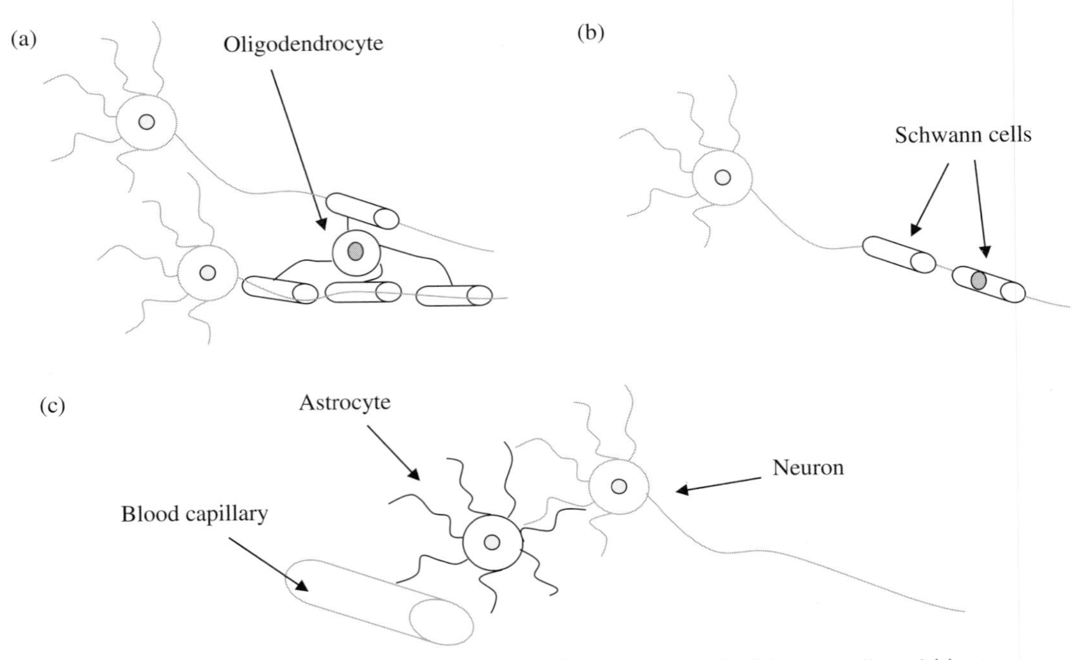

Figure 1.2 Diagram of different glial cells including (a) oligodentrocytes, (b) Schwann cells and (c) astrocytes.

function by cleaning up the debris of dying neurons. In addition, astrocytes are thought to provide a mechanism by which neurons receive nutrients and get rid of any waste material by acting as a mediator between neurons and blood capillaries. Astrocytes receive glucose from the blood, which is converted into lactate which, in turn, is taken up by neurons and is used as a source of energy. Recent research indicates that astrocytes not only perform a supporting role for neurons, but maybe involved in signal communication as well.

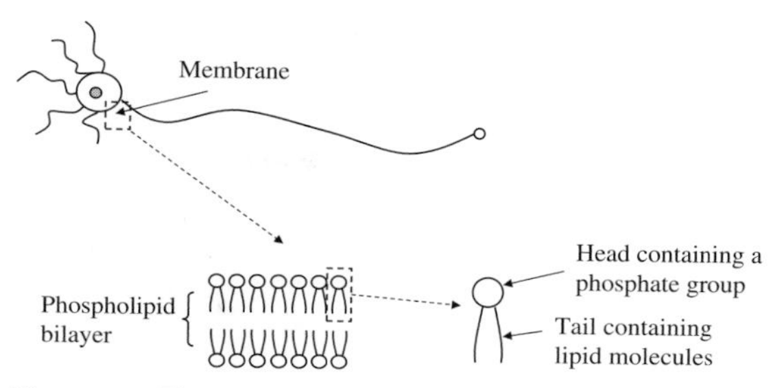

Figure 1.3 The membrane of a neuron is composed of a phospholipid bilayer, which regulates the movement of substances in and out of the cell.

Communication within Neurons

The Cell Membrane

One of the critical structures of any cell is its membrane. The membrane helps to give the cell its shape and helps keep all the internal organelles within the cell, and it also functions to regulate what goes in and out of the cell, a very important function. We can imagine cells to be like balloons, which come in different shapes and sizes. Likewise, some brain cells are spherical and some are pyramidal. It is the membrane that gives a cell its shape. If we blow into a balloon, it will expand and eventually burst; if air is removed it will shrivel. Further, the air that you have blown into the balloon is different to the air outside. When you tie an inflated balloon, nothing can come in or out. Likewise, the fluid within a cell, the intracellular fluid, is of different composition than the extracellular fluid, and must be regulated.

The cell membrane is composed of what is termed a *phospholipid bilayer* (Figure 1.3), and it is this layer that prevents fluids and other substances from entering and leaving the cell. Because the double layer is composed of lipid, a fat-like substance, water and other fluids are unable to pass through. But cells need to receive nutrients and get rid of waste material, so how can the materials get through this membrane?

Embedded within the cell membrane are large proteins, and it is through these proteins that materials can pass in and out of the cell. Proteins have the fantastic ability to change shape, and it is this property that makes them so useful and versatile. There are different types of proteins embedded within the membrane. Some proteins form a channel through the membrane, but are of a particular size and shape that only allow very specific ions (charged atoms) through (Figure 1.4a). For example, there are protein channels that only allow sodium ions (Na^+) through, and there are other channels that only allow potassium ions (K^+) through. Other protein channels, called *ligand-gated channels*, only allow ions through the channel when a molecule is attached. When a particular molecule attaches to the protein channel, the channel changes shape, forming a hole through the centre and allowing ions to flow into the cell (Figure 1.4b). A third type of channel operates as a pump, changing dimensions so that the channel pumps some ions into the cell and pumps others out of the cell (Figure 1.4c). We will meet these channel types and others as we progress through the chapter.

Resting Membrane Potential

Imagine if we were to take a very fine electrode (a small wire about the size of a single strand of hair), place it into a neuron and record its voltage or electrical charge. We would notice that the inside of the cell is more negative

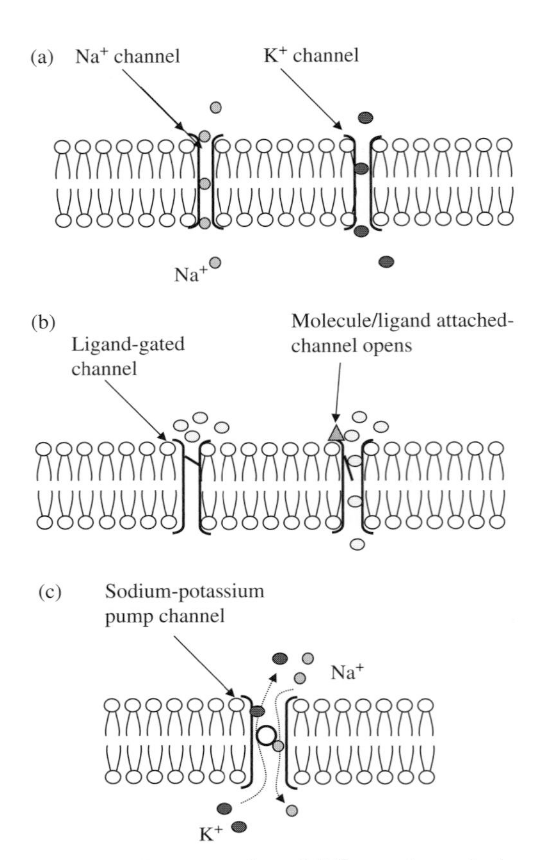

Figure 1.4 Representation of different channels that are embedded in the cell membrane. (a) Some channels have a particular shape, which regulates the flow in and out of the cell. (b) Some act like a gate and only open in the presence of a particular molecule; (c) others act as a pump, transporting ions in and out.

than the outside. The voltage difference between the inside and outside of the cell is about −65 or −70 mV (millivolts). The inside of the cell has a charge of approximately −70 mV. This is known as the *resting membrane potential*. Why is the inside of the cell more negative compared to the outside? Recall we suggested that the air blown into a balloon is different than the air outside. The neuron is similar. The composition of the intracellular fluid in a neuron is different compared to the fluid outside the cell. This difference in composition is due mainly to distribution of various ions inside and outside of the cell.

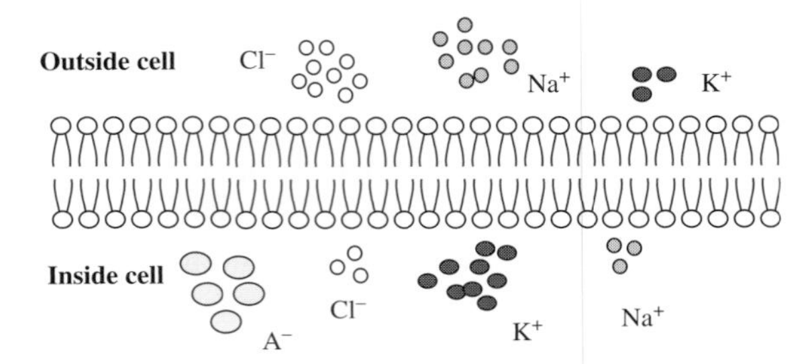

Figure 1.5 Distribution of various ions inside and outside of the cell.

Figure 1.5 shows the four main ions and their distribution on either side of the cell membrane. One of the main reasons that the inside of the cell is more negative compared to the outside is the presence of large negatively charged proteins (represented by A^-), these are unable to pass through the membrane. Although potassium (K^+), sodium (Na^+) and chloride (Cl^-) ions are found on both sides of the membrane, the distribution is not even. Extracellular fluid consists mostly of a saline-like solution, and therefore has a large presence of Na^+ and Cl^- ions, whereas there is a larger concentration of K^+ ions on the inside of the cell.

If the membrane just contained the phospholipid bilayer without any channels, as is illustrated in Figure 1.5, there would be no difficulty in maintaining the balance of composition between inside and outside of the cell. However, there are channels within the membrane that allow ions to flow in and out. How is it that the voltage inside of the cell remains fairly constant, at −70 mV without fluctuating? There are two major forces that help maintain this situation. The first is termed the *diffusion gradient* and the second is called *electrostatic pressure*. Diffusion refers to the fact that molecules will spread or diffuse from an area of high concentration to an area of low concentration. For example, if we were to put some salt into water, it would initially be concentrated in a small area, but with time it would spread evenly throughout the container, thereby spreading from an area of high concentration to areas of low concentration. The second factor, electrostatic pressure, means that particles of different charges are attracted to each other. Positively charged ions (cations; for example Na^+ and K^+) are attracted to areas that more negatively charged. Negatively charged ions (anions; for example, Cl^-) are attracted to areas that are positively charged.

When we look at Figure 1.6 we can see the forces that are working on each of the ions. As mentioned, the inside of the cell is more negative compared to the outside due to the large proteins that cannot escape; this is represented by the minus signs inside the cell in Figure 1.6. The K^+ ions are concentrated on the inside of the cell. Forces of diffusion want them to flow out of the cell, but they are attracted by the negativity on the inside of the cell. The net result is that they do not really move. Likewise, Cl^- ions want to flow into the cell (diffusion) but are repelled by the negativity on the inside. So these ions do not really move either. Finally, the Na^+ ions want to flow into the cell, both due to diffusion and by attraction to the negativity, but they remain outside, mainly because the Na^+ channels *remain closed*. There is, however, some leakage into the cell, but it is the job of the sodium-potassium channel pump to pump the Na^+ ions back out (and for balance, to bring K^+ ions back in). There are 3 sodium ions (Na^+) pumped out for every 2 potassium ions (K^+) pumped in.

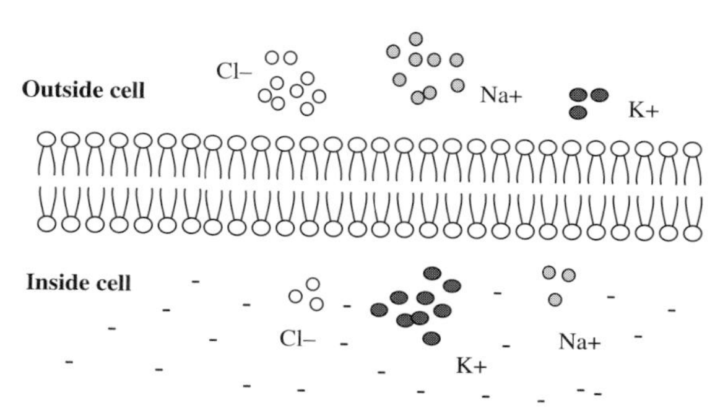

Figure 1.6 The inside of the cell is more negative compared to the outside due the presence of large negative ions.

Action Potential

Generation of the action potential

Information is sent down axons of neurons via small electrical pulses called *action potentials*, represented in Figure 1.7a by the stars. Imagine we were again to insert a small electrode into the axon and record what we would see. Figure 1.7a

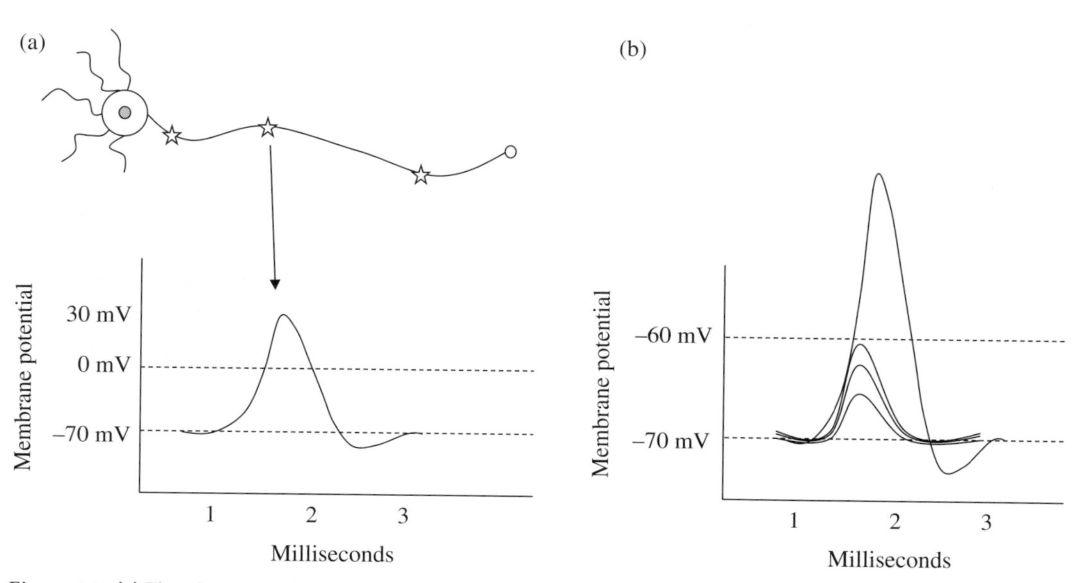

Figure 1.7 (a) The change in the resting membrane potential as an action potential is generated. (b) An action potential is not generated until a certain threshold is reached.

gives an idea of what an action potential looks like, its shape and dynamics. Essentially, what we see is a rapid change in the resting membrane potential. The inside of the cell briefly becomes more positive (termed *depolarisation*), moving from approximately −70 mV to +30 mV before returning to its normal negative baseline state (termed *hyperpolarisation*) of −70 mV. This is all completed in less than 3 milliseconds – an extremely rapid event. The generation of an action potential is, however, an "all or nothing" event. A threshold must be reached before the action potential is fired. The threshold is approximately −60 mV, that is, the inside of the cell must change from approximately −70 mV to −60 mV. If it does not, an action potential will not be generated (Figure 1.7b).

The change in the resting membrane potential during the expression of an action potential must be reflected by the change in the flow of ions in and out of the cell. The next section will provide a step-by-step account of what happens at the cellular level and how this is represented by the various phases of the action potential (Figure 1.8).

1. Once the threshold is reached, Na^+ channels open and sodium ions flow into the cell. They are attracted inwards due to negativity on the inside (electrostatic pressure) and because they are moving from an area of high concentration to a region of low concentration (diffusion gradient). As these positive ions rush in, the inside of the cell becomes more positive, and this is reflected by the upward depolarisation phase of the action potential (Figure 1.9).
2. Next, the K^+ channels start to open slowly and potassium begins to flow out of the cell. The K^+ ions want to move from an area of high

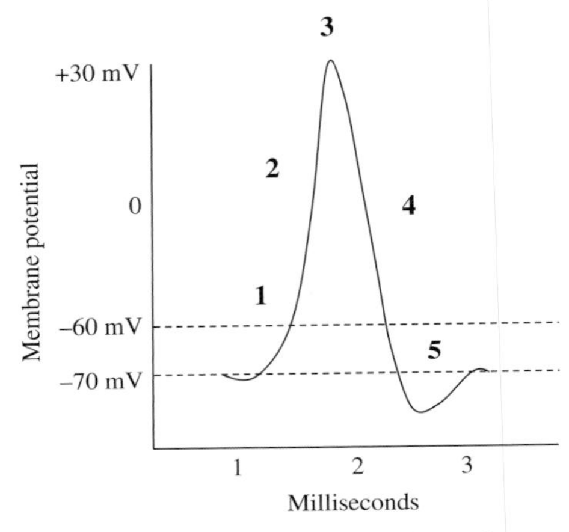

Figure 1.8 The various phases of the action potential. The cellular and the corresponding electrical changes are described for each phase (1–5) are described in the text.

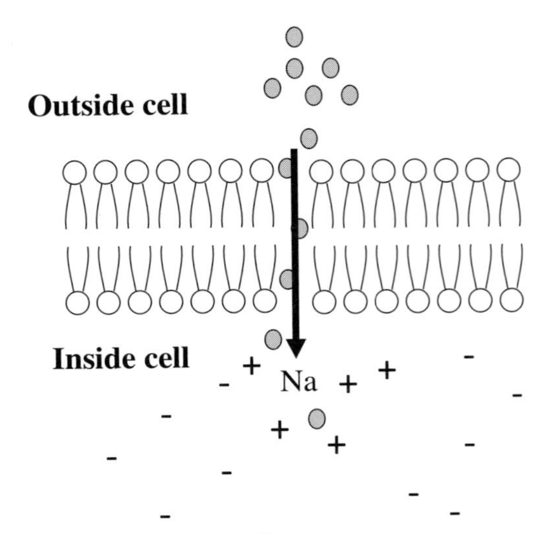

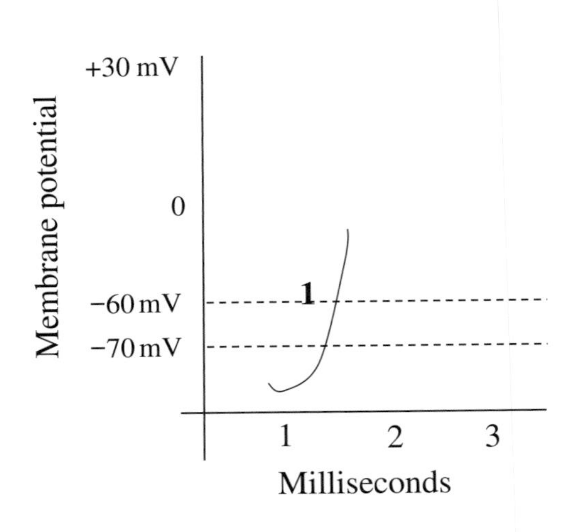

Figure 1.9 Phase 1. Na^+ ions flow in, in the upward depolarisation phase of the action potential.

concentration to one of low concentration, and because the inside of the cell is now starting to get more positive (due to the entry of sodium ions), the positive potassium ions are being repelled outwards (Figure 1.10).

3. As the action potential reaches its peak, the sodium channels close and no more Na^+ ions can come into the cell; this is often termed the *refractory period.*
4. At this stage, the inside of the cell has reached its maximum peak positivity, thereby forcing more and more positive potassium ions out of the cell (Figure 1.11, first panel). All the potassium channels are now open. As K^+ ions flow out of the cell, the inside starts to lose its positivity, and thereby begins to regain its negativity (Figure 1.11, middle panel).
5. As the inside of the cell regains its negativity, the potassium channels close. No more ions can come in or out of the cell. But often too many positive ions have left, so the membrane potential dips below the resting membrane potential of –70 mV (Figure 1.11, right panel). The sodium-potassium pump helps to restore the balance.

Conduction of the Action Potential Down the Axon

Information must go from one end of the axon to the other. All action potentials are very similar in nature; they travel in one direction from the site of generation at the soma to the synaptic bouton. All action potentials remain at the same

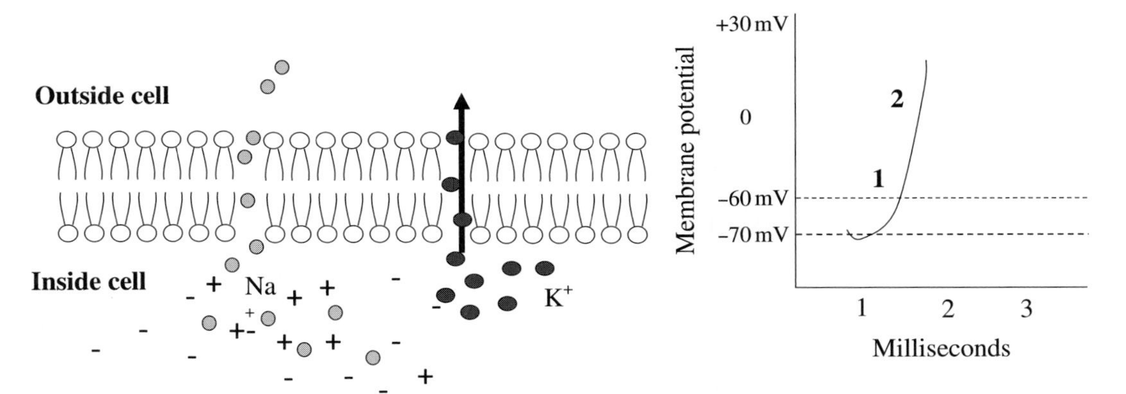

Figure 1.10 Phase 2. K^+ ions move out.

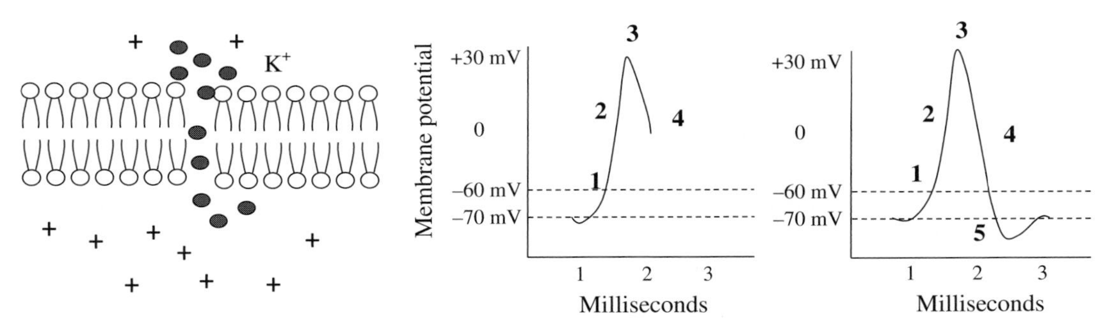

Figure 1.11 Phase 4 and 5. As K^+ ions flow out of the cell (left), the inside starts to lose it positivity (middle). However, too many positive ions have left, so the membrane potential dips below the resting membrane potential of –70 mV (right).

magnitude. Despite the long distances some action potentials have to travel, they do not diminish in size. The rate of firing does, however, change. A strong stimulus, for example, may evoke many action potentials in rapid succession, whereas a weaker stimulus may only evoke one or two.

So far we have described changes that occur in the resting membrane at a particular point along the axon. But how does the action potential 'move' along the axon? Imagine that we wanted to blow a hole in a wall. Using some gunpowder we could create a trail from the wall to a more distant, safer place. We then light the powder at one end, watch the flame move along the trail and see the wall blow up. For all intents and purposes it seems like the flame moves along the path, but in fact it does not. Each single grain of gunpowder lights up and dies out. However, the heat generated from that grain sparks the grain beside it, so the second grain lights up and dies out, and this continues on until it reaches the end. Because the action of heating up the next grain is so quick, it looks as if it is the flame is moving along the trail. If we imagine that an action potential is like a flame moving along an axon, the mechanism is very similar. As the positive ions rush into the cell, an action potential occurs at that particular point. However, these positive ions do not just remain at the point of entry – they spread along the inside of the axon. Therefore, the next segment along the axon becomes more positive, allowing the threshold to be reached and the next action potential to be triggered. As the positive ions flow towards each subsequent segment, action potentials are continually produced until the end of the axon is reached (Figure 1.12).

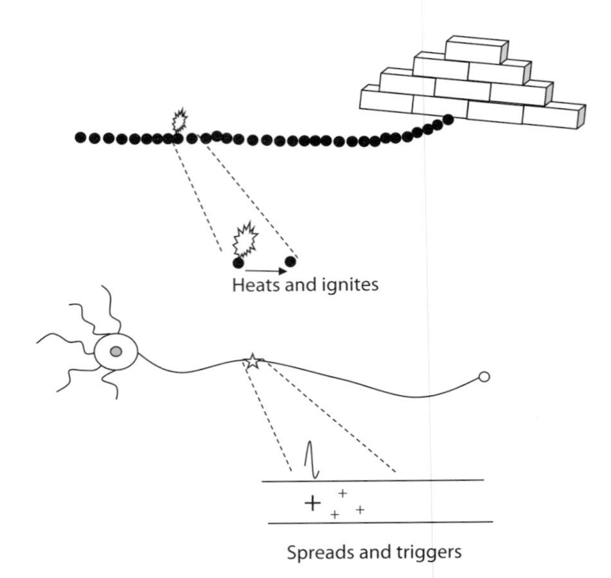

Figure 1.12 Like a flame that seems to be moving along gunpowder (top), the action potential is conducted along the axon (bottom).

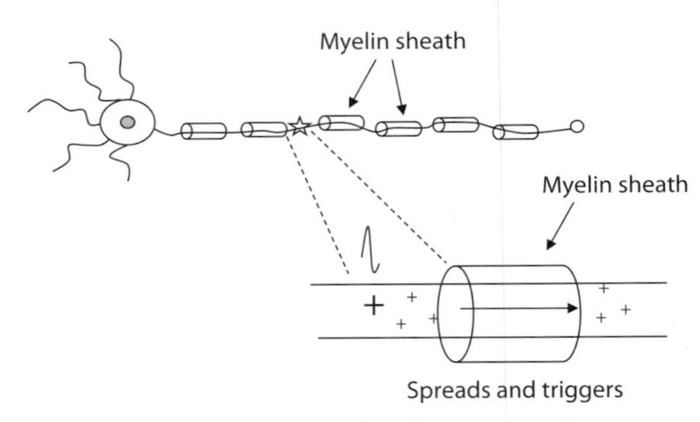

Figure 1.13 The myelin sheath helps speed up the conduction of action potentials along the axon.

Continuing with the example of the gunpowder trail, we can see that if the grains are further away from each other, the heat generated from one grain may not be strong enough to spark the next one, and so the flame dies out. The axons of some neurons are wrapped with a myelin sheath from the oligodendrocytes. Along this myelin sheath there are tiny gaps called *nodes of Ranvier*. It is the myelin sheath that allows for the fast transmission of the action potential down the axon. Rather than activating each segment along the axon, the sheath allows the positivity to jump, jumping to the next gap or node of Ranvier, located between the sheaths. The action potential gets retriggered at each gap by the passive spread of positivity through the sheath (Figure 1.13). Conduction of action potentials through myelinated axons offers two advantages over conduction through nonmyelinated neurons. First, the action potential reaches the end of the axon much quicker (up to 15 times faster).

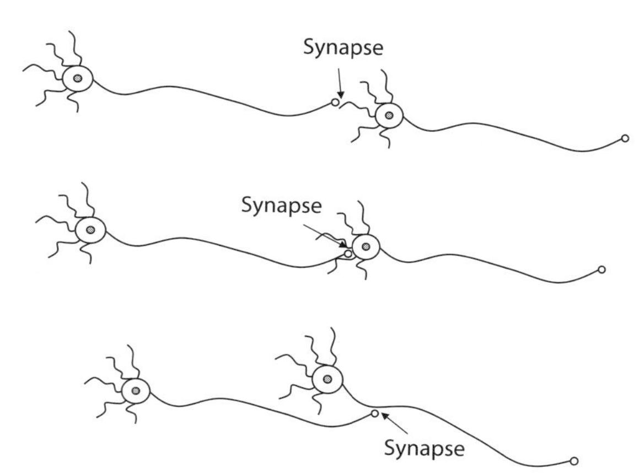

Figure 1.14 Examples of different types of synapses: axondentric (top), axosomatic (middle) and axoaxonic (bottom).

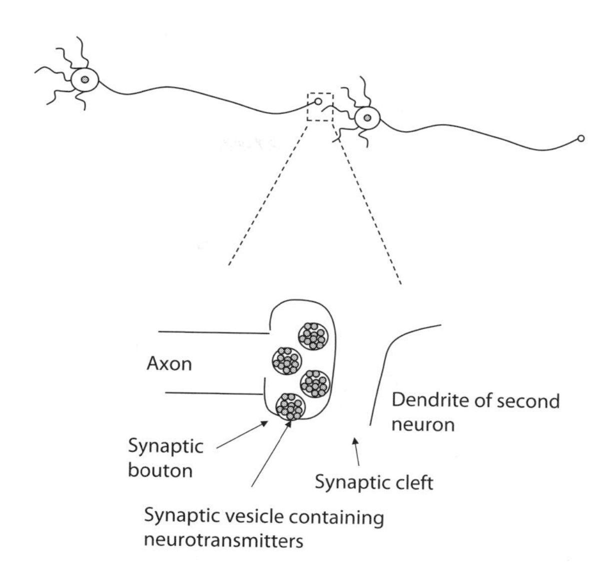

Figure 1.15 Close-up of a synapse, depicting its various components.

Second, the neuron saves energy: fewer action potentials means less use of the sodium–potassium pump to maintain the resting membrane balance at each segment. Note that in patients with multiple sclerosis, an inflammatory disease resulting in damage to the myelin sheath, neural conduction is much slower. This leaves patients with a variety of symptoms including loss of sensitivity, difficulty with movement, balance issues and visual problems.

Synaptic Transmission

As the action potential makes it way towards the end of the axon, it reaches the synaptic terminal/bouton. Information must pass from one neuron to the next, and it does so at junctions formed with a second neuron; these junctions are termed *synapses*. Synapses can occur in three places on the second neuron – on the dendrites, soma or axons, and so the synapses are referred to as axondentric, axosomatic and axoaxonic synapses, respectively (Figure 1.14). Information, however, cannot pass directly from one neuron to the other, as there is a small gap between neurons; this gap is called the *synaptic cleft*. This information must be transmitted by other means: by converting the electrical signal of the first neuron to a chemical signal that can pass across the gap before it is converted back into an electrical signal in the second neuron.

If we were zoom in on the synapse using a powerful microscope, we would see that the synaptic bouton contains a number of small oval or round organelles called *synaptic vesicles*. Each of these vesicles contains neurotransmitters (Figure 1.15). The majority of neurotransmitters fall into three categories. The first category includes amino-acid molecules, which have an amine (NH_2) and carboxyl (COOH) groups in their molecular structure. Examples of neurotransmitters that fall into this family include glutamate, the main excitatory neurotransmitter in the brain; and GABA (gamma-aminobutyric acid), the brain's main inhibitory neurotransmitter. The second category includes amine molecules. Examples of these neurotransmitters are dopamine, adrenaline, noradrenaline, serotonin and acetylcholine. The third category contains the peptides, which include neuropeptide Y and substance P.

When an action potential reaches the synaptic bouton, neurotransmitters are released from the synaptic vesicles into the synaptic cleft. The released neurotransmitters then cross the cleft, attach onto specialised receptors on the second

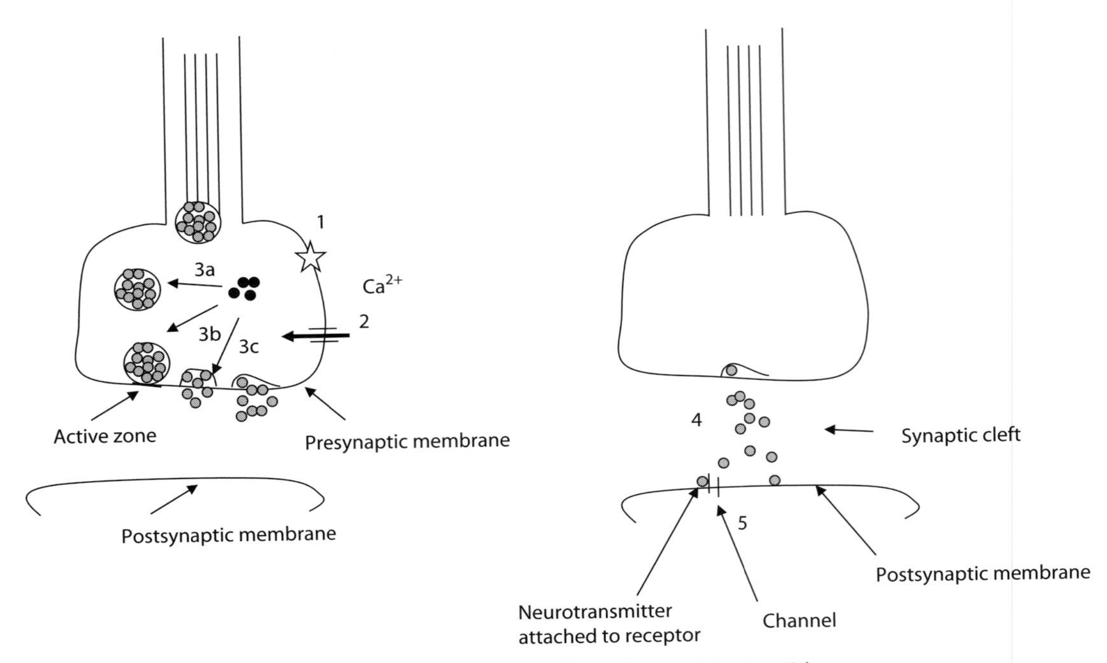

Figure 1.16 Step-by-step description of synaptic transmission (see text for details).

neuron and trigger a response. Thus, information is passed from one neuron to another. There are a number of steps that allow this to occur (see Figure 1.16): The action potential reaches the synaptic bouton (1), which opens voltage-dependent calcium channels (2). Calcium ions (Ca^{2+}), like sodium ions, are in high concentration outside the cell. So when the action potential arrives, the voltage on the inside of the cell becomes more positive, causing the channel to open. Calcium then flows into the cell via diffusion. The influx of calcium into the synaptic bouton has the effect of causing synaptic vesicles to move towards the presynaptic membrane's active zone (3a), to dock to the presynaptic membrane (3b) and to fuse with the membrane, releasing the neurotransmitter into the synaptic cleft (3c). The neurotransmitter crosses the synaptic cleft (4) and attaches onto receptors located in the postsynaptic membrane (5).

What happens when the neurotransmitter attaches onto the receptors depends upon the neurotransmitter involved. Neurotransmitters attach onto receptors that are part of the channel structure. Once the neurotransmitter is attached, the channel modifies its shape and allows either positive or negative ions to flow into the postsynaptic membrane (Figure 1.17). For example, glutamate, the main excitatory neurotransmitter in the brain, attaches onto specific receptor sites on the channel, which then open and allow positive sodium ions (Na^+) into the cell. As a result, the inside of the cell becomes more positive. This positive change in the resting membrane potential in the dendrites of the second neuron is termed an *excitatory postsynaptic potential* (EPSP). GABA, on the other hand is the brain's main inhibitory neurotransmitter; it attaches to receptors and the channel opens, but this channel only allows negative chloride ions in (Cl^-). The resting membrane potential of cell in this case becomes more negative; this transient negative change is termed an *inhibitory postsynaptic potential* (IPSP). These types of channels are termed *ligand-gated ionotropic channels*: 'ligand' because they need a neurotransmitter to activate

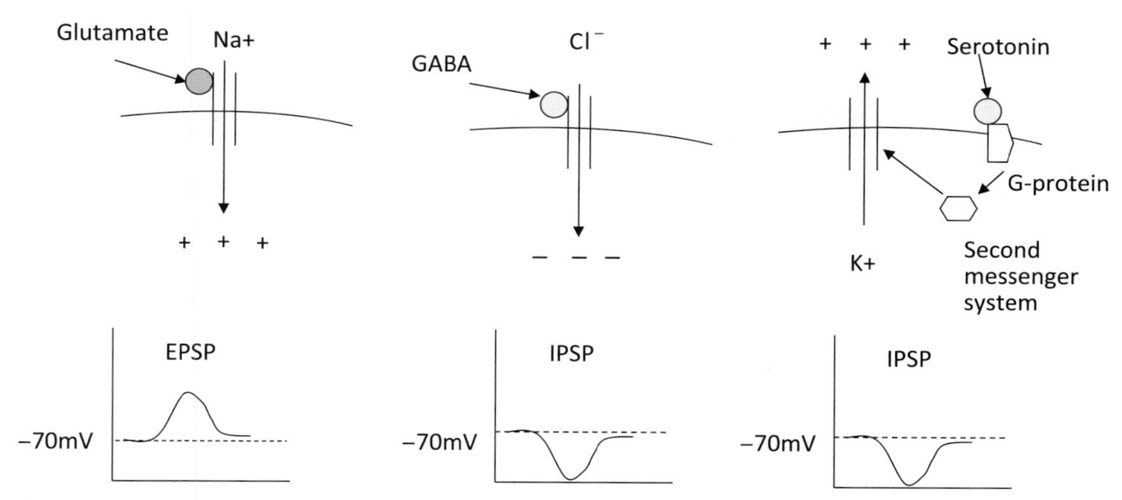

Figure 1.17 Changes in the internal electrical dynamics of the postsynaptic membrane depend on the neurotransmitter involved; e.g. glutamate (left), GABA (middle) and serotonin (right).

them and 'ionotropic' because they allow ions to flow through them. This is the most direct way of converting the chemical signal of the first neuron to an electrical signal in the second.

An indirect method of converting a chemical signal to an electrical one is when the neurotransmitter attaches onto specialised non-iontrophic receptors; these are termed *metabotropic receptors*. In this case there is no modification of the channel to allow ions to flow in; rather, the receptor activates a specialised phospholipid membrane-bound protein termed a *G protein*, which in turn activates a whole cascade of molecular events. The net result of these events is that some of the molecules can activate other channels, away from the original site, which then allow ions to flow in or out of the cell. Although the net results of activation of a metabotropic receptor and an ionotropic receptor may be the same, that is, the inside of the cell becomes transiently more negative or positive, the indirect method is a much slower process (Figure 1.17, right).

At this stage, the neurotransmitters have done their job and are no longer of any use, as such. They are then deactivated through a number of different mechanisms. Some neurotransmitters simply diffuse away. Some are deactivated and broken down by special enzymes located in the synaptic cleft. For example, the neurotransmitter acetylcholine (ACh) is deactivated by the enzyme acetylcholinesterase (AChE). Other neurotransmitters are taken back up into the presynaptic terminal and are recycled. Still other neurotransmitters may be removed via glial cells.

So far we have described the synaptic transmission between two neurons. However, the receiving neuron does not just get messages from a single neuron; it receives information from multiple neurons, which synapse on its dendrites, axon and cell body (Figure 1.18a). The receiving neuron must gather all this information and 'decide' whether to generate an action potential, thereby continuing the process, or not. Therefore, the final step in the process involves integrating all the excitatory and inhibitory inputs that it has received. The more excitatory inputs a neuron receives (i.e. if more EPSPs are produced), the greater the chance that the neuron will generate an action potential. Excitation from the dendrites spreads into the soma and reaches the axon hillock, which is where the soma and axon meet. If the excitation threshold in the axon hillock

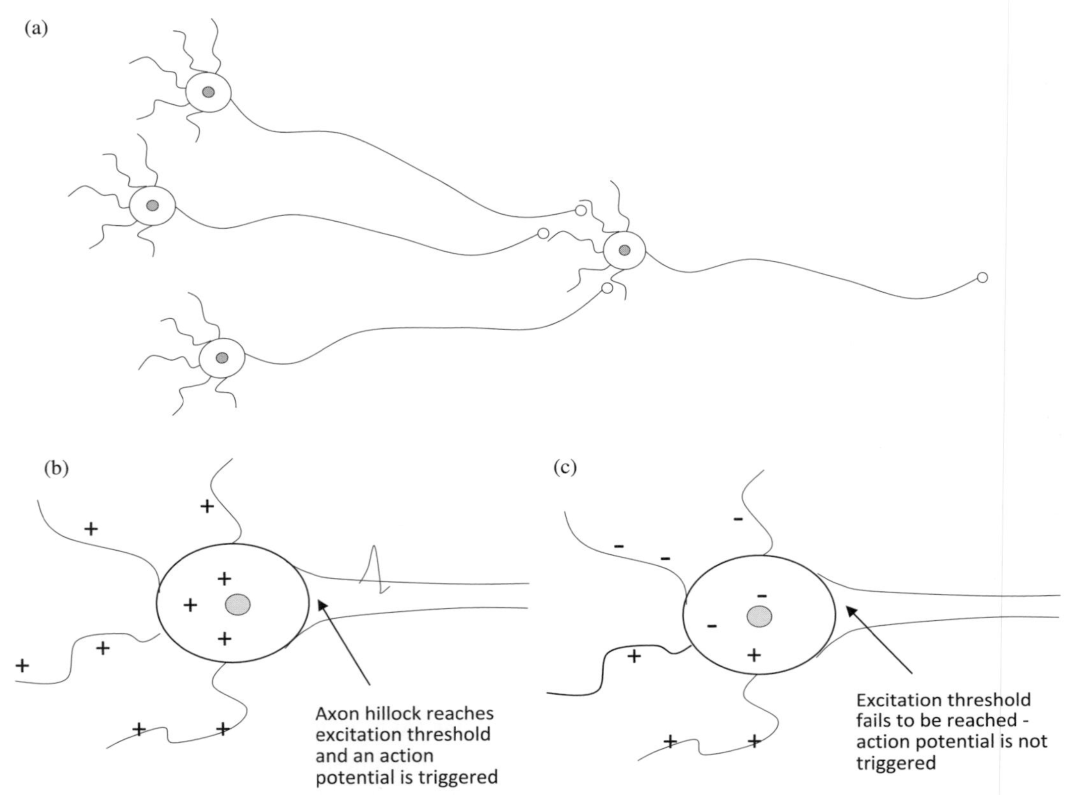

Figure 1.18 Integration of excitatory and inhibitory signals from multiple input neurons (a) determines whether an action potential is generated (b) or not (c).

is reached, an action potential is generated and the information continues to be passed along the axon of the neuron (Figure 1.18b). However, if the dendrites receive inhibitory inputs (i.e. if more IPSPs than EPSPs are produced), this serves to dampen the spread of excitation, the threshold for the generation of an action potential is therefore not reached and so the process stops (Figure 1.18c). It is important to note that neural inhibition does not necessarily produce a behavioural inhibition. Imagine a group of neurons in the brain that are in charge of inhibiting a behaviour, e.g. stopping an arm from moving. If this group of neurons is itself inhibited, the net result is a movement of the arm.

Summary

The composition of the cell is key to understanding how it functions. Due to the cell membrane and the proteins embedded within this membrane, there is an imbalance between the concentration of ions on the inside of the cell compared to concentration on the outside. The inside of the neuron is more negative (–70 mV) compared to the outside. This is known as the *resting membrane potential.*

An *action potential* is the means by which information is passed along the length of a neuron. The action potential is a change in the electrical charge at a particular point along the axon. The change occurs as a result of positive ions flowing into the cell, making the inside of the cell temporarily more positive. Then ions flow back out of the cell, returning it to a normal resting state. The change at one point along the axon triggers a change at the next point and so forth, allowing the action potential to propagate along the axon.

As the action potential reaches the end of the neuron, it triggers the release of chemical molecules called *neurotransmitters.* These neurotransmitters cross the gap between neurons (synaptic cleft) in a process called *synaptic transmission.* The neurotransmitters then attach onto receptors on a second neuron, which changes the electrical dynamic in the dendrites. This change may then trigger the initiation of a second action potential in the second neuron. Information is therefore passed from one neuron to another.

2 An Introduction to the Brain

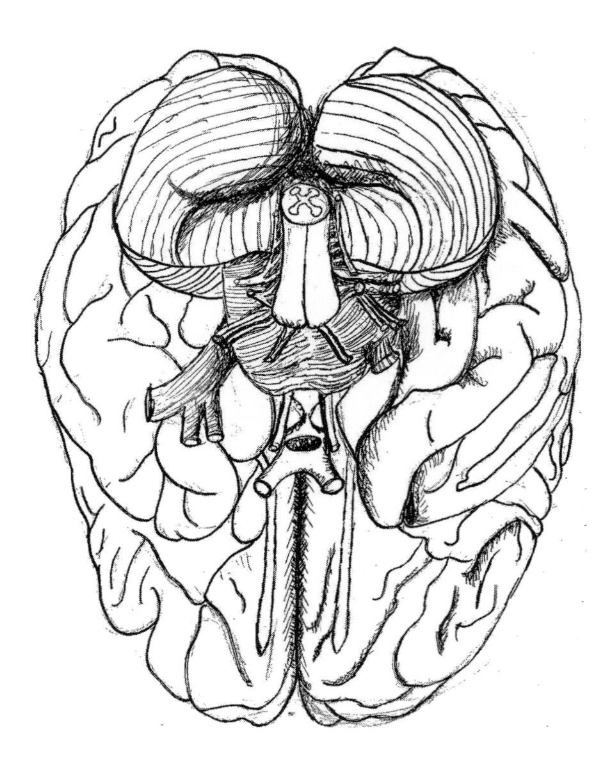

Protecting the Brain

In the previous chapter we looked at single neurons and how they communicate with each other. In Chapter 2 we want to look at the brain as a whole. The human adult brain weighs approximately 1.5 kg, is very soft and feels like jelly. As such it is quite susceptible to damage if it receives a direct blow of any kind. However, there are a number of different features surrounding the brain that provide protection. The first layer of protection is the skull, which encases the brain and allows the brain to keep its shape. The skull is a very hard bone structure, which itself can cause damage to the brain if a connection is made. To prevent this from happening there are a number of soft-tissue layers that cushion the brain and can absorb any shock that might arise following trauma (Figure 2.1). These layers are collectively known as the *meninges*. The meninges consist of three main layers, the first is termed the *dura mater*, which is derived from the Latin meaning 'hard mother'. This layer is a tough, thick layer that stretches around the brain and lies directly under the skull. The next layer is termed the *arachnoid membrane*, which derives its name from the Greek word for 'spider'. This is a web, meshlike layer that is soft and spongy. Lying below this layer is a space (subarachnoid space) that is filled with cerebrospinal fluid (CSF). The final layer is termed the *pia mater* or 'gentle mother' and, as its name suggests, it is a soft tissue layer. Meningitis, a potentially fatal disorder, arises when these layers become inflamed through a bacterial or viral infection.

As the brain is surrounded by CSF, it essentially floats. This not only serves to reduce the weight of the brain and therefore relieves the pressure against the skull, but the fluid also absorbs any shock if the head is suddenly hit. The brain, similar to the heart, is not completely solid but comprises a number of hollows, which are termed *ventricles* (grey areas in Figure 2.2). There are four ventricles. Two large ones in the middle of the brain, one in each hemisphere, are termed the *lateral ventricles*. Located below these is another ventricle, termed the *third ventricle*. Then, the final ventricle, called simply the *fourth ventricle*, is located between the cerebellum and the brain stem at the base of the brain (see Figure 2.2 and 2.3).

The ventricles are filled with CSF, a clear liquid substance that is produced from blood. The CSF is produced in all four ventricles, at locations where the blood supply is very rich. These locations are

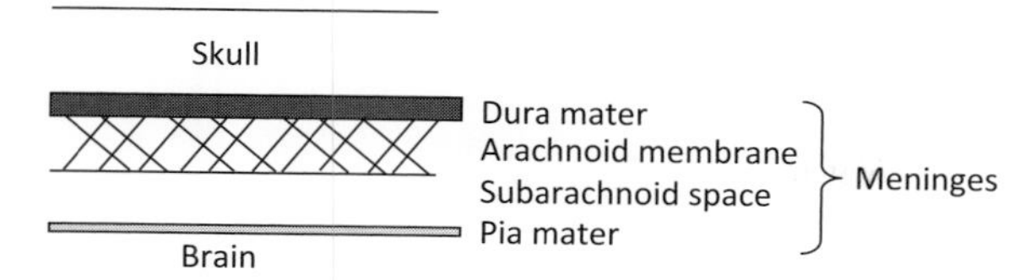

Figure 2.1 Multiple layers that protect the brain from the overarching skull.

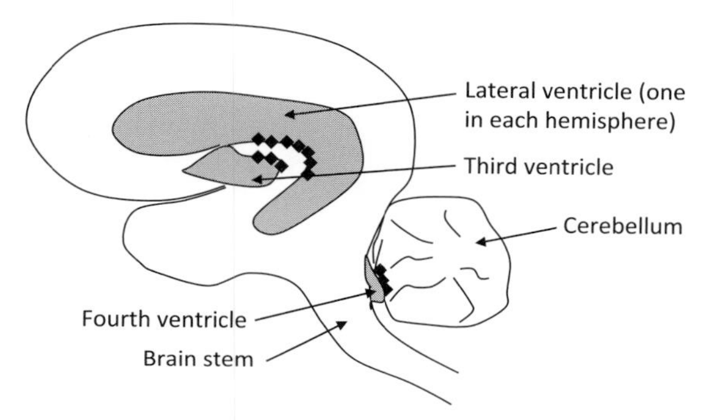

Figure 2.2 A side view of the brain depicting the location of its ventricles (grey) and the choroid plexus (black diamonds).

called the *choroid plexus* and are represented in Figure 2.2 by the small black diamonds in each ventricle. CSF is continuously produced and is circulated around the ventricles before being removed and reabsorbed into the bloodstream by specialised structures in the arachnoid membrane. Apart from providing buoyancy and protection for the brain, the CSF is also involved in removing waste materials from the brain tissue. Sometimes, unfortunately, a build-up of CSF can occur in the ventricles due to a blockage of the system (e.g. a brain tumour), resulting in a condition known as hydrocephalus. In this condition, the walls of the ventricles expand and may cause brain damage. The pressure can only be released by implanting a tube into the ventricle to remove the excess CSF.

Orienting Yourself Around the Brain

Before delving into the brain itself and its various structures, it is important to become familiar with different terms that are often used to indicate the location of brain regions. When we talk about structures that lie towards the front of the brain, the term *anterior* is used. The term *rostral* may also be used. An easy way to remember this is to think of a rostrum, which is a podium that is located at the front of the stage, from which people can deliver speeches. Structures that lie towards the back are referred to as *posterior*. Likewise the term *caudal* is sometimes used. This term is derived from the Latin word *caudum*, meaning 'tail'. Structures that are located towards the top of the brain are referred to as *dorsal*; think, for example, of the dorsal fin of a shark, which lies on its back. The term *ventral* refers to structures located towards the bottom of the brain. (*Venter* means 'belly' in Latin.) Structures that are located towards the centre or middle of the brain are known as *medial* brain regions, and the term *lateral* describes those lying towards the side of the brain. Related to this are the terms *contralateral* and *ipsilateral*, which refer to structures located on the opposite or on the same side, respectively. The word *contra* comes from the Latin meaning 'against', while *ipsi* derives from the Latin for 'self' or 'same'.

As the brain is a three-dimensional structure, it can be viewed along three different planes or axes. Imagine that we wanted to slice the brain into multiple sections. Moving the blade up and down, we can start at the eyes and gradually travel along to the back of the head. Each one of the slices along this orientation is known as a *coronal* section. Next, if we wanted to slice the brain into sections running from one ear to the other, each of the resultant sections are along what is termed the *sagittal* plane. Finally, if we wanted to take a slice off the top of the head and gradually keep making sections until we reached the base of the brain, slices along this orientation are known as *horizontal or transverse* sections.

Overview of Human Brain Structures and Functions

Anatomically, the brain is not a single spherical mass but has distinctive features. It is from these features that neuroscientists are able to map out the various regions and structures.

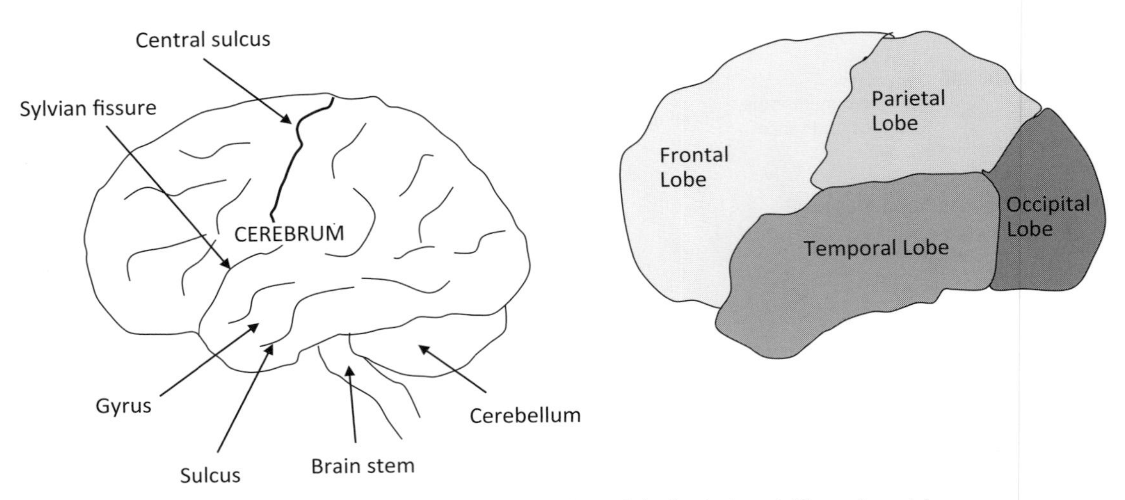

Figure 2.3 Side view of left hemisphere of brain. The surface of the brain is wrinkly and contains many anatomical features (left); each hemisphere can also be divided into four distinct regions termed lobes (right).

There are three main structures of the brain, the cerebrum, the cerebellum and brain stem. The cerebrum refers to the part of the brain that is observable (known as the cortex), as well as structures deep within the brain (known as subcortical structures). The word *cortex* comes from the Latin for 'bark'. Just as the bark surrounds the tree and is observable, the cortex is the outer, observable part of the brain. The first thing you notice when looking at the human brain is that it is wrinkly (Figure 2.3, left). It has grooves and folds. The grooves are called *sulci* (the singular is *sulcus*), a word that is derived from the Latin for 'furrow'. The folds or ridges are called *gyri* (the singular is *gyrus*). Although the human brain has such folds, brains of other species (e.g. the rat) are completely smooth. As the human brain evolved, it grew larger. However, the skull places a limit to this expansion. To circumvent this limitation and to increase its surface area, the brain started to fold in on itself. In addition to the folds, you also notice that the brain is divided into two parts which are termed *hemispheres*. (Note: a side view of the left hemisphere is depicted in Figure 2.3.) The left and right hemispheres are connected together by a group of neural fibres called the *corpus callosum*; this allows the two hemispheres to communicate with each other. If the corpus callosum is damaged or severed, communication can be disrupted and the hemispheres can seem to act independently, commanding one side of the body to do something without the other side 'knowing'. This results in very dramatic behavioural changes, observed in so called split-brain patients. In addition to the cerebrum, comprising most of the neural tissue, the brain also contains the cerebellum and the brain stem (Figure 2.3, left). The *cerebellum*, whose name is derived from the Latin for 'little brain', is located at the back of the brain. The brain stem is located at the bottom of the brain and runs directly into the spinal cord.

Each half of the brain is traditionally divided into four main regions, called the *frontal*, *temporal*, *parietal* and *occipital lobes* (Figure 2.3, right). The frontal lobes (left and right), as the name suggests, are located to the front of the brain and comprise all the tissue that lies anterior to (in front of) the central sulcus. The temporal lobes are located ventral to the Sylvian fissure and the parietal lobes. The occipital lobes are the smallest regions and are located at the back of the brain.

Exploring the Four Lobes

Rather than dividing the brain by anatomical features, the brain can be divided based on function. Korbinian Brodmann, a German neurologist working in the early twentieth century, was one of the first to describe the brain in this fashion, mapping the cerebral cortex into 52 distinct areas. Although descriptions of many of the original areas have changed over the years, the division of the brain by function is still very useful. The following sections describe the main functions of various regions within each lobe and, where appropriate, also indicate Brodmann areas (BAs).

The frontal lobes comprise approximately 40% of all cortical tissue. The frontal lobe deals with three main functions: motor movement, planning of movement and cognition. Motor function is carried out by a small strip of cortex lying just anterior to the central sulcus, known as the *motor cortex* (Figure 2.4, BA 4). Each part of our body is mapped along this narrow strip of cortex, from our lips and eyelids to our big toes.

Movements of muscles on the right side of the body are controlled by the left motor cortex and vice versa. The representation of each body part along the motor cortex is proportional to the amount of movement and use. As humans are verbal and expressive, the areas devoted to our lips, tongue, mouth and hands are huge compared to the area represented by, for example, our toes.

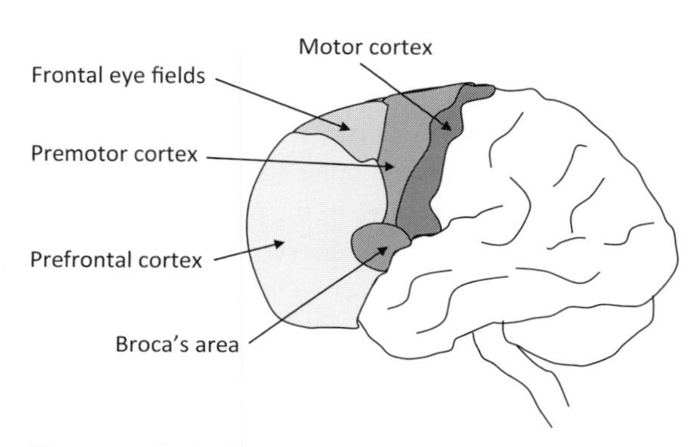

Figure 2.4 Sagittal view of left brain, showing subregions of the frontal lobe.

Lying anterior to the motor strip is the premotor cortex (BA 6), which is involved in the preparation and planning of motor movements. If we want to raise a hand, this involves selecting the correct movement to be carried out – therefore, our hand moves, not our leg. Broca's area, part of the premotor cortex, is also involved with the planning of motor movements – the movements related to speech.

As humans produce language, the functioning of the muscles of our throat, lips and tongue must be planned and coordinated before words emerge. Damage to this region, termed Broca's aphasia, results in an impairment in speech production. Vision also is particularly important for humans. We also need to plan our eye movements, directing them towards different stimuli. This function seems to involve the frontal eye fields. The rest of the frontal lobe (prefrontal cortex) deals with higher order cognitive functions, including planning, working memory, abstract thinking and aspects of personality. It is this region and the associated functions that have become particularly evolved in humans, compared to other species.

The parietal lobe comprises tissue lying posterior to the central sulcus. It is bordered by the occipital lobe caudally and the temporal lobe ventrally. The strip of cortex lying beside the central sulcus is called the *somatosensory cortex* (Figure 2.5, top; BA 1, 2, 3) and is concerned with sensations related to touch. Very similar to the situation in the motor cortex, each part of the body is represented along the somatosensory cortex contralaterally. However, this region detects touch, rather than movement. If someone was to touch your face, neurons in the 'face' region of the somatosensory cortex would fire. As on the motor cortex, the body areas are not equally represented. There is a greater representation of our hands and tongue compared to our knees. This reflects the amount of touches we make with our hands and fingers and the importance of taste on our tongues.

Posterior to the somatosensory cortex lies the *superior parietal lobule*, which is concerned with visual–spatial perception, the detection of

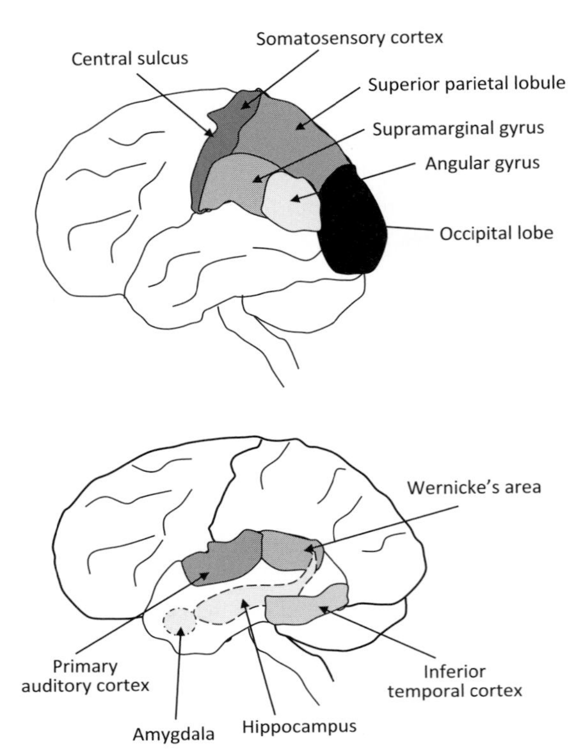

Figure 2.5 Subregions of the parietal lobe (top illu.) and temporal lobe (bottom illu.).

movement and visual attention. Damage to this region of the brain often leads to symptoms of spatial neglect, a condition in which a patient can completely ignore one side of space. For example, the patient, when shaving, may neglect the right side of his face, or only dress one side of his body. The two regions of the inferior parietal lobule, the supramarginal gyrus and the angular gyrus, are involved in reading, writing and arithmetic, particularly in the language-dominant hemisphere (usually the left hemisphere of the brain).

The *occipital lobe* is generally concerned with function of vision. This region contains neurons that respond to light, the movement of light, the orientation of light and colour. For example, damage to area V4, an area specialised in colour processing, leads to the loss of colour perception (known as *achromatopsia*). Damage to other areas of the visual system can lead to loss of the perception of motion or indeed to blindness.

The *temporal lobe* comprises tissue lying below the Sylvian fissure and bordered caudally by the occipital lobe. The *primary auditory cortex* (BA 41, 42) deals with sound information. The auditory cortex is tonotopic in organisation, meaning that different regions of the auditory cortex respond to different frequencies. Related to this is Wernicke's area, which is a higher order speech-processing area. Damage to this brain area leads to Wernicke's aphasia. With this condition, patients fail to put meaning onto sounds, leading to speech comprehension problems.

The *inferior temporal cortex* processes visual information coming from the occipital lobe. Damage to this region often leaves patients with an inability to recognise forms and even faces. These conditions are referred to as visual agnosia and prosopagnosia, respectively.

Deep within each temporal lobe are two important structures: the *hippocampus* and *amygdala*, which we will explore in more detail later. The hippocampus is particularly involved in memory, especially memory for facts and events, and memories of personal events. The hippocampus is also critically involved in spatial memories (see Chapter 14). The amygdala is a small almond-shaped structure and is mainly involved in emotional responses, particularly the emotions of fear and anger (see Chapter 10).

Comparison of Brains Across Species

Although it is useful to have an overview of the human brain, most of the work described in the subsequent chapters will relate to behaviours studied in rats, birds and other species. Figure 2.6 shows the main structures in the human, rat and bird brain. As can be observed, many of the structures are conserved across species; however, there are differences in the relative size of some brain regions. This difference is thought to reflect the importance of the region and function for a particular species. For example, the cerebellum plays an important role in the coordination

(a)

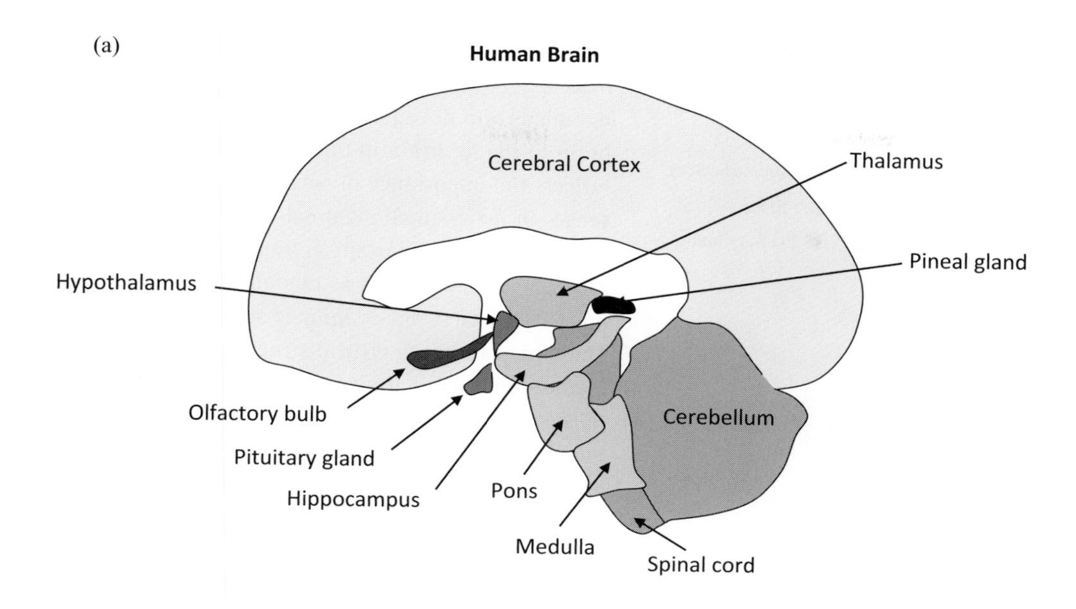

(b)

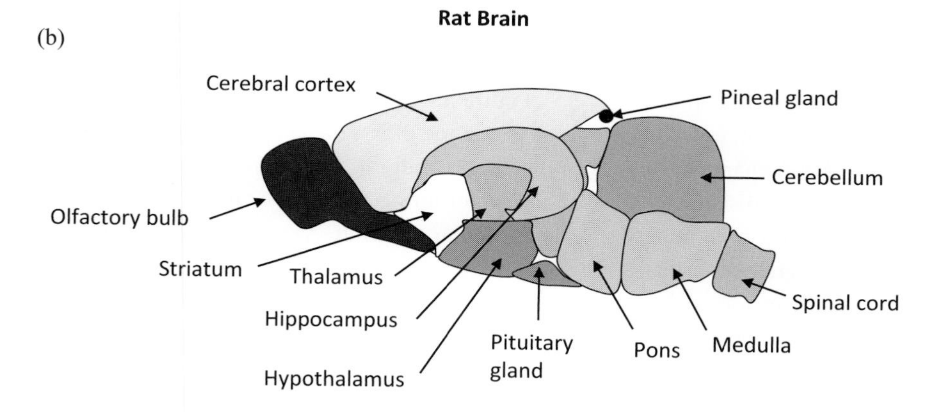

(c)

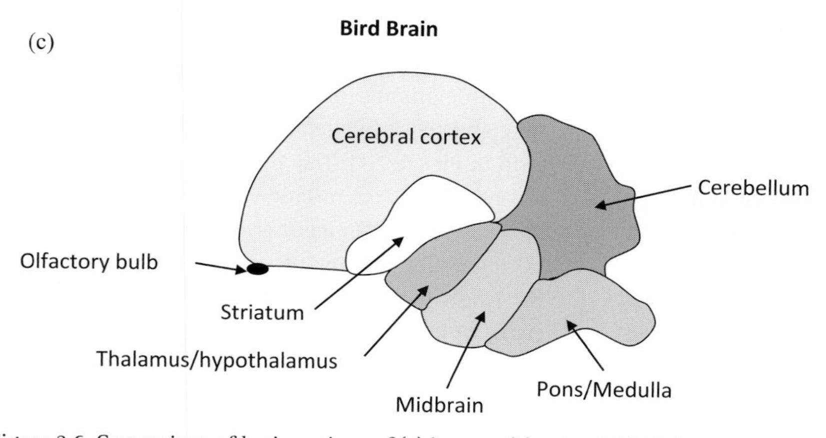

Figure 2.6 Comparison of brain regions of (a) human, (b) rat and (c) bird.

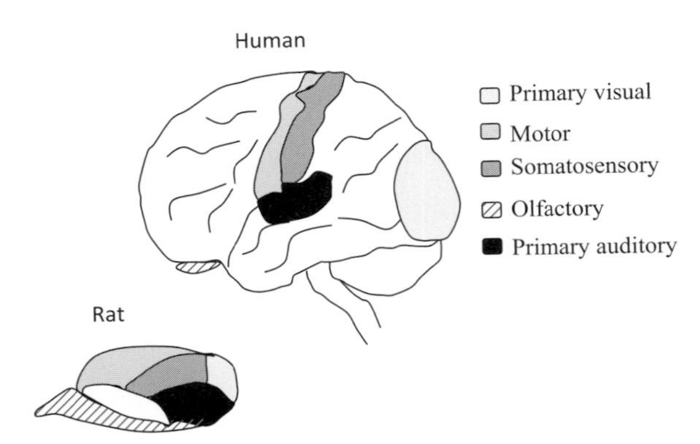

Figure 2.7 Representation of the location of different functions in the human brain compared to the rat brain.

of motor movement. A bird's cerebellum is significantly larger relative to its total body size compared to other species. This is thought to reflect the greater complexity of motor coordination needed by birds operating in three-dimensional space, compared to what is needed when simply walking on the earth's surface. In contrast, the cerebral cortex in the mammalian brain, especially in humans, is much larger, with an expanded cortical surface that creates wrinkles (discussed earlier). These are not observed in lower mammals (e.g. rats and mice) or in birds, where the cortical surface is smooth. Interestingly, there was a reclassification of the avian brain by the Avian Brain Nomenclature Consortium in 2005. The major change has been to reclassify many areas originally thought to be part of the striatum (e.g. neostriatum and hyperstriatum) as cortical. The realisation that birds have a large cortical region that processes sensory and motor information similar to the brains of mammals has lead researchers to explore the ability of birds to perform many higher order tasks. This has proven to be very successful, as many birds show very complex behaviours, such as tool-making abilities, episodic memory, sophisticated sound localisation (Chapter 12) and vocal learning (Chapter 15). Figure 2.6c shows the large cortical region of the bird brain.

The anterior part of the brain contains the olfactory bulb, which is very large in some mammals and not in others. Contrast the human brain to the rat brain in this respect. Again, this reflects the importance of smell above other senses in many small mammals. In contrast, the visual sense, so important in humans, is poor in small mammals such as rats and mice. The majority of these small animals are nocturnal. The hippocampus, a critical structure for spatial memory and navigation, is also relatively large in the rat brain, compared to its size in the human brain. The thalamus, located in the centre of the brain, acts as a coordinating centre for sensory and motor information. The thalamus also serves as a relay station for many sensory signals. For example, information from the eyes passes through the thalamus before reaching the visual centres of the occipital lobe. The thalamus is also well-conserved across species.

Figure 2.7 provides an excellent demonstration of where the different senses are represented in the human and rat brain. In addition it also gives an idea of how much of the cortex is devoted to a particular function. As mentioned earlier, the olfactory sense in lower mammals plays a much more prominent role than it does in humans and it is therefore represented by a larger proportion of cortex. Although motor movement is clearly an important function in humans, allowing us to interact with the environment, the motor area is confined to a small strip of cortex. There is a large area of the human brain that is taken up with planning for those movements in both time and space (white area anterior to the motor strip in Figure 2.7). This planning function does not have the same prominence in rats. Although vision is predominately associated with the occipital lobe in humans, as much as 55% of the whole cortex is devoted to visual functions. Areas of both the parietal and temporal lobes are specialised in various aspects of visual processing.

Summary

The brain is one of the most remarkable organs of the body. It allows us to think, remember, learn, perceive, feel and move. Containing billions of neurons, the brain is also one of the most complex structures known. Surrounded by a hard skull and cushioned by multiple layers of tissue, the brain is well protected, as it should be. Despite the brain's complexity, neuroscientists have over many decades tried to understand what role various regions of the brain play. Each of the four lobes (frontal, temporal, parietal and occipital) has a different function, yet they integrate well with each other to allow us to act coherently. Although the functions and anatomy of the human brain compare well to the brains of other species, it is important to note that the brain of any given species has evolved so that that species is best suited to the environment in which it operates.

3 Neuroscientific Methods

Neuroscientists use a multiplicity of techniques in order to probe the inner workings of the brain. Some scientists work at a very minute level, examining genes, molecules and dynamic changes in cellular function in order to relate these to a particular behaviour. For this, many different methodologies and techniques may be employed. Other scientists prefer to work at the whole brain or behavioural level and therefore require a very different set of techniques and skills. Increasingly, a combination of methods is needed to fully understand a behaviour, and this often requires close collaboration between scientists of very different disciplines, such as biochemists, psychologists, computer scientists and physiologists. As there is a huge array of techniques used, this chapter will only examine the most common ones. As well as providing a brief description of the techniques, we also describe the advantages and disadvantages of each particular method.

Imaging the Living Brain

Recent technological advances have allowed scientists to peer inside the living human brain. This is a remarkable achievement as the techniques used are minimally invasive and allow scientists to get an idea of what is happening in the brain at a particular moment in time. This was not always the case. In order to understand a particular behaviour, case studies of patients were (and still are) examined. If, for example, a patient had a particular disease or had a traumatic brain injury, the patient's behaviour could be categorised and examined, but it was difficult to know what brain region might be affected. It was only after death that the brain could be removed and studied, and inferences made about possible causes of behavioural change. However, even if the brain was removed very rapidly and preserved immediately, the information obtained was only related to that particular time point. The information was not in real time. The patient's disorder or trauma might have occurred many years prior to death, and the brain might have changed radically in the intervening years.

An alternative method used is to ablate (surgically remove) targeted brain regions in other species and infer how the loss of a brain region translates into a behavioural change. This is a useful method, but relating the functioning of brains of other animals to that of the human brain may not be very exact. Although, as discussed in the previous chapter, many structures are conserved across species, there are many regions that do not map readily onto the human brain.

Imaging techniques have changed things, not only allowing us to see inside a human brain but giving us techniques that can also be applied to non-human species.

Computerised Tomography (CT)

The first imaging technique developed was the CT scan. This method requires the participant to lie on a table. The table then slowly moves so that the participant's head is placed in a large doughnut-shaped machine. On one side of the doughnut is an X-ray tube and on the opposite side, an X-ray detector. X-rays are passed through the participant's head, but as the brain contains both white matter (axons and fibres) and grey matter (cell bodies), the amount of radioactivity that passes through the brain varies accordingly. The detectors pick up on this variation and construct visual images. The beam scans the brain from different angles, so a visual image of the whole brain can be constructed. CT scans are very useful for detecting brain blood clots, lesions, tumours and other brain traumas (Figure 3.1a). However, the scans are not very detailed, thereby making precise judgement regarding the size and extent of the lesion/tumour difficult. A further disadvantage is that the images are constructed along a single dimension – the horizontal plane.

Magnetic Resonance Imaging (MRI)

Similar to participants in a CT scanner, participants in a MRI scanner are required to lie on a table, which moves into a ring-shaped machine. However, rather than being X-rayed, the participant's head is placed in a strong magnetic field. This has the effect of aligning hydrogen atoms in the brain to the field. Remember, the brain consists mainly of water, H_2O molecules, so there are plenty of hydrogen atoms in the brain. However, the amount of water in the brain varies from tissue to tissue. A beam of radio waves is then passed through the brain, causing a misalignment of the hydrogen atoms. The energy released from the misalignment is detected and images of the brain then can be constructed. The images of the MRI scan are very detailed, getting down to a resolution close to that of an actual brain slice. The grey matter and white matter of the brain can be distinguished easily. Fibre tracts (such as the corpus callosum) and individual structures (such as the hippocampus) are also easily observed. Furthermore, images are constructed along three dimensions – sagittal, horizontal and coronal planes. Figure 3.1b shows an MRI scan of a stroke patient following a parietal infarct (white arrow). Notice how much clearer the MRI image is when compared to the CT scan.

Diffusion Tensor Imaging (DTI)

Using a modified MRI scanner and taking advantage of fact that most water molecules move along nerve fibres (like water in drainpipes), thousands upon thousands of nerve tracts can be imaged and visualised. DTI measures the diffusion of these water molecules and allows for the visualisation of the neural tracts. The DTI can not only determine the location but also the orientation of neural axons in the brain. This gives scientists an idea of how different areas of the brain are connected to each other. Colour is often added to help distinguish between bundles of different tracts (Figure 3.1c) and provides beautiful illustrations of the network of connections. As many brain disorders are thought to be related to issues of connectivity between structures rather than issues with a particular structure itself, this technique offers a particular advantage over other techniques.

Functional MRI (fMRI)

Although the techniques described earlier are essential for physicians and anatomists, neuroscientists and psychologists are often interested in relating behaviour to specific brain regions. One of the major advancements in MRI, in the early 1990s, was the addition of 'functionality'. By focusing on the magnetic properties of blood oxygen and by using the idea that an active neuron requires more oxygen, brain regions involved in a particular task could now be examined and

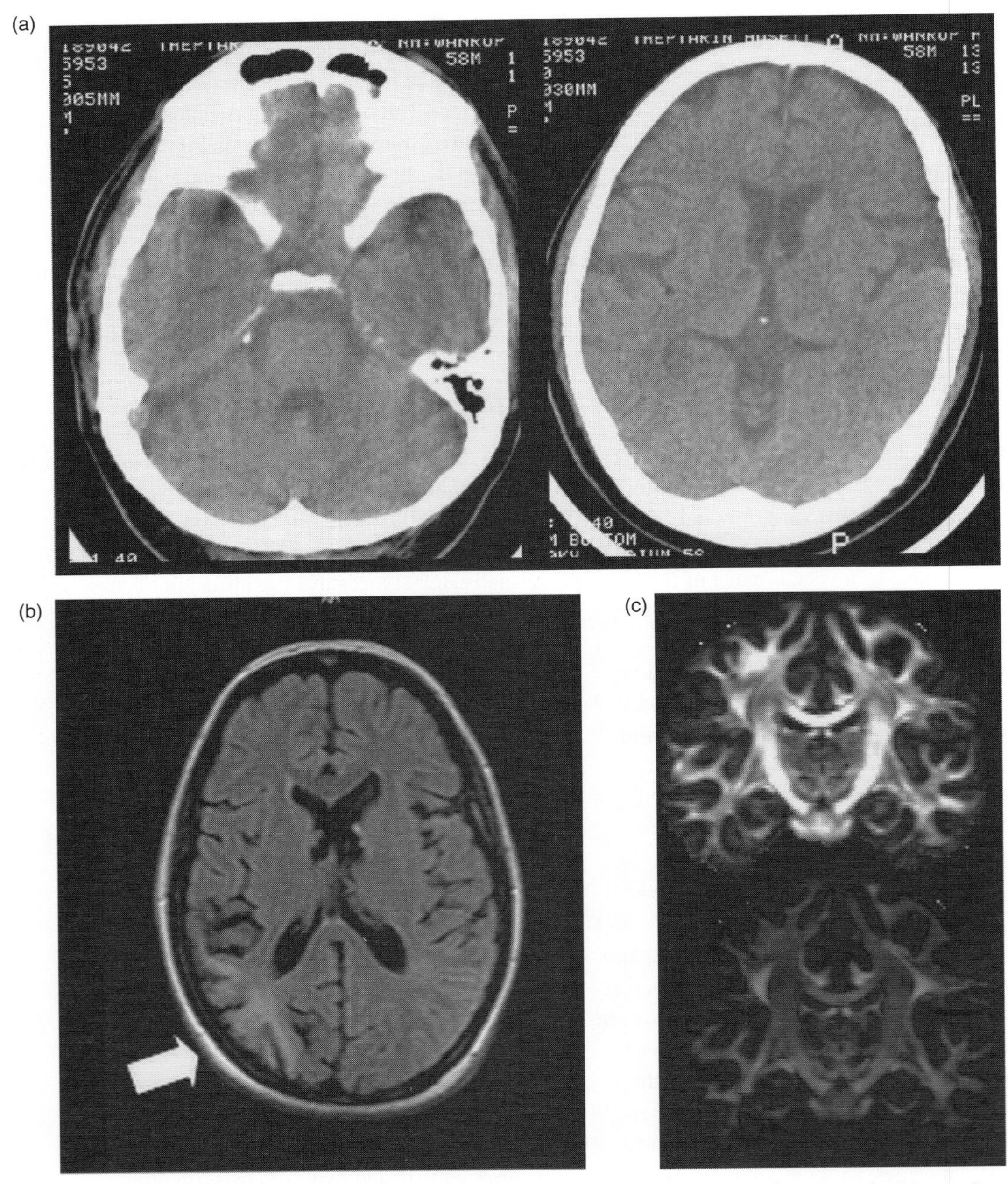

Figure 3.1 (a) Non-contrast brain CT scan. Image taken from A. Chutinet et al. (2014), Intracerebral hemorrhage after intravenous thrombolysis in patients with cerebral microbleeds and cardiac myxoma. *Front. Neurol.* 5:252. Reproduced under Creative Commons (http://creativecommons.org/licenses/by/4.0/). (b) An MRI scan of a stroke patient following a right parietal infarct (white arrow). (c) A DTI scan showing how water diffuses in the brain. Diffusion directions are RGB colour encoded. Red, left–right; green, anterior–posterior; blue, inferior–superior. Image courtesy of the WU-Minn HCP consortium. (http://humanconnectome.org). For colour version of part (c), please refer to the plate section.

imaged. Deoxygenated haemoglobin is weakly magnetic in a magnetic field, whereas oxygenated haemoglobin is much stronger. The functional MRI (fMRI) scanner measures the ratio of oxygenated to deoxygenated haemoglobin, providing a blood-oxygen-level dependent (BOLD) signal which, can be then mapped onto the anatomical scan. Although many neural events are very rapid, occurring over milliseconds, measuring blood oxygen flow to a particular region is less precise; the signal changes over the course of seconds. However, measuring brain activity combined with the excellent spatial resolution of an MRI image far outweighs any issues relating to timing of events. Figure 3.2a shows the activation of different brain regions during a working memory task.

Positron Emission Tomography (PET)

Another technique, older than fMRI, used to map activity onto brain regions is positron emission technology, or PET. Active neurons require more oxygen, but they also require more energy. Glucose provides this energy to neurons, but unfortunately it decays rapidly and therefore is difficult to measure glucose directly in the brain. For this reason, 2-deoxyglucose, a radioactive substance, is injected into the participant. This substance has a molecular structure similar to that of glucose, so it is taken up by active neurons, but it decays at a slower rate to glucose, allowing it to be measured by a PET scanner. The amount of radiation injected is not harmful. As the radioactive isotopes within the injected material decay, they emit positrons from the atoms. When positrons collide with an electron, photons or gamma rays are produced which, in turn, can be measured by special detectors in the PET scanner. From this, activity maps of the brain are produced.

Unfortunately, PET suffers from the same disadvantages as fMRI in that it does not measure the activity of neurons themselves, but rather the glucose that is taken up by neurons. In addition, the change in the signal is also slower than neural activity (seconds rather than milliseconds). A further disadvantage of PET is that it is invasive, requiring an injection into the participant. Also, the images obtained by PET are not as clear as those produced by fMRI; fMRI produces images on all three axes, but PET only provides images on horizontal plane (see Figure 3.2b). In addition, both fMRI and PET techniques produce results that sometimes are hard to interpret, and are correlational: Just because there is an increase in activity in a

(a)

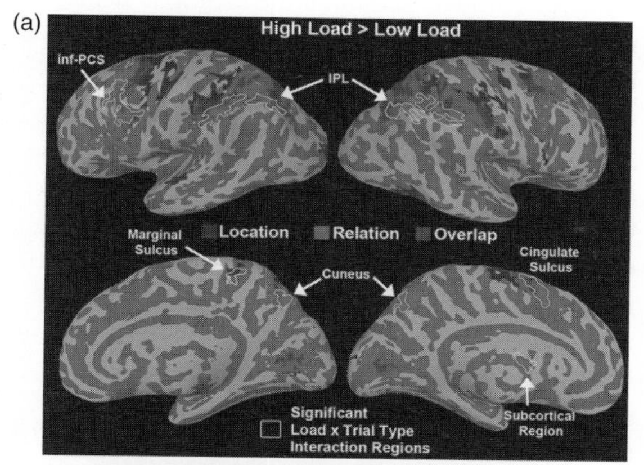

(b)

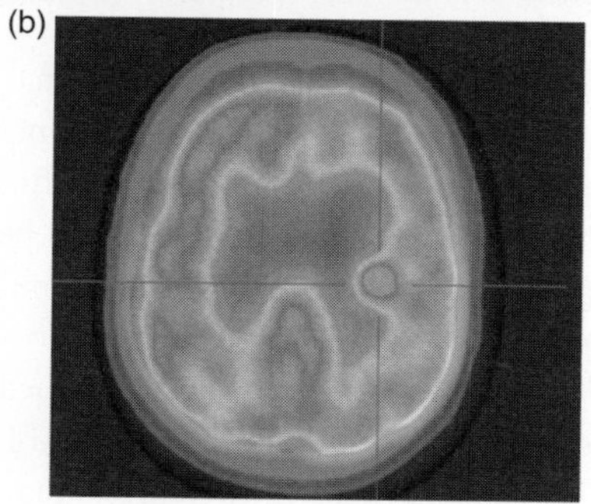

Figure 3.2 (a) An fMRI scan showing regions that were significantly more active for high vs. low load during the delay period of a working memory task. Image taken from K.J. Blacker and S.M. Courtney (2016), Distinct neural substrates for maintaining locations and spatial relations in working memory. *Front. Hum. Neurosci.* 10:594 (http://creativecommons.org/licenses/by/4.0/). (b) PET scan of a patient with cerebral lymphoma (cross hairs). Image taken from Lau et al. (2010), Comparative PET study using F-18 FET and F-18 FDG for the evaluation of patients with suspected brain tumour. *J Clin Neurosci.* 17(1):43–49 with permission from Elsevier. For colour version, please refer to the plate section.

particular region as a result of a task does not mean that this region is necessary for the task.

Magnetoencephalography (MEG)

The electrical current produced by active neurons (described in Chapter 1) produces a small magnetic field. Measuring these fields using magnetoencephalography (MEG) provides an alternative method of examining where changes in brain activity occur in relation to a particular behaviour or task. Although the magnetic fields produced are quite weak, they can be detected by special magnetometers composed of arrays of superconducting quantum interference devices (SQUIDs). As well as detecting changes in response to a task, MEG can be used in a clinical setting to help in the detection of tumours, the focus of epileptic seizures, etc. The main advantage of MEG is that it measures neural events directly and as a result detects changes in the brain very rapidly. In addition, the spatial resolution of MEG is good and superior to that of electroencephalograms (see below), because the skull and scalp distort magnetic fields less than they distort electric fields. However, in terms of spatial resolution, MEG is not as good as fMRI. Furthermore, since SQUID detects weak magnetic fields, quite often the room in which MEG is conducted should be further magnetically shielded so that other sources of magnetism (including the earth's magnetic field) do not interfere with any results obtained from the brain. A further difficulty with MEG is that it can only detect fields from neurons that flow parallel to the brain surface. So any neuron that runs perpendicular to the surface (e.g. many cortical neurons) cannot be easily measured.

Transcranial Magnetic Stimulation (TMS)

Magnetic fields generated from neurons can be measured using MEG. Alternatively, a magnetic field can be generated artificially using a well-insulated coil connected to an electrical capacitor. By placing this small transcranial magnetic stimulation (TMS) coil, slightly larger than a table tennis bat, on a participant's head, magnetic fields can be generated, which cause neurons to fire. The coil can be shaped like a figure of 8 or like a doughnut. Using the figure of 8-shaped TMS coil allows researchers to be more targeted with their stimulation, whereas the doughnut-shaped coil stimulates a broader region of cortex. Many researchers use TMS as a mapping technique. For example, by placing the TMS coil on the motor cortex and stimulating different regions along this cortical strip, muscle movements and twitches will be observed on the contralateral side of the body. From this, the researcher can observe how different body parts map onto different parts of the cortex and also how much cortex is devoted to any particular body part. For example, TMS mapping may reveal that cortical regions related to finger movements are larger in a piano player compared to those in someone who does not play.

Other researchers use TMS as a 'lesion' technique. If a cortical area is known to have a specific function, by stimulating this region with TMS the function can be disrupted. For example, stimulation of Broca's language area will temporarily disrupt a person's ability to count, recite the alphabet or verbalise out loud. As a research tool TMS is excellent; rather than correlating activity with a specific region (noted with many of the imaging techniques), disruption of a particular region allows the researcher to conclude that a region is necessary for a particular function. Further, as TMS can be used to stimulate specific regions, some researchers have shown that TMS is effective at treating neurological and mental disorders. The main disadvantage of TMS is that it not very effective at reaching subcortical structures.

Brain Staining and Imaging

Sometimes whole-brain analysis is not sufficient and more detail is required. Behaviours, diseases or disorders can affect the brain on a cellular level that might be easily missed or might not even register at higher levels of analysis. Further, some imaging techniques have very poor spatial resolution, often making individual structures and their subregions difficult to identify precisely. To overcome these problems brains can be removed (post-mortem), sliced into wafer thin sections and then stained so that individual cells may be examined. However, the brain is a very soft substance and can easily be damaged without its surrounding protective layers (see Chapter 2). The brain's jelly-like consistency makes working with the brain very difficult. The first thing that scientists need to do is to quickly remove the brain and place it into a fixative, usually formalin. This has the effect of preventing decomposition by destroying any bacteria. It also prevents enzymes from rendering the tissue into mush and serves to harden the brain. Once the brain has been fixed, it can be easily sliced on a microtome, which can produce sections of 10 to 100 micrometres thick. Before the individual slices are mounted onto glass slides and observed under a microscope, they are usually stained. Staining allows the fine detail in each brain slice to emerge, including cell structure, number of cells, and types of cells.

Golgi and Nissl Staining

Tissue staining by early anatomists was woefully inadequate. The chemicals used stained thousands upon thousands of neurons at a time. It was impossible to tell where one neuron began and another ended. In the early 1870s Camillo Golgi, an Italian anatomist, tried hundreds of different stains, varying both the timing and concentrations, and in 1872 he made an important discovery. If brain tissue was fixed in potassium dichromate for an extended time period and then soaked in a weak solution of silver nitrate, what emerged was black staining of a small percentage of neurons. As not all cells were stained, individual neurons and their components could now be identified. Figure 3.3a shows a section stained with Golgi stain; the cell body, axons, and dendrites of individual neurons can all be identified clearly. Later, Franz Nissl, a German neuropathologist, used alcohol as a

(a)
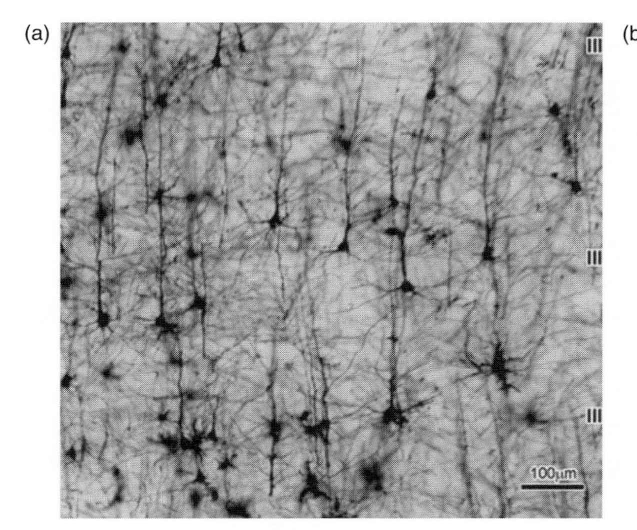

(b)
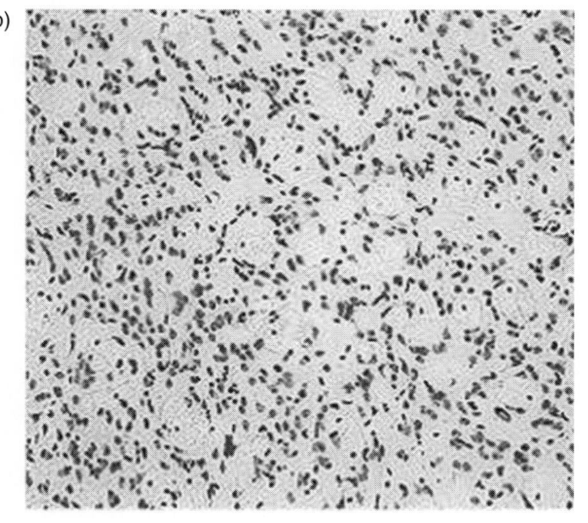

Figure 3.3 (a) Microphotography of the Golgi-stained section of the human neocortex. Image taken from Džaja et al. (2014). Neocortical calretinin neurons in primates: increase in proportion and microcircuitry structure. *Front. Neuroanat.* 8:103. (http://creativecommons.org/licenses/by/4.0/). (b) Nissl-stained section from the rat caudate-putamen. Reprinted with permission from Valdeolivas et al. (2012). Sativex-like combination of phytocannabinoids is neuroprotective in malonate-lesioned rats, an inflammatory model of Huntington's disease: role of CB1 and CB2 receptors. *ACS Chem. Neurosci.* 3(5):400–406. Copyright 2012 American Chemical Society.

fixative and stained tissue using a variety of dyes including cresyl violet. Rather than staining each component of the neuron (dendrites, axons, etc.), only the cell body is stained (Figure 3.3b). As each structure has different cell densities, the stain allows researchers to identify different structures and identify different layers or subregions within a brain area. The stains also allow researchers to identify any abnormalities within the tissue and see whether a particular region has been damaged.

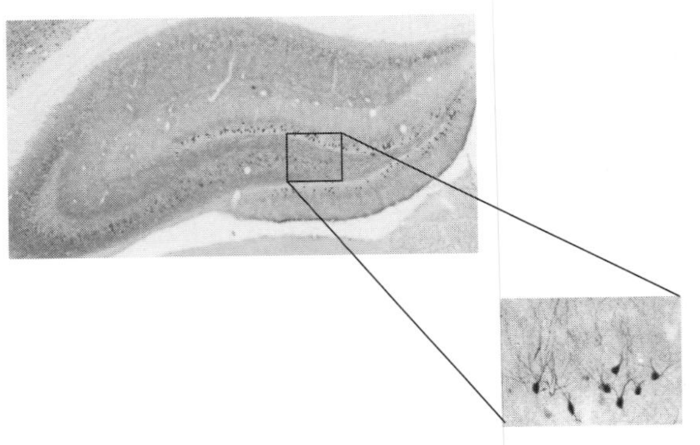

Figure 3.4 Section of a rat hippocampus stained for immediate-early gene Arc. Image thanks to Drs Francesca Farina and Daniel Barry.

Immunohistochemistry

This technique is widely used in neuroscience to identify different proteins that are expressed in cells of tissue. As such, this technique can be used as an imaging tool, identifying activity of different brain structures in response to a task or behaviour. This can be achieved by monitoring proteins located within the neuron, which are expressed when the cell is active (c-Fos is one such protein). The technique is also very useful to examine how a tissue responds to an immune challenge by monitoring proteins such as IBA-1 (ionised calcium-binding adapter molecule 1) that are expressed in microglial cells. More recently immunohistochemical techniques have been useful in a clinical setting whereby proteins known to be involved in certain cancers or diseases can be targeted and examined, allowing for more accurate diagnosis. Immunohistochemistry, as its name suggests, applies techniques used in immunology, i.e. the use of antibodies to tissue (*histos* comes from the Greek meaning 'tissue'). There are a number of major steps that are involved in this procedure. The first step is to apply the primary antibody that binds specifically to the targeted protein (e.g. c-Fos). The next step involves the application of secondary antibodies that target the primary antibodies. This serves to amplify the staining signal. Next, to increase the chance of the protein being detected, enzymes (such as horseradish peroxidase or avidin-biotin-peroxidase) are added, which bind to the secondary antibody. The final step requires the addition of a chromogen, which provides colour to the protein, allowing for visualisation under the microscope, For example, chromogens such as nickel-DAB stain blue/purple. Fluorescent tagging can also be used for visualisation. Figure 3.4 shows staining of the immediate-early gene Arc in neurons of the hippocampus following a spatial learning task using this technique.

The major advantage of this technique is its spatial resolution, by which a targeted protein that is expressed can be identified precisely. The major disadvantage of this technique is that quantification is not precise. Typically the researcher manually counts each stained neuron or uses computer software to do so (in a particular brain region). Both techniques are not ideal and may lead to errors. By comparison, a related technique, such as immunoblotting, a technique also used to detect proteins from sample tissue, does not have as good spatial resolution but does allow for better quantification and ensures that the stained protein is actually the targeted one. This technique is done by comparing the targeted protein against its known molecular weight.

In Situ Hybridisation

Rather than targeting proteins, *in situ* hybridisation is used to examine the expression of messenger ribonucleic acid (mRNA). RNA, unlike DNA, is a single-stranded structure. As many genes have been identified, single-stranded probes that can match a particular RNA strand of a gene that a researcher wishes to study can be artificially created. The probe can then be introduced to the tissue sample, where it targets the RNA stand. Once located, the probe then conjugates/joins with it. A hybrid has been created, hence the name *in situ hybridisation.* The probe can be tagged to allow for visualisation under a microscope. This tag can be a radioactive label or, similar to the immunohistochemical techniques described in the previous section, can be detected by antibodies. Alternatively, the probe could have a tag that fluoresces under ultraviolet (UV) light. The main advantage of this technique is that the spatial resolution is very good. Similar to immunohistochemistry, targeted mRNA can be identified at the cellular level. In addition, the many uses described above, for example, the monitoring of an immune response and activity of particular regions, can also be ascribed to this technique. The main disadvantage of in situ hybridisation is that the probes can be very sensitive and often difficult to obtain. As a consequence the probes can be costly. In addition, the technique targets the mRNA, and mRNA may or may not be translated into protein products, so why a particular mRNA is expressed in a particular location may be open to interpretation.

Invasive Techniques

Sometimes it is important to perform surgery and probe the brain itself. For example, if a person has a brain tumour that is malignant but is easily accessible, the neurosurgeon may make the decision to remove that piece of tissue. Likewise, in cases of temporal lobe epilepsy, it may be necessary to remove the focus or site of origin of seizures. Surgeons must compare the consequences of brain surgery and the damage that this may cause to what might happen in terms of quality of life if surgery is not performed. Likewise, many disorders may require a brain implant that may help regulate normal brain activity. All of these procedures require the opening of the skull to provide access to the brain.

Further, in order to gain an understanding of the function of a particular brain region, how it works or how it is connected to other regions, it may be necessary to purposefully target brain structures in animals. Depending on nature of the experiment, the neuroscientist may remove tissue or implant an electrode in the targeted region. How does a neurosurgeon or neuroscientist know where to target? Neurosurgeons make use of the many imaging techniques described earlier, but neuroscientists typically use *brain atlases*. These are maps of the brain on which every structure in the bird, rat, mouse or other species is mapped. Every brain structure has a coordinate that is based on a reference to 'landmarks' on the skull (e.g. points where the various plates of the skull fuse together). Once the researchers know the structure that is to be removed, they can go to the atlas and find the coordinates, and then, using the skull markings, can pinpoint the exact region of interest in the brain.

Clearly, for any type of brain surgery, the head and brain must be as stable as possible. For this a metal frame in which the head can be placed is used. This is called a *stereotactic frame.*

Lesion Techniques

A lesion can refer to a damaged part of the brain that has resulted from brain trauma, a tumour, disease or an injury of any type. The word 'lesion' is derived from the Latin *laesio,* meaning injury. Knowing the location of a particular lesion provides invaluable information about the function of that particular brain region. Case studies of patients with damage centred on specific sites can prove very useful. For example, the case study of HM, who had his hippocampus removed, has contributed enormously to our understanding of memory (see also Chapter 7). However, case studies with well-defined damage are thankfully rare, so targeted lesions in animals have provided this much-needed information.

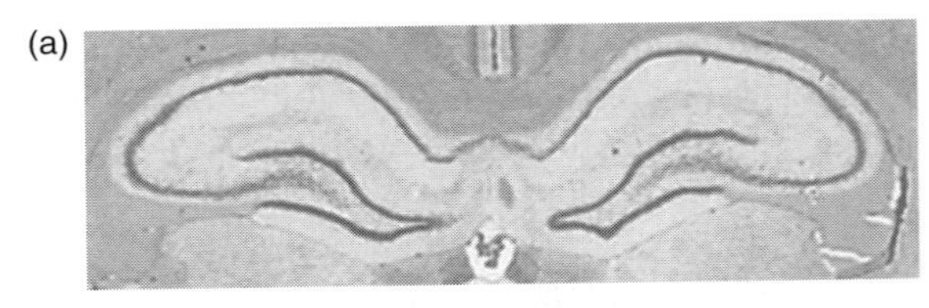

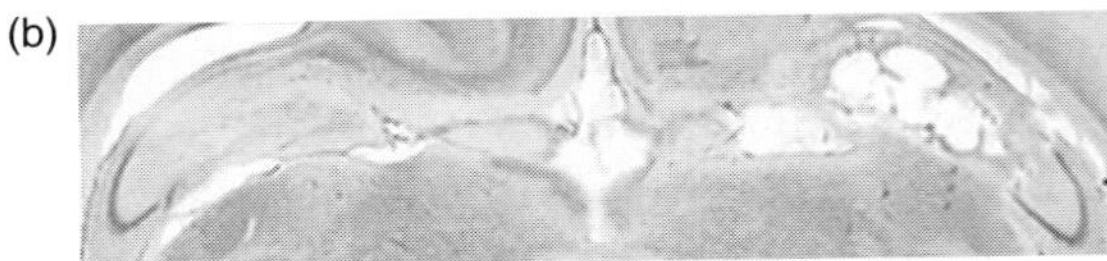

Figure 3.5 Coronal sections of rat brain, comparing a non-lesioned (a) to a lesioned hippocampus (b). Images thanks to Dr Mairead Diviney.

There are five main lesion techniques. An *aspiration lesion* involves the use of a small suction tube. This technique is mainly used to remove tissue from the cortex. A *radiofrequency lesion* involves the insertion of a thin hair-sized electrode into the targeted region, through which a current is then passed. The area at the tip of the electrode is destroyed by the heat of the current. This technique can be used both cortically and subcortically. The intensity and duration of the current determines the extent of damage. The main disadvantage of this technique is that it destroys both cell bodies and axons of the neurons, so if there are axons passing through the targeted region these too may get destroyed. Results may, therefore, be open to interpretation. For example, is the change in behaviour due to the targeted region or due to the disconnection of two separated brain regions? A *neurotoxin lesion* involves the insertion of a small cannula so that a neurotoxin (e.g. ibotenic acid) can be injected into the region. Like the radiofrequency lesion, this technique can be used on cortical and subcortical structures, but has the advantage of only targeting the cell bodies (see Figure 3.5).

A *knife cut* is often used to section a group of fibres or destroy a particular region. Here a small metal probe is inserted and when the targeted site is reached, a blade is released. Small movements are then used to cut the region. The final lesion type is called a *cryoprobe*. As its name suggests, this technique uses a small probe that passes a coolant through to the tip. Any neurons that are located at the tip stop firing. The advantage of this technique, unlike the others, is that it is reversible. Once the temperature increases, neurons fire again.

Tracing Techniques

As there are millions of neurons within the brain and trillions of connections between them, clearly each neuron or brain structure does not operate in isolation. However, each structure does not connect with *all* structures of the brain. Each structure receives information and then passes this information on to the connected region within a network. How do we know which structures are connected and which are not? To identify connecting structures, tracing techniques are used. There are two main types of tracers: anterograde and retrograde. *Anterograde* means moving forward. If we want to know to where a particular structure is sending information, an anterograde tracer can be used. For example, if we wanted to know how the hippocampus is connected and to what regions it is sending information, we would inject an anterograde tracer such as phytohaemagglutinin (PHA-L) or biotinylated dextran amines (BDA) into the hippocampus. This anterograde tracer is taken up by the neurons and follows the axons, which can then be visualised. From this we know in what structures the hippocampal axons end (Figure 3.6a left). Figure 3.6a right shows the tracing of a sample of PHA-L labelled neurons under increasing magnification.

A *retrograde tracer* is used if we want to know from where a structure is receiving its information. For example, if we want to know what structures project to the hippocampus, we would inject a retrograde tracer such as fast blue or Fluoro-Gold into the hippocampus. This is taken up by the axons in the hippocampus and transported back to the cell bodies. Unlike the anterograde tracers where everything (cell bodies, axons, etc.) is visualised, only the cell bodies of the projecting neurons are visualised. Figure 3.6b shows cell bodies of the

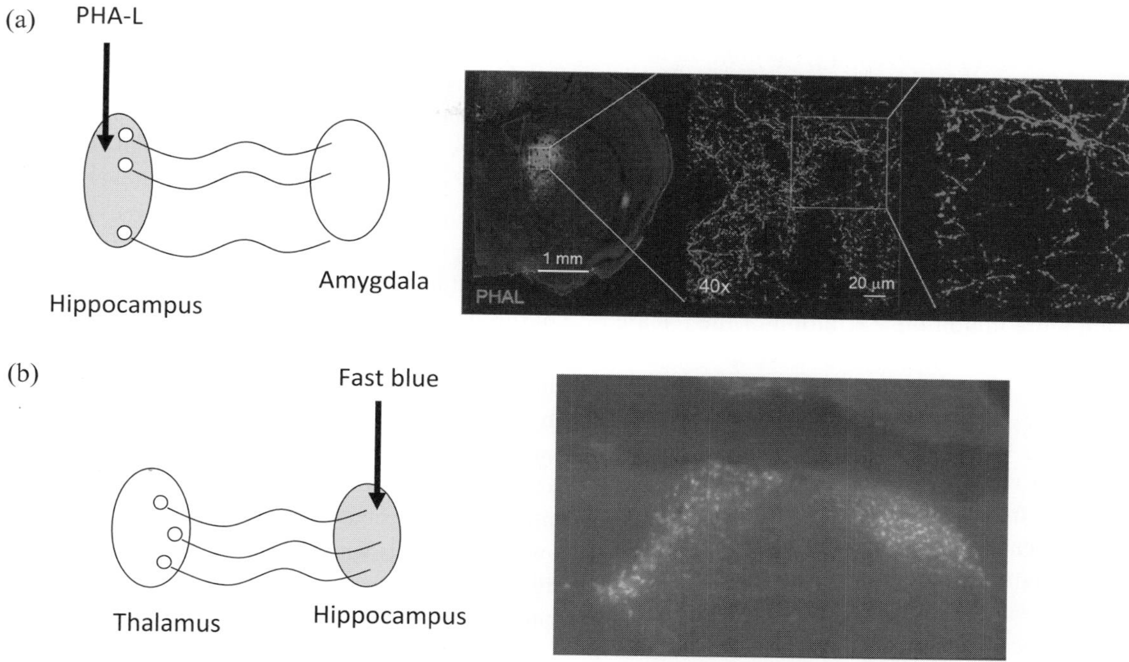

Figure 3.6 (a) Representation of an anterograde tracer injected into the hippocampus, which traces its projections to the amygdala (left). Increasing high- resolution confocal imaging at 40x magnification showing PHA-L-labelled axons (right). Image reprinted by permission from Macmillan Publishers Ltd: Nature Neuroscience, from Hintiryan, H. et al. (2016), The mouse cortico-striatal projectome. *Nat Neurosci.*19(8):1100–1114), copyright 2016. (b) Representation of a retrograde tracer injected into the hippocampus, which traces its projections back to the thalamus (left). Example of cells in the thalamus labelled with fast blue following injection into hippocampal region (right). Image thanks to Drs Maria Cabellero-Bleda and Miroljub Popović, University of Murcia, Spain. For colour version, please refer to the plate section.

thalamus light up and fluoresce with fast blue following injection of the tracer in the hippocampal region – the thalamus projects to the hippocampus.

Optogenetics

The brain contains thousands of neurons that are of different types, of different chemical composition and of different connectivity. Even within a specific brain region, neurons are not distributed equally. For example, the hippocampus region of the brain contains different layers. Pyramidal cells, which are mainly excitatory, are located within a very specific layer of the hippocampus. Other layers contain different types of cells such as basket cells, which are mainly inhibitory. A technique that allows researchers to turn on and off different types of neurons in specific brain regions has recently been developed. This technique is known as *optogenetics*. Researchers have discovered a special light-sensitive protein called *channelrhodopsin*, found in certain green algae. When blue light is directed onto the channelrhodopsin protein it is activated, allowing sodium and calcium to flow into the cell, causing excitation. Other light-sensitive proteins have been found that are activated not by blue light but by yellow/orange light. Such proteins, for example halorhodopsin, upon activation allow chloride ions into the cell, which in turn causes inhibition. Therefore, by inserting the DNA that encodes for channelrhodopsin into, for example pyramidal cells of the hippocampus, light-sensitive proteins are then synthesised and inserted into the cell membrane of this structure.

This set of neurons (only pyramidal neurons) can now be controlled by simply turning on or off a blue light directed towards these neurons. The blue light is delivered to the region via the implantation of optical fibres. This technique is very powerful as it gives researchers control over a specific region, thereby controlling the function pertaining to this region. In addition, it allows for the control over specific types of neurons, causing more inhibition, excitation or the release of specific neurotransmitters in a region.

Microdialysis and Biosensors

In Chapter 1 we learned that neurotransmitters are released from neurons into the extracellular fluid and then act on receptors in the postsynaptic terminal. Brain regions contain many different types of neurons, each releasing neurotransmitters in different concentrations. Dopamine, acetylcholine, adrenaline, glutamate and GABA are just a few amongst many neurotransmitters that are released. Tasks and behaviours are often linked to a particular neurotransmitter being released in a specific brain location (although often behaviours can release many neurotransmitters). For example, amphetamines can increase the concentration of dopamine in the frontal lobes, while patients with Parkinson's disease show lower concentrations of dopamine in the striatum region of the brain.

One way of measuring the concentration of a particular neurotransmitter is by using microdialysis. With this technique, a probe is lowered into the region of interest. Artificial cerebrospinal fluid (aCSF), made up to have a similar osmotic balance to natural extracellular fluid, is pumped into an inner cannula and reaches a membrane-covered tubing located at the cannula's tip. As the fluid is infused through the probe, material from the extracellular fluid of the brain is absorbed back into the probe across a concentration gradient (neurotransmitter concentrations will be higher in the brain than in the aCSF so will readily enter the microdialysis probe), where it is taken up by a second outer cannula and collected for chemical analysis. The analysis can characterise the absorbed substance, identifying the concentration of different neurotransmitters in the brain region of interest.

A related technique used to measure the neurochemical composition of a brain region is via *biosensors*. Rather than collecting a sample of brain fluid and analysing its constituents, a probe specific to the neurochemical of interest is lowered into the brain. For example, a glutamate biosensor can be implanted in the brain to measure the concentration of this neurotransmitter only. A biosensor is typically composed of a platinum electrode with a biological component attached (e.g. an enzyme). The biological component interacts with the substance of interest in the brain and is then measured. For example, the enzyme glutamate oxidase can be attached to a probe; when this is lowered into the brain it interacts with glutamate (the substance of interest) in the presence of oxygen, resulting in the production of hydrogen peroxide which, in turn, can then be measured electrochemically. The level of enzymatic interaction and hydrogen peroxide produced are determined by the amount of oxygen in the region. By changing the biological component on the probe, other chemicals can be detected. This technique offers the advantage of measuring real-time changes in a specific neurochemical.

Further, very small, minimally invasive microbiosensors may prove very useful in a clinical environment. For example, measuring levels of lactate and glutamate in a patient's brain may prove useful in detecting and determining the level of brain trauma that he or she has experienced.

Electrophysiological Techniques

Ions flow in and out of nuerons/brain cells in the form of action potentials or postsynaptic potentials, creating electrical changes across the membrane that provide a direct measure of neural activity. There are a number of ways to measure this electrical activity. Electrodes may be implanted directly into the brain to give exact readings from a particular region. Alternatively, electrodes can be placed on the scalp, providing a more general readout of the electrical activity.

Electroencephalogram (EEG) and Event-Related Potentials (ERPs)

Using multiple electrodes that are placed across the scalp, full-brain recordings can be made. Typically, a cap containing multiple holes and plastic electrode holders is placed onto a participant's head (Figure 3.7a). Caps come in different sizes and can contain up to 256 recording locations, especially for a high-density array in a research environment. However, most clinical EEG recordings use caps with just 20 to 32 sites. Then, each electrode, a small flat doughnut-shaped metal disk attached to a wire, is placed on each of these sites. To ensure contact is made between the electrode and the scalp, a small bit of conductive gel is injected into each site. The wire from the electrode is then attached to an amplifier, which serves to enhance the signal coming off the brain. To ensure that potentials associated with eye blinks do not interfere with the brain signal, two more electrodes are placed just below the eye.

The EEG signal that comes from each brain site is in the form of a continuous squiggle or wave pattern and is recorded and displayed on a monitor. This pattern can be measured in terms of the height of each wave (the amplitude, measured in microvolts) and the frequency (how often the wave repeats itself in a single second), which is measured in Hertz (Hz). An EEG is often used in a clinical setting for patients with epilepsy, to help identify the type of seizure and where in the brain the seizure might originate. For example, a generalised seizure is characterised by large-amplitude waves occurring simultaneously across multiple electrode sites. EEG is also very useful in sleep studies, helping with the diagnosis of sleep disorders and identifying how patterns of electrical activity change across the sleep cycle. Sometimes EEG is also used to help identify the location of lesions, although the imaging techniques described earlier in this chapter provide better spatial resolution.

Instead of continuous free-recordings, sometimes it is important to see how an EEG pattern changes in relation to a particular event. For example, how does the brain react when it hears or sees something? This technique, used widely in cognitive neuroscience, is termed *event-related potentials* (ERPs). With this technique, the EEG pattern is locked to a certain stimulus, giving the experimenter more control.

For example, a participant may be required to respond to a particular stimulus (e.g. a green circle) presented on the screen. After multiple trials, the researcher can examine the averaged EEG pattern that is related to the stimulus presentation and make inferences about what area of the brain might be involved, the timing of response and how some people may respond to compared to others. For example, Figure 3.7 shows the electrical activity over a posterior site on the scalp (indicated by the red pattern on the scalp map) comparing the responses of males and females upon the presentation of a spatial scene. A larger positive (upwards) response occurring approximately 400 ms after the stimulus onset can be observed with females when compared to males.

EEG/ERP offers many advantages. It is a relatively cheap technique compared to many of the other imaging techniques and can be set up in an area requiring a minimal amount of space. Care is needed to ensure that electrical interference from the room does not contaminate the brain signal, but many EEG systems come with electrically shielded components to reduce the interference effect. The temporal resolution with EEG is excellent; electrodes record responses in the milliseconds range. Compare this to the seconds range of fMRI and PET. The technique is noninvasive and does not require participants to be injected or lie in a claustrophobic tunnel. However, the main disadvantage of this technique is its spatial resolution. As the electrodes are on the scalp, they are good at picking up cortical signals, but signals generated deeper in the brain are hard to detect and differentiate. EEG recordings represent the average activity of thousands of neurons and so typically are only able to pick up postsynaptic potentials. Action potentials are too fast and are not synchronised enough to be recorded with this technique.

Invasive Electrical Recording Techniques

Sometimes it is necessary to record the electrical activity from specific regions located deep in the

(a)

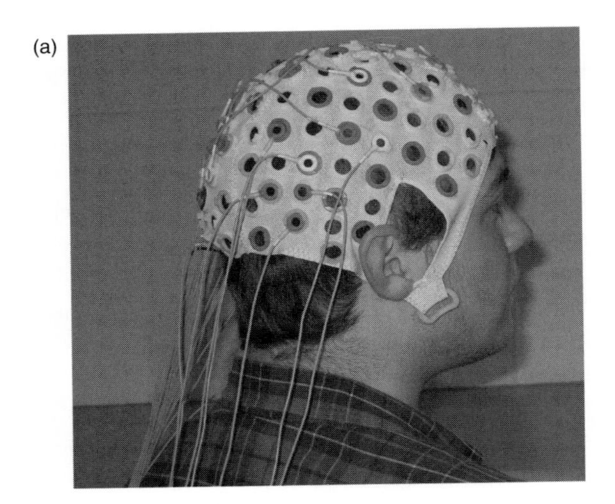

(b)

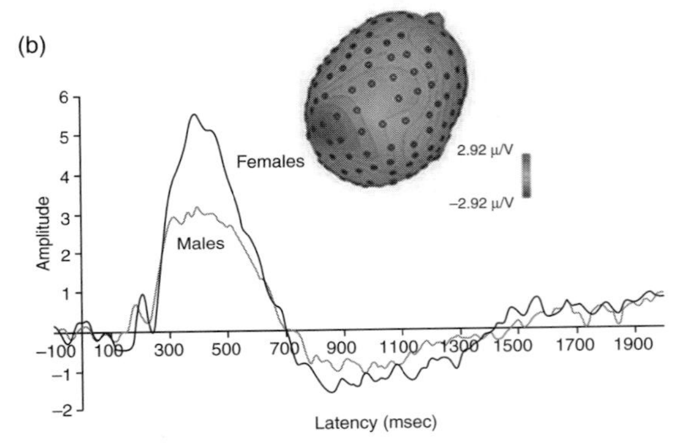

Figure 3.7 (a) Example of high-density EEG array. Image thanks to Mr Jamie Cummins. (b) Averaged ERP waveforms over occipital regions for male and female participants following a spatial memory task. Reprinted from Murphy et al. (2009), High-resolution ERP mapping of cortical activation related to implicit object-location memory, *Biol. Psychol.* 82(3):234–245, with permission from Elsevier. For colour version, please refer to the plate section.

brain. Surface electrodes are not sufficient to pick up this activity. A fine wire (microelectrode) no thicker than a single strand of hair can be implanted stereotactically into the brain (see section of this chapter on Invasive Techniques). The electrode is attached to a small amplifier and has the capability of recording action potentials from single cells (termed single units). Such precision has brought neuroscientists a wealth of information. Using such techniques, specialised neurons have been discovered across many areas of the brain. For example, *place cells*, neurons that fire in response to a single location, have been recorded in the hippocampus; *face cells*, cells that respond to faces, have been found in the inferotemporal cortex. Cells that respond to the movement and orientation of light have been recorded in the occipital lobe. There are even cells that respond to not only to your own movement but also to the movement of others. These cells are termed *mirror neurons* and may have implications in how we learn from others and show empathy.

Sometimes, rather than recording from a single neuron, it is desirable to record the activity from a brain region. For this, macroelectrodes can be used. For example, neurosurgeons implant macroelectrodes directly into the human hippocampus to try to detect the source of abnormal electrical activity during a seizure. As the hippocampus (an area vulnerable to seizures) is located deep in the temporal lobe, such activity may be missed using surface EEG. Rather than picking up action potentials, macroelectrodes are similar to EEG electrodes in that they record postsynaptic potentials of many thousands of neurons.

Instead of recording the electrical activity from a particular brain structure, many neuroscientists may wish to stimulate it. By passing a small current into a region, researchers can ‘activate’ it and determine its function (somewhat similar to transcranial magnetic stimulation, described earlier in the chapter). In the clinical setting, deep brain stimulation (DBS) is a technique used to help relieve patients with movement and affective disorders that cannot be treated otherwise. In the case of Parkinson’s disease, for example, electrodes are implanted into the globus pallidus or the subthalamic nucleus, and the electrical impulses that are passed can try to regulate the abnormal electrical or chemical activity in the region. The amount of stimulation is controlled by a type of pacemaker, usually placed under the skin, which is connected to the electrodes via a wire. Although deep brain stimulation does not cure Parkinson’s disease or other disorders, it may improve the quality of life of the patient.

Other types of research may require the simultaneous stimulation of one brain region and recording from another. In such cases a fine wire, the stimulating electrode, is implanted into one brain structure and another wire, the recording electrode, is placed into a second brain region. The stimulating electrode passes a small electric current to the underlying tissue, essentially creating action potential(s) in this region. The recording electrode picks up the excitatory postsynaptic potential (EPSP) response from the adjoining structure. Typically, the electrode records the extracellular changes from a large number of neurons and this is therefore termed a *field potential.* This technique is very useful in determining whether two areas are communicating with each other. Although tracing techniques will tell if two areas are projecting to each other, this technique will determine whether there is an active signal that passes between them. In addition, by injecting certain drugs into a recording area, it is possible to see how this affects the response, and the researcher may be able to infer which neurotransmitter and receptors are important for creating the response. This technique is also very useful for determining the plasticity or strength of a particular synapse. For example, by applying a high-frequency stimulation pattern, the field potential response can increase in magnitude. Importantly, this response can remain enhanced over a long period of time, even if the researcher applies a much lower stimulation pattern. Memories are thought to be encoded by this process (see Chapters 7 and 8 for more details).

The suggestion that synapses are able to change their responses and that they are not fixed or set in stone has important implications. The brain can be changed and is constantly changing. Experience changes the brain, learning changes the brain, living changes the brain! Furthermore, neurons can adapt and change, even if subject to severe trauma, giving hope for rehabilitation to the many people with brain damage and disorders.

Summary

From the development of the microscope to the ability to look inside the human brain, neuroscientists have always used the latest technology to make new and exciting discoveries. However, all technologies and methods come with their advantages and disadvantages. These have been discussed in this chapter and should be weighed up carefully before experimentation. It is also of critical importance to apply the appropriate method for each experiment; using technology just for the sake of it is counterproductive. As technology develops further in the future, even greater insights into the workings of the brain await us.

References

Blacker, K.J., and Courtney, S.M. (2016). Distinct neural substrates for maintaining locations and spatial relations in working memory. *Frontiers in Human Neuroscience*, 24, 10:594.

Chutinet, A., Roongpiboonsopit, D., and Suwanwela, N.C. (2014), Intracerebral hemorrhage after intravenous thrombolysis in patients with cerebral microbleeds and cardiac myxoma. *Frontiers in Neurology* 5:252.

Džaja, D., Hladnik, A., Bičanić, I., Baković, M., and Petanjek, Z. (2014). Neocortical calretinin neurons in primates: increase in proportion and microcircuitry structure. *Frontiers in Neuroanatomy*, 8:103.

Hintiryan, H., Foster, N.N., Bowman, I., Bay, M., Song, M.Y., Gou, L., Yamashita, S., Bienkowski, M.S., Zingg, B., Zhu, M., Yang, X.W., Shih, .JC., Toga, A.W., and Dong, H.W. (2016). The mouse cortico-striatal projectome. *Nature Neuroscience*, 19(8), 1100–1114.

Lau, E.W., Drummond, K.J., Ware, R.E., Drummond, E., Hogg, A., Ryan, G., Grigg, A., Callahan, J., and Hicks, R.J. (2010). Comparative PET study using F-18 FET and F-18 FDG for the evaluation of patients with suspected brain tumour. *Clinical Neuroscience*, 17(1), 43–49.

Murphy, J.S., Wynne, C.E., O'Rourke, E.M., Commins, S., and Roche, R.A. (2009). High-resolution ERP mapping of cortical activation related to implicit object-location memory. *Biological Psychology* 82(3):234–245.

Valdeolivas, S., Satta, V., Pertwee, R.G., Fernández-Ruiz, J., Sagredo, and O. (2012). Sativex-like combination of phytocannabinoids is neuroprotective in malonate-lesioned rats, an inflammatory model of Huntington's disease: role of CB1 and CB2 receptors. *ACS Chemical Neuroscience*, 3(5), 400–406.

4 Examination of Animal Behaviour: General Principles and Techniques

Behavioural Neuroscience, a Historical Context

Connecting Brain and Behaviour

Linking brain to behaviour has a long and distinguished history that stretches back to Roman times. However, it is only in the last 200 years that our knowledge of brain–behaviour interactions has blossomed. One of the major breakthroughs came in the late nineteenth century with the idea of localisation. This idea suggests that function is localised in specific regions of the brain. So rather than memory, language, perception, and motor or sensory functions being dealt with in all parts of the brain, functions are compartmentalised: different areas of the brain specialise in particular functions. Paul Broca (1824–1880), a French physician, studying a patient who was only able to utter the word 'tan', was one of the first to notice this localisation effect. Upon the death of the patient, Broca noted damage to the patient's brain was confined to a small region in the frontal lobe in the left hemisphere, and concluded that language production was a function of this brain area. Other physicians began to observe similar findings. The German physician Carl Wernicke (1848–1905) noted very different language problems (comprehension difficulties) in patients who had brain damage in a region separate to that identified by Broca. The picture began to emerge that behaviour such as language could be broken into different aspects, which are dealt with by the brain in distinct regions. These regions do not operate in isolation, but are connected to each other, somewhat similar to a circuit, to give a complete experience.

Over the following decades, patients with damage to specific brain regions and who demonstrated other particular behavioural deficits were identified and researched. One such famous patient, Henry Molaison (HM, 1926–2008), who had his hippocampus removed in 1953 to prevent his ongoing seizures, was unable to form any new memories. Following many years of research, it was found that the hippocampus is a critical structure in the formation of very specific types of memories – memories for facts and events – and other brain structures are involved in other aspects of memory, such as habit formation.

In parallel to such discoveries, great strides were also made in the understanding of neurons. Camillo Golgi (1843–1926), the Italian anatomist who we met earlier, developed a staining method so that individual neurons could be identified. Using this technique, Santiago Ramon y Cajal (1852–1934), a Spanish anatomist, went a step further and suggested how neurons communicated with each other. He promoted the idea that neurons were separate cells, rather than a large continuous network. Charles Sherrington (1857–1952), a British neurophysiologist, helped with the understanding of neural communication by developing the concept of the synapse, the small gap between neurons, thereby confirming Ramon y Cajal's idea of neurons being separate entities.

Throughout the twentieth century, a deeper understanding of the mechanisms by which communication both within and between neurons occurs was elucidated, giving us a fuller picture of the mechanics of the brain cell (see Chapter 1). However, it was only in the latter half of the twentieth century that direct links between neurons and behaviour started to emerge, giving impetus to the study of behavioural neuroscience. Pioneers such as David Hubel (1926–2013) and Torsten Wiesel (1924–) placed very fine microelectrodes into the visual cortex and were able to record action potentials from a single neuron. When the scientists shone a light onto a screen, the recorded neuron became active and fired action potentials. Here a direct link between a neuron and a specific behaviour had been made. Hubel and Wiesel went on to identify many different types of neurons within the visual cortex; for example, some neurons responded to light oriented at a particular angle while others responded to light moving in a specific direction. For these findings and for an in-depth description of the workings of the visual system, both men were awarded the Nobel Prize in Physiology or Medicine in 1981.

More specialised neurons were soon to be discovered in other brain regions. In 1971 John O'Keefe discovered cells in the hippocampus that responded to a particular location in space. These place cells, coupled with cells that respond to head direction (discovered by James Ranck in 1984) and grid cells, discovered by Edvard and May-Britt Moser in 2005, led to a greater understanding of how mammals navigate through their environment. For this, O'Keefe and the Mosers were awarded the 2014 Nobel Prize in Physiology or Medicine.

Although discovery of such cells represents an important step, behaviours like how we see or make our way in the world are very complex and will take many years or decades to understand completely. In addition, there may be many different brain areas that are involved in multiple aspects of the same behaviour, as observed above with respect to language. Connecting the activity of neurons within a behavioural circuit and trying to integrate experiences pose great challenges for the field of neuroscience. However, these challenges are being met with the establishment of large-scale international collaborative projects. For example, the Human Brain Project was set up in 2013 to provide a computational model of the brain, and the 5-year Human Connectome Project, established in 2009, attempted to understand how different areas of the brain are connected with each other within a network.

Behaviour, Learning and Its Underlying Principles

The study of behaviour itself has an equally distinguished history, dominated primarily by the work of psychologists rather than physiologists. Given the complexity of behaviour, its personal nature and variability, it was thought that behaviour could never be studied scientifically or objectively. Despite this, many early philosophers argued against this view and attempted to explain the nature of behaviour. One of the first was the French philosopher René Descartes (1596–1650), who divided behaviours into two categories: voluntary and involuntary. Voluntary behaviour was thought to be governed by reason and not subject to physical laws, whereas involuntary behaviours such as the knee jerk or the eye-blink reflex could be understood through mechanical laws. In contrast to Descartes, the British philosopher Thomas Hobbes (1588–1679) believed that the activities of the mind could also be explained via mechanical laws. Although it is true that many behaviours are indeed inborn and reflexive in nature, many are not, and these behaviours can be learned. The first systematic study of behaviour, and particularly of the principles underlying learning, was conducted by Ivan Pavlov (1849–1936, Figure 4.1a). Pavlov, a Russian physiologist, won the Nobel Prize for Medicine or Physiology in 1904 for his work on the digestive system. Following this work, he noticed that dogs would readily salivate when food was placed into their mouths, a reflexive response. However, he also observed that dogs would also salivate at the sight of other objects – for example, the white coat of the lab assistant. Dogs could not possibly have a reflexive salivary response to white lab coats. Pavlov concluded that an association had been learnt between objects that normally do not produce a reflexive response and one that does. With this, Pavlov had discovered a learning behaviour that could be studied and explained scientifically and objectively. Pavlov termed this type of learning *conditioning* and devoted the rest of his life to studying this phenomenon.

From this work a number of important principles emerged. For example, the *unconditioned stimulus* or US (in this case, food) and the *conditioned stimulus* or CS (lab coat) must be presented very close together in time in order for the *conditioned response* or CR to occur (salivation upon presentation of the lab coat). Indeed, in a standard conditioning paradigm the CS is presented just before the US. The time between the pairing maybe as short as 0.5 second, as seen with eye-blink conditioning. In other paradigms the interval may be longer. However, if the interval is too long, conditioning will not occur. Likewise if the US is presented before or simultaneously with the CS, conditioning will not occur. For example, if the tone is sounded after food is given, the tone will not, on its own, produce a salivary response. Interestingly, if the US is presented repeatedly at a regular time interval, in the absence of a CS, a conditioned response will be made just before the occurrence of the US. Therefore, animals are able discriminate and learn on the basis of time.

However, not all learning could be properly explained through this mechanism. Around the same time of Pavlov's experiments, the American psychologist Edward Thorndike (1874–1949, Figure 4.1b) sought to explain how a cat in a puzzle box could learn to open the cage door and retrieve food that had been placed outside it. Initially the cat would try to reach the food through the cage. After many unsuccessful attempts, the animal would give up and start wandering around the cage. Then, by chance, it would hit upon a lever in the cage and the door would open. The next time the cat was placed in the cage, it would be quicker to hit the lever and open the cage door. With each successive trial, the cat would become quicker and quicker. Here was a series of complex behaviours that were goal-driven. The gradual decrease in time suggested to Thorndike that the animals did not solve the problem through insight or reason (otherwise there would be a sudden drop in escape time) but rather by a process of trial and error.

Thorndike (1911) suggested that if a behaviour is followed closely in time by a reward, then the response is strengthened and the likelihood of the behaviour recurring is increased. This idea became known as the *law of effect*. Likewise, if a behaviour is followed by punishment, this decreases the chances of the particular behaviour recurring. The strengthening process is gradual. This type of conditioning is very different from that observed by Pavlov and is known as *instrumental conditioning* or *operant conditioning.*

There were a number of issues with the methods used by Thorndike. One such issue was how to measure the escape behaviour. For example, was the escape behaviour the time to reach the lever or the number of times the lever was pressed? B. F. Skinner (1904–1990, Figure 4.2a) took on some of these methodological issues and developed an operant chamber, the Skinner Box, in which he attached a recording device to the lever. The frequency of lever pushes could now be recorded. More than this, the environment was kept constant and really well controlled. Skinner expanded on Thorndike's idea and developed what became known as *radical behaviourism*. This suggested that an individual behaviour arises from a history of reinforcement, reinforcements that can be either positive (reward) or negative (punishment or removal of reward). He suggested that some very complex behaviours can arise from a sequence of relatively simple responses that have been reinforced. Indeed, Skinner was able to train animals to perform very complicated behaviours using these principles

Figure 4.1 (a) Ivan Pavlov (b) Edward Thorndike. Images from Wikimedia Commons.

(e.g. he taught pigeons to play ping-pong (Figure 4.2b) and taught chickens how to dance.

Unlike earlier behaviourists, Skinner did not deny thoughts, emotions and other inner events; however, he felt that there was no point in looking at these as they could not be accurately measured. As Pavlov did for classical conditioning, Skinner examined the underlying rules of operant conditioning; for example, he examined the different schedules of reinforcement and under what conditions reinforcement arose. Skinner's work is still very influential today, particularly in the classroom setting where many educators set up a *token economy*, whereby a child's action is rewarded with tokens that can be used to buy a treat. Likewise, applied behavioural analysis is an effective technique used to modify behaviour in many different aspects of life, including work with children with autism.

In the 1940s and 1950s a debate arose between psychologists on the nature of learning. Some believed that learning was simply a series of stimulus–response connections, in which some of these connections will be strengthened and others will be weakened. For example, in maze learning, where animals have to run down a series of alleyways to obtain food in a goal box, animals are simply responding to succession of stimuli; e.g. turn left, then turn right, then turn right again. The correct responses will become strengthened at the expense of the incorrect ones, and these correct responses will be more likely recur. This results in the animal making less errors and obtaining the food quicker on each subsequent trial.

Other psychologists, known as *field theorists* at the time and now referred to as *cognitive psychologists*, believed that learning was more flexible than this. Edward Tolman (1886–1959, Figure 4.3a), an American psychologist, demonstrated that animals could use cues in the environment such as the window and lights to find their way to a goal. This behaviour is more flexible. Tolman suggested that animals develop what might be considered a cognitive

(a)

(b)

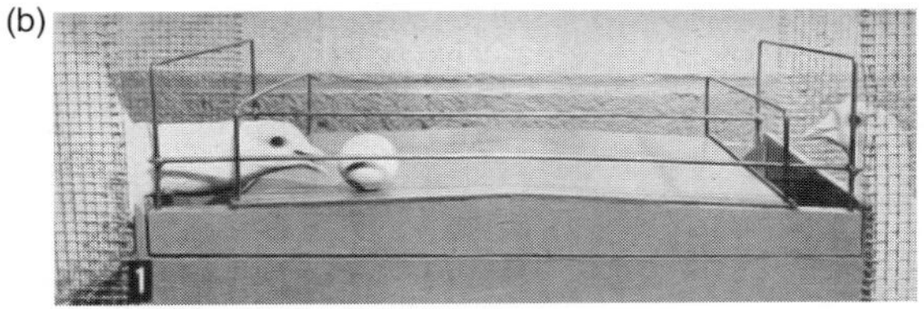

Figure 4.2 (a) B.F. Skinner (b) Pigeons playing ping-pong. Image provided courtesy of www.all-about-psychology.com

map. Tolman demonstrated this idea in an important experiment. Animals were placed on an initial starting platform and were trained to go down a series of alleyways, turning left and right, before obtaining food at a goal location (G in Figure 4.3b left). Following multiple trials, animals were then tested on a 'starburst maze' in which the initial pathway was blocked. Tolman then observed which alley, if any, animals would go down. If animals had learned the task as a sequence of stimulus–response behaviours, they would be impaired and go nowhere. However, Tolman observed that the majority of animals picked arm 6 (Figure 4.3b right), which was in the correct direction of the food relative to the starting position. This suggested that animals used the various cues in the room to navigate, rather than relying on a specific path.

A second important finding reported by Tolman concerning the flexible learning behaviour of animals is what is termed *latent learning*. Animals were trained to find food at the end of a maze with multiple alleyways. There were three groups. The first group, HR, was rewarded food when they reached the target (Figure 4.3c). A second group, HNR, was allowed to explore the maze but was not provided with a reward. The third group, HNR-R, was also allowed to explore the maze without a reward for the first 11 days; then from day 11 onwards a reward was provided. Examination of the number of errors made (i.e. going down a blind alley) demonstrated that the group that had been provided with a reward on every trial quickly learnt the task, especially compared to the unrewarded group. Interestingly, the group that received a reward from day 11 onwards initially made many errors, but once they started receiving food, the number of errors made rapidly decreased to the level of the rewarded group. This demonstrated that animals, even without obtaining a reward, had learned something about their environment. This learning was hidden (latent) and was only revealed when a reward was provided after day 11.

Through the work of Tolman and others, the focus of research shifted to the examination of the cognitive mechanisms underlying behaviour. In addition, with improvements in technology, particularly computer technology, these putative cognitive mechanisms could be both modelled and tested experimentally. Everything was now open to research, including topics that were considered by the early behaviourists as being beyond the realm of study. Topics such as memory, decision-making, emotion, imagery, personality and attention could all be analysed. Furthermore, as brain imaging and recording techniques improved, measurements of such inner workings could be done in an objective and scientific fashion.

The dreams of the early philosophers were and are currently being realised. Although the rise of cognitive psychology and neuroscience in general did push traditional behavioural psychology to the periphery, there was also a realisation that many of the laws and behavioural principles unearthed using traditional learning paradigms, such as classical conditioning, could be also be examined at a neural level. Furthermore, well-controlled traditional techniques and paradigms such as mazes and conditioning chambers could be used to elucidate higher-order behaviours

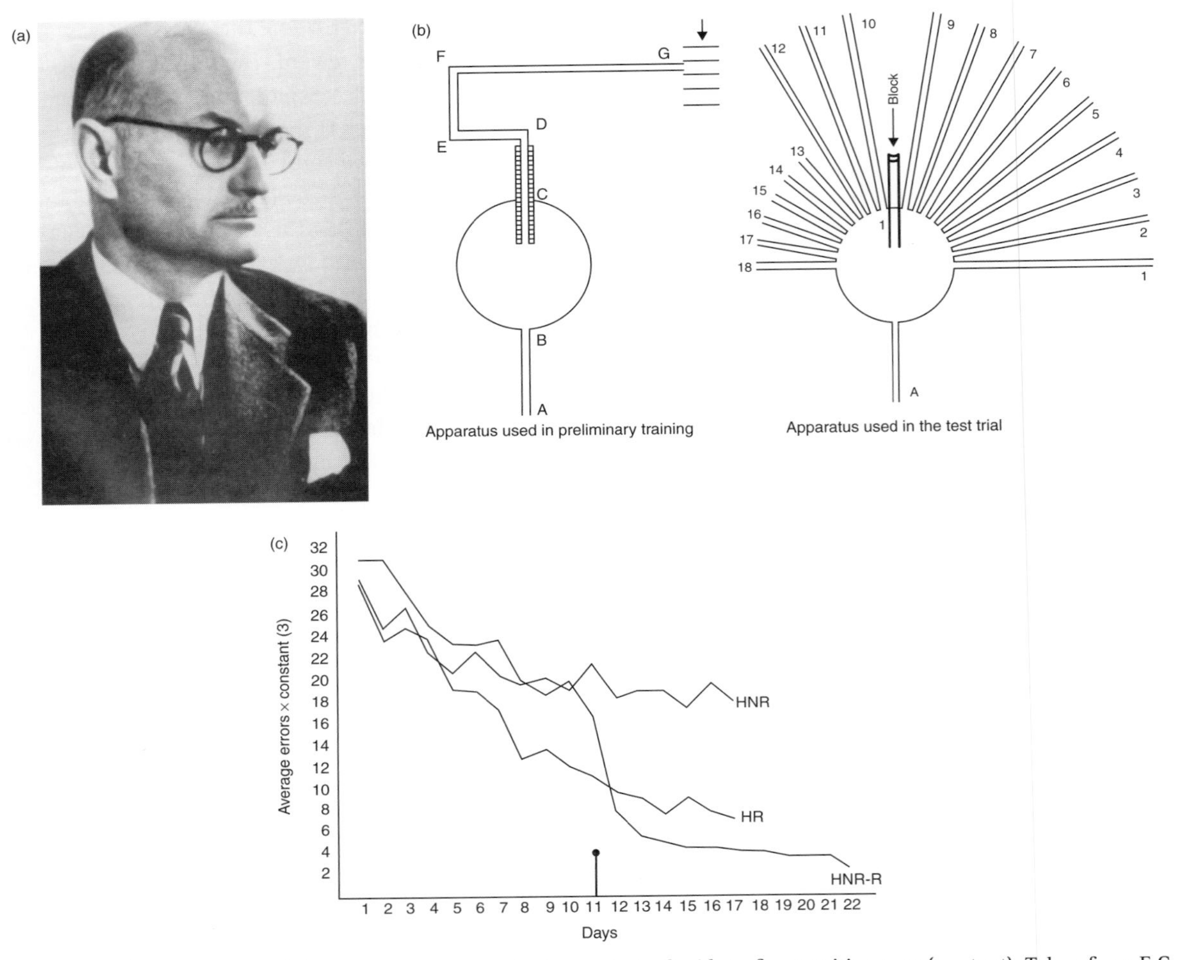

Figure 4.3 (a) E.C. Tolman. (b) Starburst maze used to demonstrate the idea of a cognitive map (see text). Taken from E.C. Tolman, B.F. Ritchie, and D. Kalish (1946), Studies in spatial learning. I. Orientation and short-cut. *J. Exp. Psychol.*, 36:17. (c) Learning curves for HR, HNR and HNR-R demonstrating latent learning (see text). Taken from E.C. Tolman and C.H. Honzik (1930), Introduction and removal of reward, and maze performance in rats. *Univ. Calif. Publ. Psychol.* 19: 267.

such as memory and fear as well as more basic behaviours, such as the eye-blink response. We will explore some of these topics later in the book.

Measuring Behaviour: Methods and Techniques

Chapter 3 described many of the techniques used to examine the brain and measure its activity. This section will briefly look at the different methods used to measure behaviour. Although the book will explore the behavioural repertoire of many species, the tasks and tests typically used in behavioural neuroscience have been developed primarily for rats, mice, non-human primates and birds (mainly pigeons).

Mazes: Tests of Learning, Memory and Exploration

Mazes have a long history. Willard Small, a young American graduate student, was one of the first to develop a maze for rodent learning.

In 1901 he developed a maze (Figure 4.4a) that was based on a hedge maze at Hampton Court constructed in the 1690s for King William III of England. Animals would be released into the maze and Small would record every turn and behaviour the animal would make, would time how long it took to reach the centre and the number of errors the animal would make. John Watson (1878–1958), considered to be the founding father of behaviourism, continued to use Small's maze with minor modifications. However, it was Tolman who really elaborated the maze, using a wide variety of different shapes and sizes. Some mazes were very similar to those used by Small and Watson, with multiple alleyways, some of which were blocked (Figure 4.4c). Some were quite complex, like the starburst maze, with arms radiating from a central platform. Others were very simple, shaped like a cross (Figure 4.4d). Using such mazes, Tolman demonstrated that

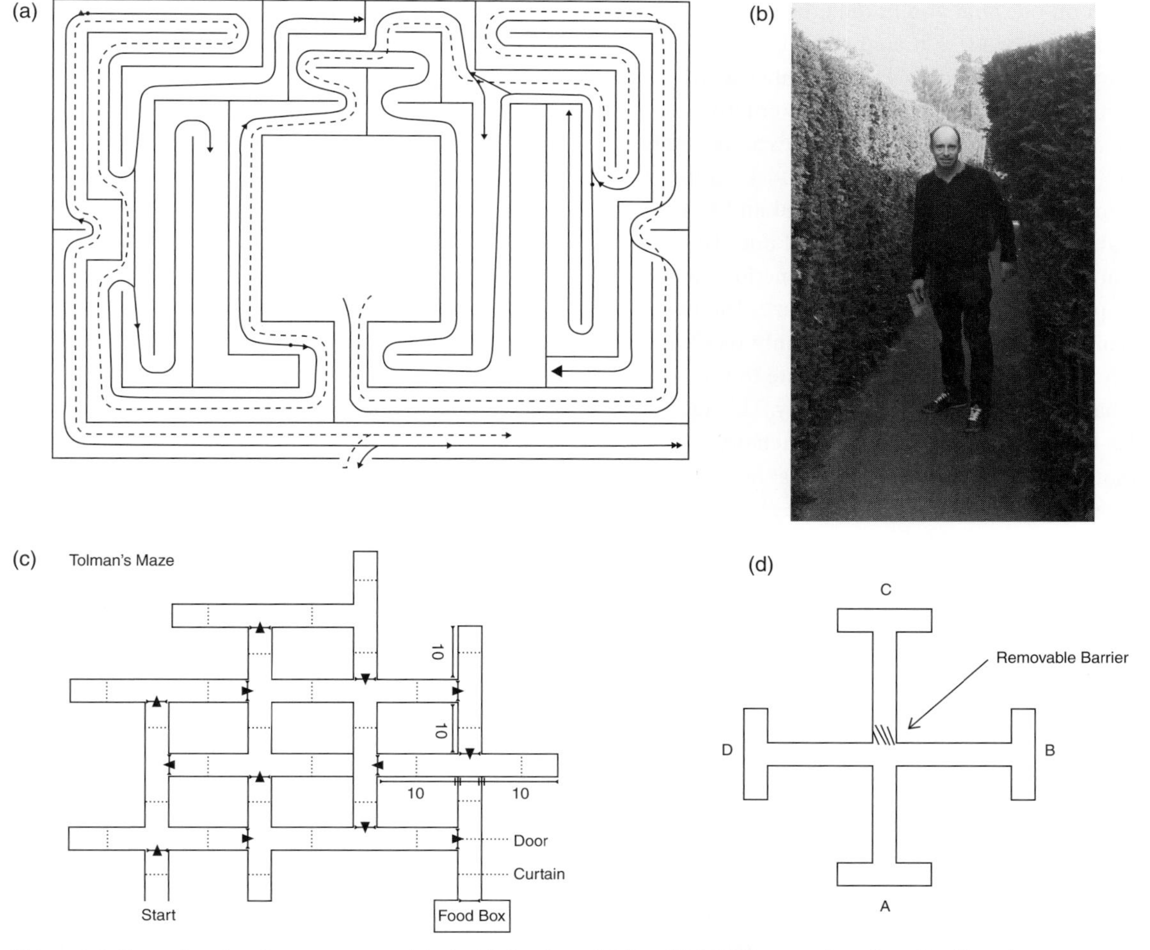

Figure 4.4 Example of various mazes used to test behaviour. (a) Small's (1901) maze, based on the Hampton Court maze. Image taken from Willard S. Small (1901), Experimental study of the mental processes of the rat, II, *American Journal of Psychology*, 12 (2): 206–239. (b) The author trying to find his way in the maze at Hampton Court, UK. (c) and (d) Complicated and simple mazes designed by Tolman. Reprinted with permission from E.C. Tolman (1948), Cognitive maps in rats and men. *Psychol Rev.* 55(4):189–208; published by APA.

animals navigated by using environmental cues rather than a learnt motor response.

One popular maze used today is the *T-maze*. As its name suggests, it is in the shape of a T and is a variant of the plus or cross maze. Animals are placed in the bottom stem with food located at the end of either of the top arms. Animals are simply required to locate the food. Time taken to find the food or the number of errors (running down the arm that does not have the food) are indicators of memory. In an unrewarded T-maze, animals tend to alternate their searching, search in one arm on one trial and in the other in the next trial. This behaviour is thought to rely on working memory, as the animal has to remember which arm was previously visited. As an alternative to this spontaneous alternation, many experimenters use a 'forced-sample' version of the task. In this case, one of the maze arms is blocked and so the animal is 'forced' into the unblocked one. The animal is then taken out for a short period and is again placed into the stem of the maze. The two arms are now open but the animal only receives the reward if it picks the arm opposite to the one that it had been down previously. This task is also thought to rely on working memory.

Another standard maze used to test learning and memory today is the *water maze*, which was developed by Richard Morris in 1981. The maze (Figure 4.5) is a large circular pool filled with water. Somewhere within the pool, located just below the surface of the water, is a small platform. Animals, particularly mice and rats, are very good swimmers, but they do not like water too much and try to escape it. Thus, the water provides a good motivation for animals to search for an escape, and they can do so by finding and climbing onto the hidden platform. Once the platform has been found, animals must remember the platform's location, which can be done by using the landmarks in the room (note they cannot see the platform directly when swimming, as the water is usually coloured with a nontoxic dye). The time taken to find the platform on a particular trial is used as a measure of learning and memory. This task

Figure 4.5 Photograph of the Morris water maze. Image thanks to Drs Anne-Marie McGauran and Deirdre Harvey.

is excellent as it is easy to set up and relatively cheap. Further, the water eliminates any odour cues that may linger in other dry mazes, thus controlling for rodents' keen sense of smell.

Another maze used to test memory was originally described by David Olton and Robert Samuelson in 1976 and is known as the *radial arm maze*. This maze has a large circular central platform, extending from which are 8 equally spaced arms. At the end of each arm, a small food pellet is placed. A hungry rat or mouse is initially placed on the central platform, and it runs down each of the arms to collect the food reward. Once the animal has retrieved the food, it tends not to go down that arm again. This is an excellent test of working or short-term memory; the animal must store the visited arms in memory on a particular trial. To test spatial reference memory, food may be placed in just some of the arms, and the animal has to remember the location of the food relative to cues in the environment. However, care must be taken to ensure odour trails are not left behind, and the maze must be cleaned between trials to prevent animals from following the scent left from previous trials.

A variant on the water maze was developed by Carol Barnes. This maze consists of a large circular platform. Surrounding the circle, on the periphery, are 18 holes. All the holes are blocked apart from one, which leads into a dark tunnel and back to the animal's cage. The circular platform is bathed in a bright light. Rodents do not like bright exposed spaces and prefer to find somewhere dark. Therefore, once placed into the centre of the arena, the animal tries to escape the light. Once it finds the appropriate hole leading back to its cage, it has to remember this location. Similar to their behaviour in the water maze, animals tend to remember the correct location relative to landmarks in the environment; however, the task may be used to test other aspects of memory and other behaviours.

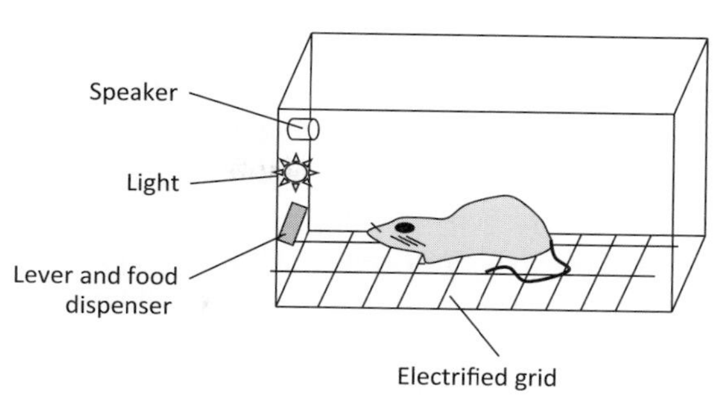

Figure 4.6 Graphical representation of a rat in a Skinner Box.

Conditioning Chamber: Operant and Classical Conditioning

Mazes make use of the natural inclination of animals, particularly rodents, to explore and to avoid bright, exposed places. However, as we saw above, Skinner did not think the maze was particularly useful for examining basic learning processes. He developed a chamber now called the Skinner box, where the learning behaviour of many animals, mostly pigeons or rodents, could be examined under very strict conditions. The basic chamber (Figure 4.6), usually soundproof and lightproof, consists of a cage with a lever that is connected to a food dispenser. The floor of the chamber can used to deliver small shocks to the animals' feet. Included within the chamber are speakers and a light source. Skinner originally showed that pigeons' behaviour could be reinforced through the presentation of a reward or eliminated through the use of punishment (e.g. withdrawal of the reward). When an animal pecked at a disk or pressed a lever (in the case of a rat), food was given. This, in turn, would lead to an increased behavioural response (disk pecking or lever pressing). This learning is referred to as *operant conditioning*. The chamber can also be used for classical conditioning. For example, if a small shock is applied to the feet of the animal through the electrified grid on the floor, the animal will produce a startle response. If every time the shock is delivered a light also comes on or a tone is sounded, after a number of such pairings, presentation of the light or tone alone will produce a startle response.

Tests of Stress, Anxiety and Depression

Other behavioural repertoires can be examined using other techniques. For example, one of the primary tests that is used to assess stress and anxiety in animals is called the *elevated plus maze*. This apparatus consists of four arms in the shape of a cross, each arm emanating from a central square. The width of each arm depends on the animal being tested. For mice, a typical arm would be 25–30 cm in length and 5 cm wide. Two of the four arms are open and two arms are enclosed. The enclosed arms have high sides (again with mice, approximately 15 cm in height). Arms of the same type are typically placed opposite to each other. Each animal is placed in the central square and the animal's behaviour is recorded over a set period of time.

Another test that takes advantage of rodents' propensity to avoid open spaces is the *open field test*. Here animals are placed in a large arena and are allowed a set time to explore the environment. The less anxious animal will readily explore the entire arena; the more anxious one will stay at the side and will be less mobile.

More invasive tests that are thought to measure depressive-like behaviours include the forced swim test and the tail suspension test.

In the *forced swim test* animals (mice or rats) are placed in a container of water from which they are unable to escape. During a 5- to 6-minute test, the time spent immobile is taken as a measure of behavioural despair. For mice the *tail suspension test* can be used. Here the animal is attached to a support by the tail, and the animal is raised above a table. The time spent immobile is an indicator of the animals 'giving up' and depressive-like behaviour.

Many animals react to a strong stimulus such as a loud sound. However, if a weaker stimulus is presented before the strong stimulus, the animal's response is much reduced compared to the response if given the strong stimulus alone. This phenomenon is referred to as *prepulse inhibition* and it is a procedure often used to examine sensorimotor reactivity and the ability of an animal to integrate and inhibit sensory information.

Social Interaction Tests

Many animals enjoy the company of others and often huddle together for comfort and warmth. Furthermore, animals such as mice, rats, cats and dogs are quite sociable, often making contact with each other by way of sniffing and other means. In the laboratory, social interaction can also be tested by using a three-chambered apparatus with the chambers separated by clear Plexiglas. The clear dividing wall contains a small hole to allow animals to freely move between chambers. The experimenter places one mouse, contained in a small wire mesh container, into the left side chamber. The experimenter then places a second mouse, also contained in a small wire mesh container, into the right side chamber. The third mouse is placed in the middle chamber. This mouse is free to go from one chamber to another without restriction. The experimenter records the amount of times the mouse makes nose contact with the other two mice (or the time spent at each chamber). Greater interaction suggests greater sociability. To test the animal's reaction to a familiar mouse or novel one, the experimenter may replace one of the mice in a wire container with a complete stranger. Then the experimenter records the number of nose contacts made with the new animal, compared to the familiar one.

Summary

Scientists cannot go forward without looking back. How neuroscience and psychology have developed over the last 100 years has profoundly influenced the direction that behavioural neuroscience now takes. Knowledge is slowly built on the hard work of previous generations; behavioural neuroscience is no different in this regard. Traditionally, scientists have stuck to their own discipline, but there has been an increasing trend over the last few decades to work collaboratively with others. This has borne fruit, and has been particularly important for behavioural neuroscience. Although we can look at individual neurons and brain regions, we would also like to understand their functions and how they relate to behaviour. But behaviour is complicated, and we have to understand the rules and principles that govern behaviour. This chapter tries to set the context in which behavioural neuroscience finds itself and outline the general principles and the techniques used to unravel these rules.

References

Small, W.S. (1901). Experimental study of the mental processes of the rat. II. *The American Journal of Psychology*, 12 (2): 206–39.

Tolman, E.C. (1948). Cognitive maps in rats and men. *Psychol Rev*. 55(4):189–208.

Tolman, E.C., and Honzik C.H. (1930). Introduction and removal of reward, and maze performance in rats. *University of California Publications in Psychology*, 19: 267.

Tolman, E.C., Ritchie B.F., and Kalish D. (1946). Studies in spatial learning. I. Orientation and short-cut. *Journal of Experimental Psychology*, 36:17.

5 Habituation and Sensitisation in the *Aplysia*

Habituation

Introduction and Background

We all tend to get bored; something that we found interesting and novel may, after awhile, no longer hold the same interest as before. Essentially we get used to things; objects, people or situations, we habituate to them all. This is good; we cannot be constantly on the alert and attentive all the time. Such hypervigilance can have long-term consequences on our health and well-being. A good example of habituation is if a lecturer suddenly and without warning clapped his or her hands in the middle of class. Immediately students would jump and heart rates would increase; the students would generally become very attentive to the stimulus. However, if the lecturer were to continue to clap, students would no longer jump, and attention levels and heart rates would return to normal levels. The students would have essentially habituated to the stimulus.

A formal definition suggests that habituation is a form of learning in which a response to a stimulus weakens with repeated stimuli presentation. Habituation is thought to be the simplest form of learning. There are at least two ways in which a response can recover from the habituation effect. The first is if the stimulus is withheld for a period of time. For example, after clapping, the lecturer might return to the talk, then after some time might clap again. The students in this situation will again become attentive to the stimulus. A second method to recover an habituated response is to present a new and perhaps a stronger stimulus.

The first attempt to study the neural basis of habituation was made by the British physiologist Sir Charles Sherrington. Sherrington noticed that stimulation of a cat's skin produced a withdrawal response in the animal's limb, but that over time, with continued stimulation, this reflex got weaker. As early as 1908, Sherrington suggested that such a change in behaviour was due to a change at central synapses (a term he had coined) rather than because of muscle fatigue or other causes. However, trying to localise where such a neural change occurred was considered to be next to impossible.

Many scientists turned to simple organisms to study physiological responses, including the electrical properties of neurons in the giant squid (Chapter 1), and they looked at the fruit fly *Drosophila* in the examination of genetics. The study of behaviour, particular learning and memory, at the neural level was considered a step too far. Behaviour was thought to be too complicated, and besides how could you translate what was found in a 'lower' animals to humans? In the 1950s and 1960s, through the work of B. F. Skinner (see Chapter 4), it was shown that behaviour could in fact be reduced to constituent parts, and furthermore behaviour could be controlled. In addition, through the work of ethologists such as Konrad Lorenz, Niko Tinbergen and Karl von Frisch it was found that humans share many behaviours with simpler animals. A change in attitude was emerging – perhaps behaviours could be examined. It was around this time that Eric Kandel felt that even learning and memory could be studied at the neural level, if he could just find a suitable animal.

Aplysia and Habituation

After much searching, Kandel settled on a small marine animal called the *Aplysia californica* (Figure 5.1a). The *Aplysia* is a sea slug that is found in many parts of the world, including the Atlantic, Pacific and Indian oceans. *Aplysia* tend to feed off algae and typically reach the size of 30 to 40 cm. The word *Aplysia* comes from the Greek word meaning 'sea-hare', because its tentacles are thought to closely resemble the ears of a hare. The *Aplysia* has a number of gross anatomical features (see Figure 5.1b) that help the animal to survive. The siphon, for example, is a fleshy tube that expels waste and seawater, and is also used to circulate water over the gill. The gill is a respiratory organ and is used for breathing purposes. The mantle is a covering that protects the gill and can move in and out, somewhat like wings of a bird, which helps the animal to move. The mantle can also fold over the gill and siphon to protect these structures. Although fairly primitive, the *Aplysia* has a number of behavioural repertoires, including defensive responses such as the expulsion of ink in some species (Figure 5.1a) and the secretion of toxins in others.

The choice of this creature may seem very strange, but there are a number of advantages that make *Aplysia* a good working model in which to study the neural basis of learning. The animal has a relatively simple nervous system that contains only about 20,000 central neurons (Kandel et al., 1995; Kandel and Hawkins, 1992). This compares to the millions of neurons contained in the mammalian brain. Although the *Aplysia* does not have a brain, the nervous system is composed of about 10 different clusters of neurons called *ganglia*, and each cluster comprises approximately 2,000 cells. In addition to this, many of the neurons are quite large and easily identifiable (Beggs et al., 1999). The cell body of a neuron, for example, can be 1 mm in diameter, making it visible to the naked eye and relatively easy

(a)

(b)

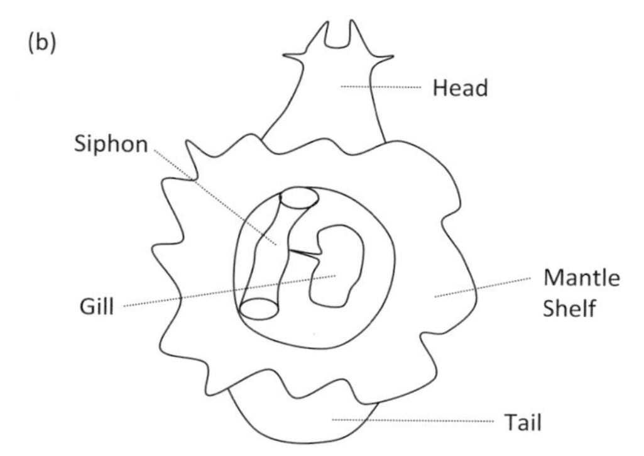

Figure 5.1 (a) *Aplysia californica* emitting an ink cloud. Image taken from G. Anderson, Marine Science, Santa Barbara City College, http://creativecommons.org/licenses/by/4.0. (b) Representation of the anatomy and key internal structures of the *Aplysia*. For colour version of part (a), please refer to the plate section.

for a researcher to place an electrode and to stimulate and record its electrical properties.

Behaviour of *Aplysia*

The *Aplysia* can move, feed and defend itself, but can this small, simple creature without a brain also learn? Can it perform the simplest form of learning – habituation? Most of us, as children, may have come across common garden snails, and many of us have even gone as far as picking them up and letting them slide across our hands. How many of us were fascinated by their protruding tentacles and observed what would happen if we were to gently touch one? Upon the slightest touch, the snail would immediately withdraw those tentacles. Imagine if we were to continue to touch the tentacle. We might see a withdrawal response, but this time not as strong as the previous one. If we were to continue with this, we would notice that the withdrawal response would become weaker and weaker. The snail has become bored with us. It has habituated to our touch.

Kandel took an *Aplysia* and found that if its tail was touched, it would withdraw it as an act of defence. If the tail continued to be touched, the *Aplysia* would gradually stop contracting it. He realised that this marine creature could habituate; it could learn. In the laboratory Kandel and colleagues went further and examined

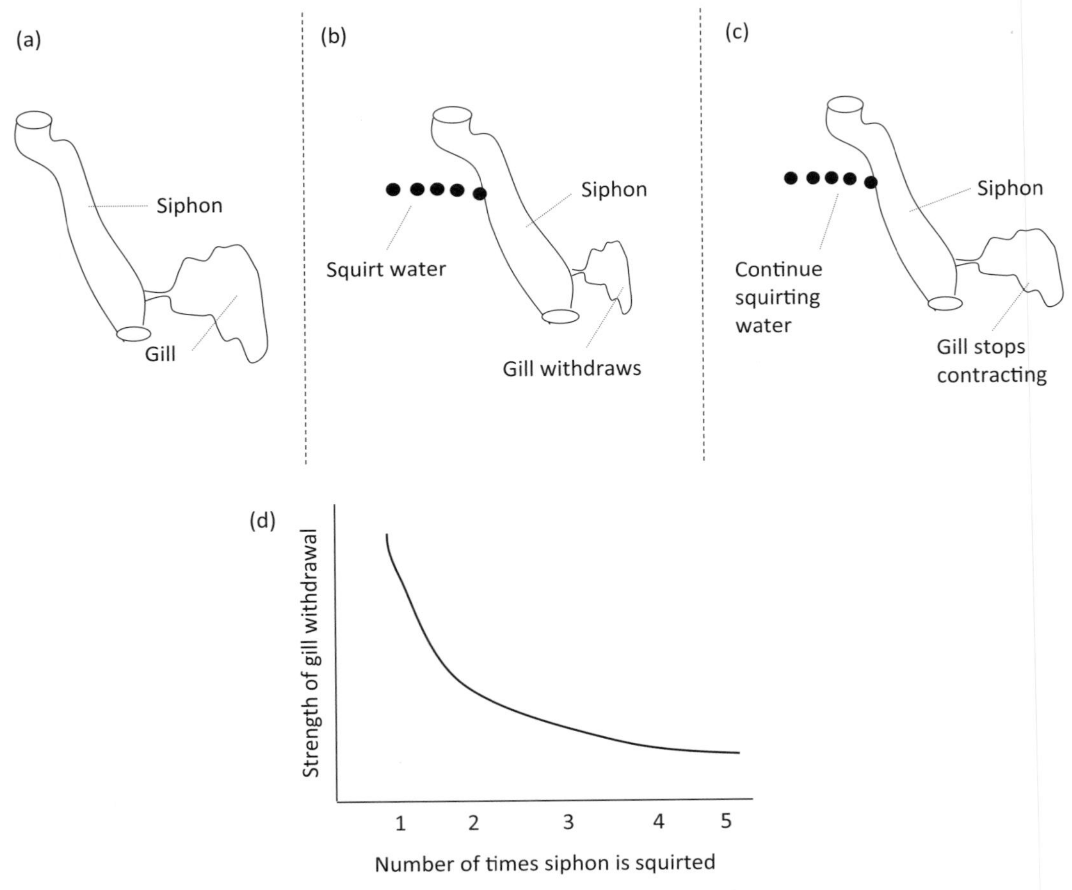

Figure 5.2 Change in the strength of gill withdrawal response in *Aplysia*, following continuous stimulation of the siphon: habituation. See text for details.

two of the *Aplysias'* internal organs; the gill and siphon, which are connected to each other (Figure 5.2a). He found that if he stimulated the siphon, by squirting water on it, it would immediately contract (Figure 5.2b). However, if water was continuously squirted at the siphon, the withdrawal response would gradually decrease in magnitude (Figure 5.2c). Here Kandel was able to demonstrate that the gill response had habituated to the stimulation of the siphon. Figure 5.2d shows a typical habituation curve, whereby continuous stimulation leads to a decrease in the gill withdrawal response.

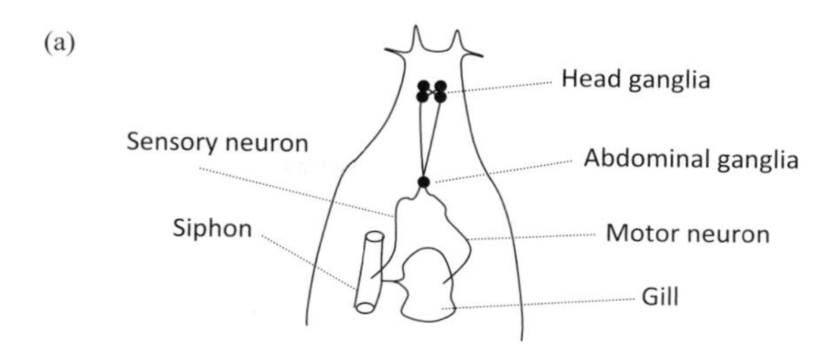

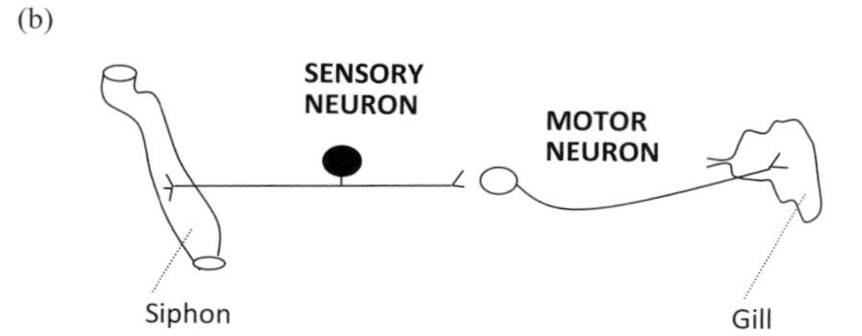

Figure 5.3 The neural connectivity of the *Aplysia*.

Circuits and Electrophysiological Mechanisms Underlying Habituation in *Aplysia*

Having found that this invertebrate could habituate, the next question that Kandel and his colleagues wanted to know was where, in the neural circuitry, did this form of learning take place? As the researchers were able to identify the various cells in the *Aplysia*, they were then able to give them names and return to the same one each time. Furthermore, the circuitry was exactly the same for each *Aplysia* they examined. With this knowledge, they were able to identify the various components and exact circuitry. The circuit was composed of 24 sensory neurons that innervated the siphon and in turn made direct connections to 6 gill motor neurons. Figure 5.3a shows a simplified wiring diagram whereby information from the siphon travels along sensory neurons, enters the abdominal ganglion and then continues to motor neurons. The motor neurons then make contact with the gill. Figure 5.3b simplifies this circuit further.

After the researchers had identified the circuit involved with the withdrawal response, the question remained: Where, along this circuit, did habituation take place? There were three possible locations: (1) at the start of the circuit, where the siphon connected with the sensory neuron; (2) the end of the circuit where the motor neuron connected with the gill; or (3) in between both, where the sensory neuron connected with the motor neuron. Kandel and his colleagues tested each of these options thoroughly. Did habituation occur at the input stage? To test this, the researchers squirted the siphon with water and inserted a fine wire electrode into the axon hillock of the sensory neuron (Figure 5.4a). The authors reasoned that if habituation was to occur at this junction, then any action potential that was recorded should diminish as stimulation continued. Figure 5.4a, right, clearly shows no decrease in action potential response, therefore the researchers were able to rule out the idea of habituation occurring at the input stage. Next, Kandel and colleagues tested whether habituation occurred at the output stage. This time they inserted a stimulating electrode into the motor neuron and recorded the response from gill muscle (Figure 5.4b). Again, they reasoned that, should habituation take place at this junction, as the stimulation of the motor neuron continued this would be accompanied by a gradual decrease in gill contractions. As can be observed in Figure 5.4b, right, no decrease in muscle contractions was observed. Habituation occurring at the output junction was thus ruled out. Finally, the researchers looked at the intermediate stage and stimulated the axon hillock of the sensory neuron and recorded excitatory postsynaptic potentials (EPSPs, see Chapter 1) from the motor neuron (Figure 5.4c).

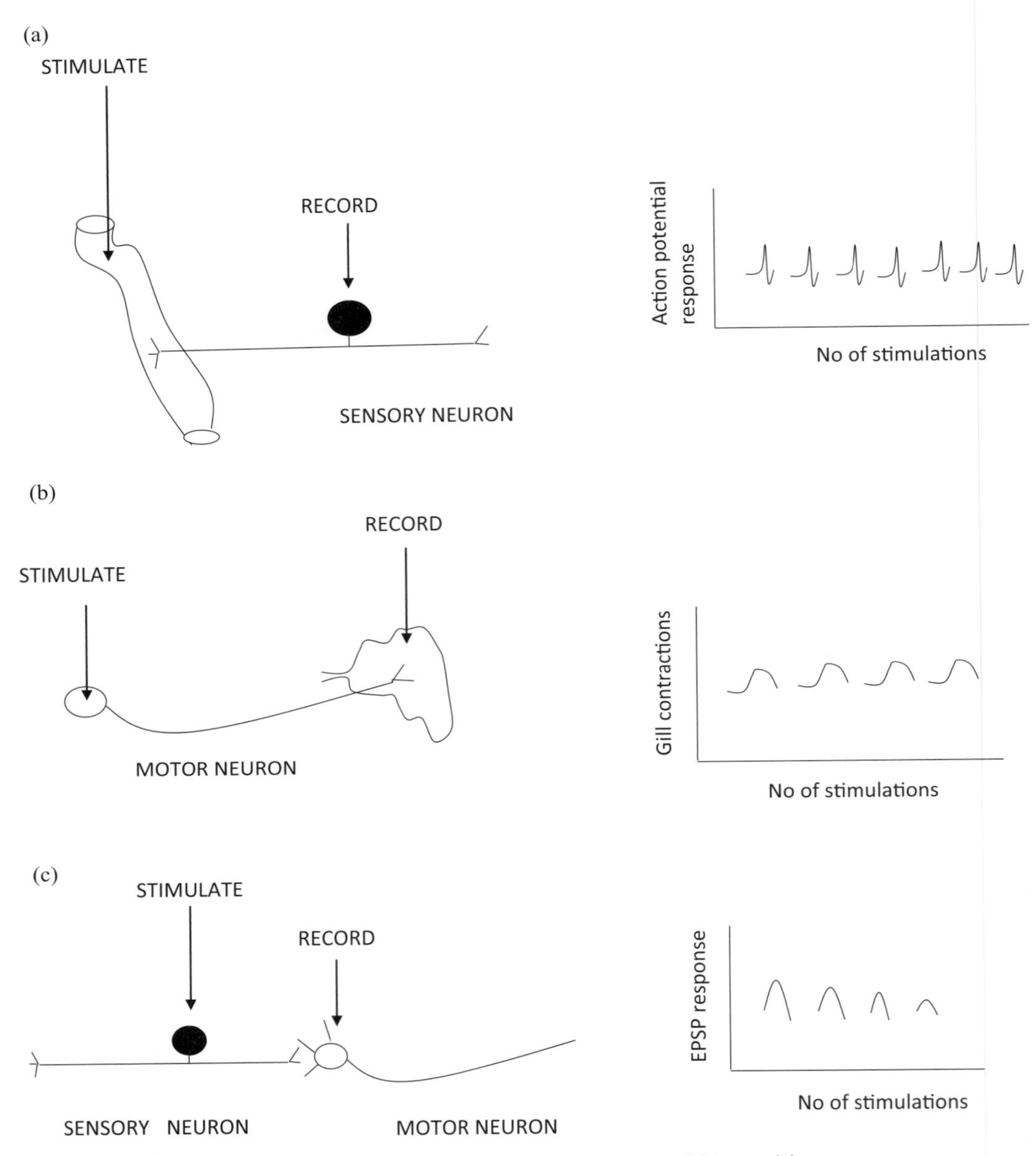

Figure 5.4 Different responses to stimulation along the neural circuit at (a) input, (b) output or (c) intermediate stage reveal where habituation takes place in *Aplysia*.

Here at last they found a change. With continued stimulation of the sensory neuron, the evoked EPSP response in the motor neuron diminished. Therefore, they realised that the critical site for habituation of the gill response to siphon stimulation occurs at the synapse between the sensory and motor neurons.

Biochemical and Molecular Mechanisms Underlying Habituation

After discovering the circuitry and the exact location where habituation took place, the next step was to identify the biochemical mechanisms underlying habituation. In 1978 Kandel, Castellucci and colleagues suggested that the decrease in the evoked EPSP response at the sensory–motor junction came about as a result of the decrease in neurotransmission between the two neurons. As we have seen in Chapter 1 during normal transmission, an action potential arrives at the synaptic terminal, causing calcium channels to open. Calcium then rushes into the presynaptic terminal (of a sensory neuron, in this case) and causes some synaptic vesicles filled with the neurotransmitter to move towards the membrane. Calcium also allows other vesicles already docked at the membrane to release neurotransmitters into the synaptic cleft by a process termed *exocytosis*. The neurotransmitter then crosses the synaptic cleft and attaches onto receptors on the postsynaptic neuron (motor neuron). This results in the opening of channels, allowing the entry of ions into the postsynaptic neuron, which in turn causes EPSPs (Figure 5.5a) and a large gill withdrawal response. However, during habituation, the calcium channels in the presynaptic terminal gradually become inactive (Figure 5.5b). This has a major consequence. The inactivation results in less calcium entering the cell with each consecutive action potential.

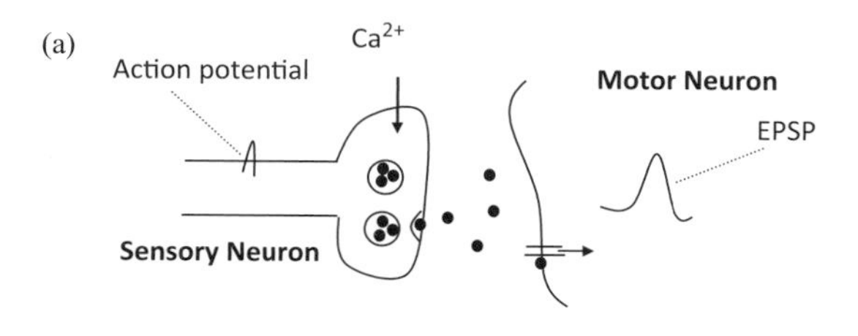

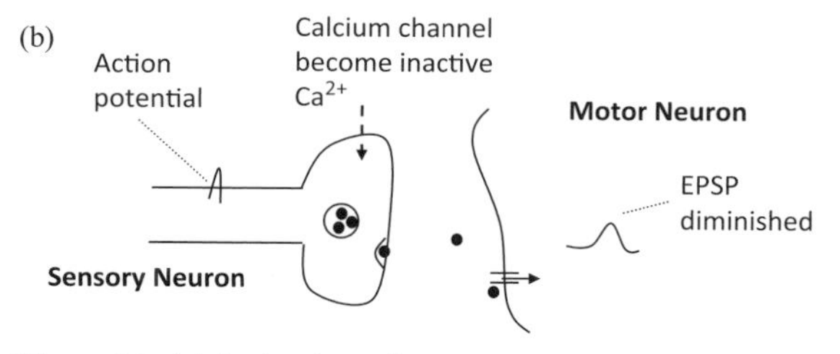

Figure 5.5 (a) Activation of a sensory neuron normally produces an EPSP response in the motor neuron.
(b) Habituation at the molecular level is caused by the inactivation of calcium channels in the sensory neuron, causing a decrease in EPSP response in the motor neuron.

Therefore, less neurotransmitter is released into the synaptic cleft, and less neurotransmitter crosses the cleft, resulting in a decrease in EPSP magnitude in the postsynaptic motor neuron. This ultimately causes a diminution of gill withdrawal response.

Short- and Long-Term Habituation

Although habituation is relatively simple, it is probably one of the most widespread of all forms of learning, and is observed in simple invertebrates right through to humans. All, at times, must learn to ignore stimuli, allowing attention to be directed elsewhere. Another interesting aspect of habituation, and one that is very relevant to us as humans, is that it can give rise to both long-term and short-term memory. Depending on the training regime, memories can remain with us for a long time or can simply fade away. Many researchers (including me, Commins et al., 2003) have shown that our memories are worse when we are subjected to mass training (that is, multiple training trials over a short period of time) compared to spaced training, which uses multiple trials but allows time between each trial. This finding has major implications for our education system, whereby multiple breaks during each learning session are often strongly recommended.

A similar effect can be seen with habituation and the *Aplysia*. After a single training session using 10 stimuli, a short-term habituation effect is observed. The memory for the stimulus is short-lived and a partial recovery of the response is observed within an hour, with full recovery after a day. However, with 4 training sessions spaced over time (with 10 stimuli per session), habituation can last for at least three weeks (Castellucci et al., 1978; Bailey and Chen, 1983, 1988; Carew et al., 1979). Therefore, memory for the stimuli is long-lived. Using knowledge of the location and biochemistry of habituation allows researchers to examine both short- and long-term memory. How does this occur and what are the differences between the memory types?

Short-term habituation was previously thought not to involve structural changes, but some type of morphological changes in the synapse is now considered. For example, according to Bailey and Chen (1988) short-term habituation seems to lead to a depletion of synaptic vesicles in the terminal of the sensory neuron immediately beside the active zone (an area on the membrane where vesicles dock). These authors report that the ratio of the vesicles ready to release their transmitters to the total number of vesicles in the terminal decreases from 28% for controls to 11.5% for habituated synapses (Figure 5.6a). However, long-term habituation leads to even more structural and

(a)

Short-term Habituation

Control synapse

Habituated synapse

Synaptic vesicles

Neurotransmitter

Docked vesicle

Active zone

Reduced number of vesicles
at active zones

(b)

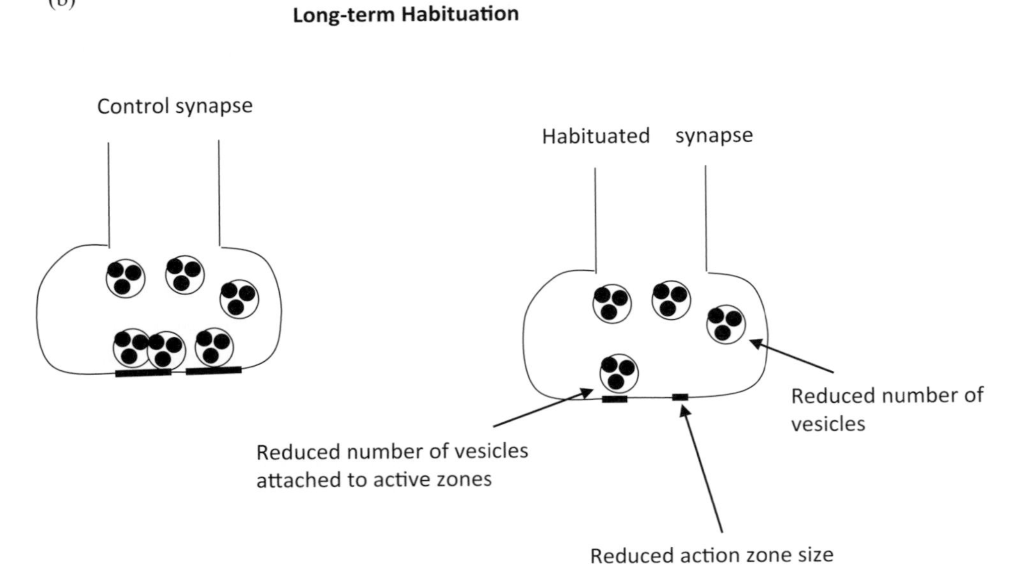

Figure 5.6 (a) Short-term and (b) long-term changes occurring at the habituated synapse compared to the control nonhabituated synapse.

morphological changes to the synapse. One structural change that seems to occur is at the active zones themselves. For example, Bailey and Chen (1983) examined the number of active zones in control and habituated synapses of sensory neurons. In control animals 41% of the presynaptic terminals had active zones compared to 12% in habituated animals. The authors also examined the total surface membrane area of the active zone in both groups and found that in the control animals the mean area of active zones was 0.195 μm^2 with an average of 13 vesicles attached. This compared to habituated animals, which had a mean active zone area of 0.13 μm^2 with an average of 6 vesicles attached. These results indicate that the number, size and vesicle complement have all decreased as a result of long-term habituation (Figure 5.6b). Therefore long-term memory brings about more radical structural changes to the synapse compared to short-term memory.

Sensitisation

Another very simple form of learning is sensitisation. In general, sensitisation occurs when we become more responsive to a situation or to the environment, usually following some disturbing occurrence. Here is a good example of sensitisation: You are walking along a well-lit street at night. You hear dogs barking, cars going past, people laughing, etc. Suddenly there is a power cut, a complete blackout. Following this, more dogs bark, other cars go by. How do you react? You now become much more edgy and jumpy. The stimuli that previously did not elicit a response now do, solely as a result of some occurrence. You have essentially become more sensitive to the surrounding environment. A more formal definition suggests that sensitisation is a process by which the response to a stimulus increases with repeated presentation of a stimulus following some intervention.

This behaviour is often very necessary; at times we do need to be alert to our surroundings. Likewise, there are times when the *Aplysia* needs to be alert to its environment. Its defence reflexes, including the gill withdrawal response, become sharper in preparation for an attack. In the laboratory, sensitisation can be studied more formally using the gill withdrawal response. As previously described, when the researcher squirts water onto the siphon of the *Aplysia*, the creature withdraws its gill (Figure 5.7a). Let's say this continues for a few trials. Next, the researcher applies a shock to the tail or some other region of the *Aplysia* (Figure 5.7b). Now, the researcher returns to squirting water at the siphon. What we now observe is that the gill withdrawal response is much stronger than before the shock (Figure 5.7c). The animal has become hypersensitive: a stimulus that had previously elicited a small response now produces a much stronger one, solely as a result of some sort of intervention – in this case, a shock to the tail. Figure 5.7d demonstrates this change in the response of the gill contractions.

Circuits, Electrophysiological and Biochemical Mechanisms Underlying Sensitisation

Interestingly, Kandel and colleagues found that sensitisation occurs at the same location as habituation, the junction between the sensory neuron and the motor neuron. As described previously, information from the *Aplysia*'s siphon travels along sensory neurons and then makes contact with motor neurons, which in turn causes the gill to contract. Sensory neurons also receive information from the tail or head. This information then travels along interneurons (neurons that are neither sensory nor motor), and finally makes contact at the sensory-motor synapse (see Figure 5.8). There are two groups of interneurons, depending upon where the shock is applied, L28 and L29. L28 is activated by shocks to the head; L29 is activated by shocks to the tail (Hawkins et al., 1981). The cell bodies of both groups of interneurons are located in the abdominal ganglion (Figure 5.3 and Kandel and Schwartz, 1982). Critically, the interneurons make contact with

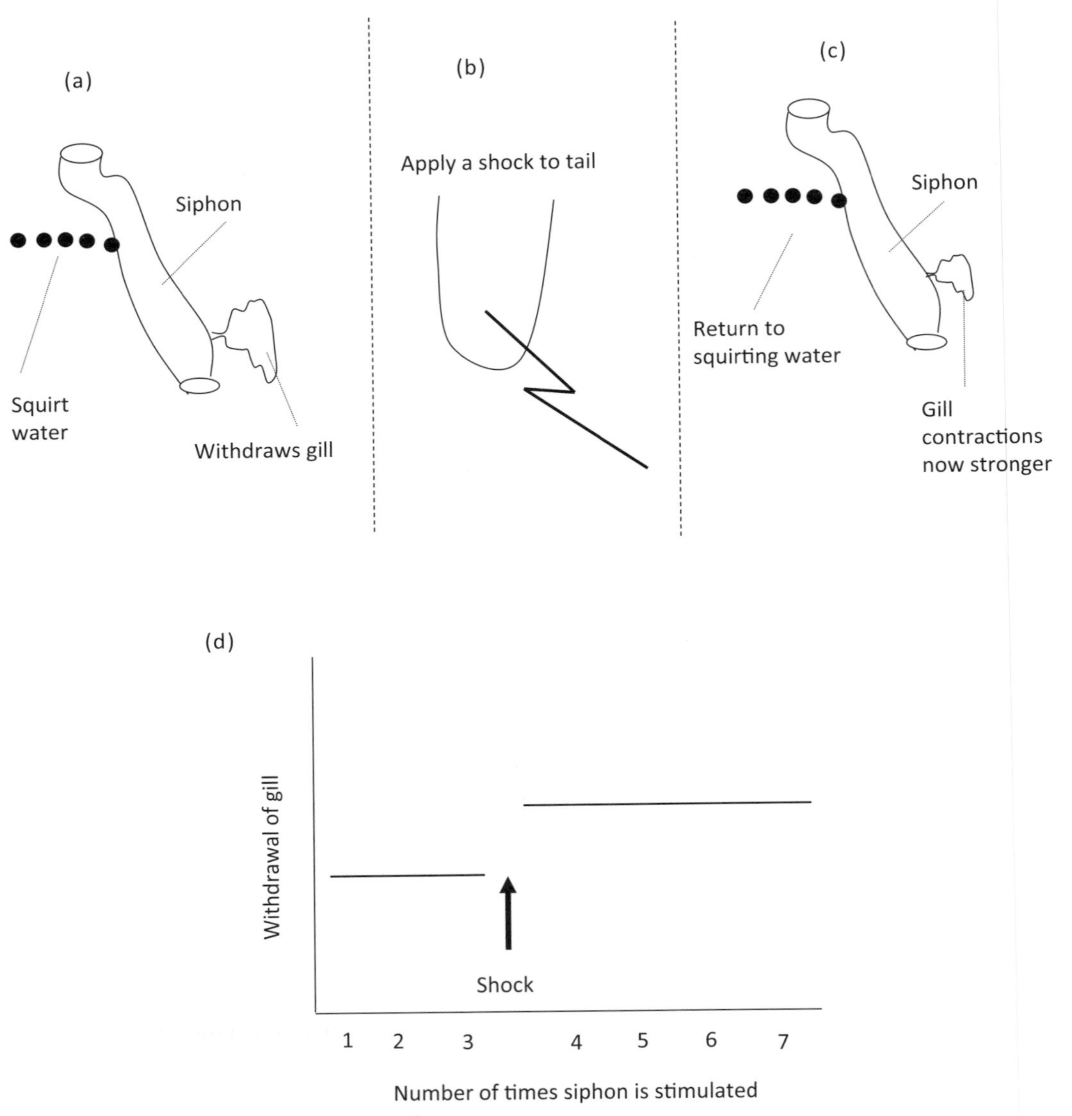

Figure 5.7 Change in the strength of gill withdrawal response following stimulation of the siphon and with a shock intervention sensitisation. See text for details.

the terminals of the sensory neuron, very close to the site where the sensory neuron releases its neurotransmitter (square box in Figure 5.8; Bailey et al., 1981). Both interneurons (L28 and L29) release serotonin, the neurotransmitter that seems to be critical for sensitisation (Brunelli et al., 1976).

The following section goes step by step through the processes that allow for sensitisation to occur.

Pre-shock

Under normal circumstances (pre-shock) a squirt of water to the siphon leads to a generation of an action potential in the sensory neuron. This action potential reaches the synaptic terminal, depolarising the membrane and causing an influx of calcium and a subsequent release of glutamate. The neurotransmitter crosses the synaptic cleft and attaches to receptors which, in turn, open channels on the postsynaptic membrane, allowing ions to enter the motor neuron. This generates an EPSP, resulting in the eventual contraction of the gill (Figure 5.9).

During the Shock

The shock at the *Aplysia*'s tail activates sensory neurons, which in turn leads to the generation of action potentials in the interneurons (L29). This action potential reaches the synaptic terminal, depolarising the membrane and causing an influx of calcium and subsequent release of serotonin (Figure 5.10a, 1). Seotonin crosses the cleft and attaches to a G-protein-coupled receptor on the sensory neuron, activating a second messenger system (2). G proteins are guanine nucleotide-binding proteins.

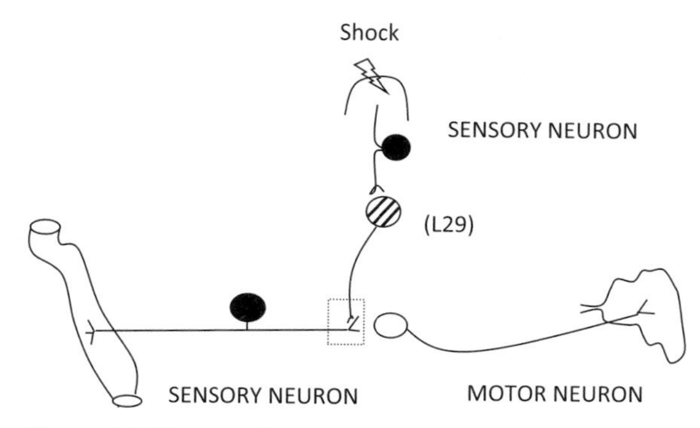

Figure 5.8 The neural connectivity of the *Aplysia* for sensitisation.

Figure 5.10b shows in more detail the action of serotonin and the subsequent (quite complicated) process involving multiple molecules of the second messenger system. Serotonin (3) activates receptors that are coupled to G proteins. Stimulation of G proteins (4) leads to production of cAMP (cyclic adenosine monophosphate). Cyclic AMP is produced from ATP (adenosine triphosphate) by the converting enzyme adenylyl cyclase (5, 6). Cyclic AMP then activates another protein, cAMP-dependent protein kinase A (PKA) (7) from its inactive state. PKA in turn phosphorylates (adds phosphate groups to) other proteins, causing different effects. One of the key actions of PKA is on the potassium channel (8a). PKA changes the conformation of the channel, closing it slightly, thereby decreasing the potassium current (see Chapter 1). The net effect of this is to prolong any further action potential that may come along this neuron. Prolonging an action potential causes a greater influx of calcium into the presynaptic terminal. This, in turn, would *increase* synaptic vesicle mobilisation and subsequent neurotransmitter release. A second effect of PKA is on calcium channels, opening these further. Again this would have the effect of increasing synaptic vesicle mobilisation and neurotransmitter release (8b). A third effect of PKA is to directly mobilise vesicles towards the active zones to be released (8c). With the modification of these channels, the sensory neuron is now 'primed'.

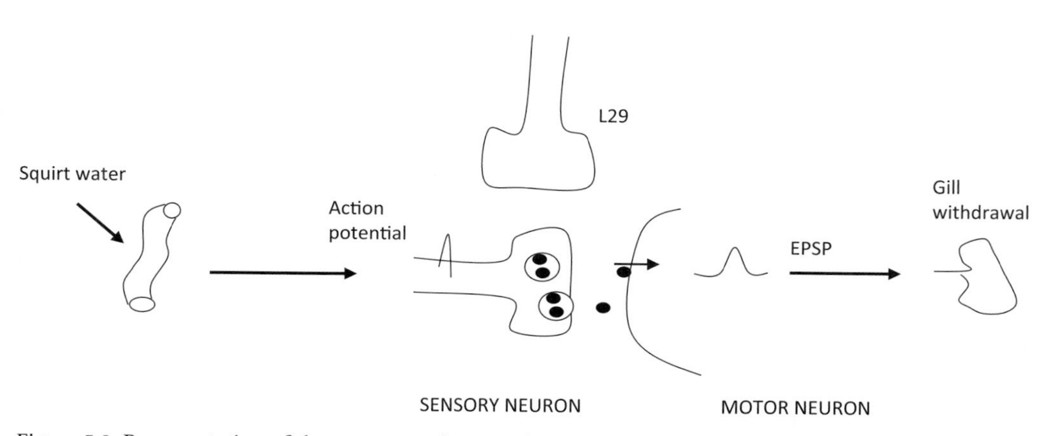

Figure 5.9 Representation of the sequence of events that occur at the pre-shock (baseline) phase of sensitisation.

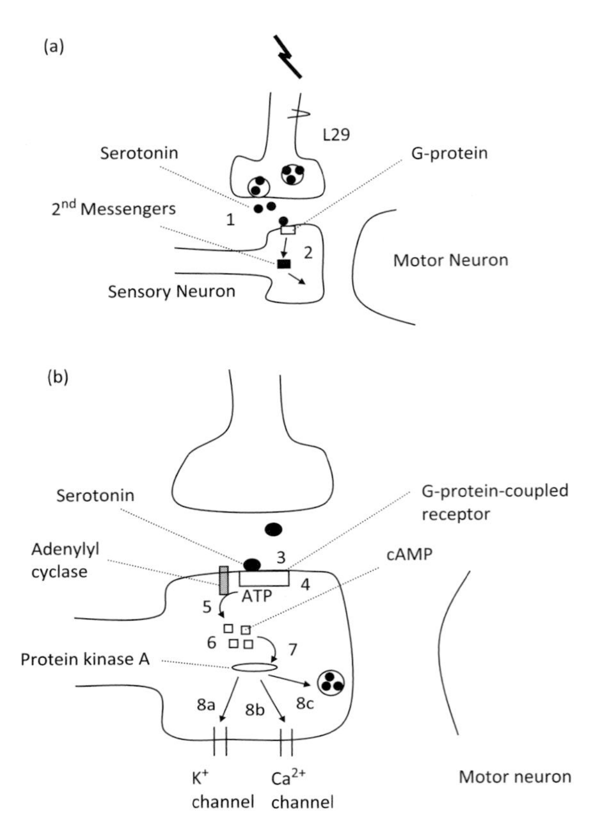

Figure 5.10 The sequence of events that occur during the shock phase of sensitisation. See text for details.

Note too that serotonin is thought to act through another set of receptors (not shown) that leads to an increase of diacylglycerol (DAG). DAG in turn activates a different protein, protein kinase C (PKC), which also affects the potassium channel, prolonging the action potential further.

Post-shock

The shock causes a change in the channels of the sensory neuron. Now, if you return to squirting water at the *Aplysia*'s siphon, this leads to a generation of an action potential in the sensory neuron. This action potential reaches the synaptic terminal, depolarising the membrane. However, due to the fact that the K^+ channel has been modified and the action potential is prolonged, as described earlier, a greater amount of calcium enters the presynaptic cell. In addition to this, calcium channels themselves are modified, allowing even more calcium to enter the cell. With so much calcium flooding the terminal, more synaptic vesicles are mobilised and even more neurotransmitter is released. The increased glutamate release crosses the cleft and attaches to receptors, which in turn open channels on the postsynaptic membrane, allowing ions to enter the motor neuron. This generates a larger EPSP, resulting in even greater contractions of the gill. The response to the squirting of the water has increased: the animal has become hypersensitive (Figure 5.11).

Long- and Short-Term Sensitisation

As with habituation, sensitisation can be divided into long- and short-term effects and therefore is also a useful paradigm in understanding the mechanisms of long- and short-term memory in general. A single shock stimulus produces an increase in the gill-withdrawal reflex that can last several hours. With four consecutive shock stimuli, the memory can last up to 24 hours. With 16 spaced stimuli (4 per day for 4 days), the withdrawal reflex can last up to several weeks (Kandel and Schwartz, 1982). Short- and long-term sensitation have a number of properties in common: both are associated with changes at the sensory-motor synaptic junction. Both involve an increase in synaptic strength due to the increase in neurotransmitter released from the sensory neuron. Both involve serotonin as the modulatory transmitter (Kandel et al., 1995). The biochemical changes that underlie short-term sensitisation follow the sequence described earlier.

Long-term sensitisation, however, requires the synthesis of new proteins, which leads to morphological and structural changes to the sensory neuron that are more permanent. This is more complicated and is still under much investigation. In summary, repeated stimulation of G-proteins by serotonin leads to production of cAMP, which in turn activates PKA from its inactive state. But rather than having a direct action on channels

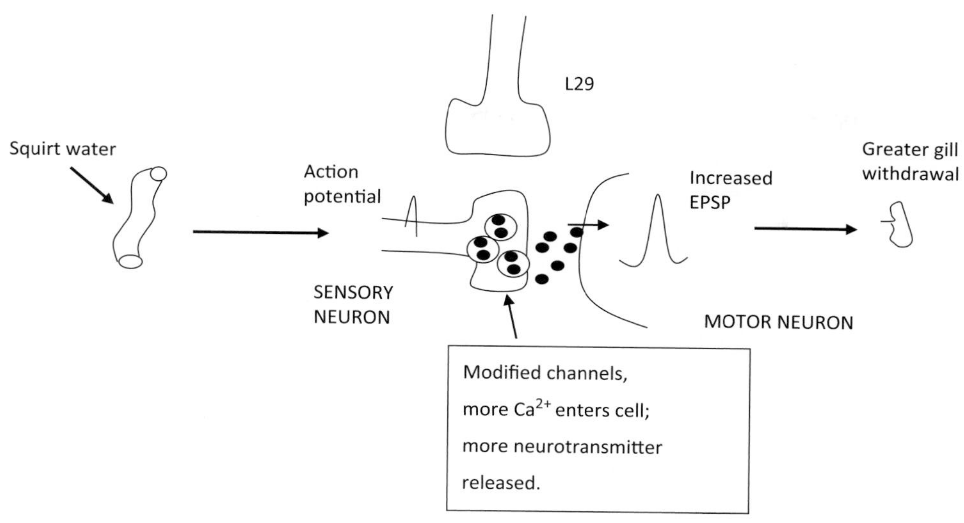

Figure 5.11 Sequence of events that occur at the post-shock phase of sensitisation.

as described earlier, with long-term sensitisation PKA translocates to the nucleus of the cell, where it phosphorylates proteins such as cyclic AMP-response element binding protein (CREB; Kaang et al., 1993). This, in turn, activates a number of genes, which encode a number of different key memory proteins. One protein (ubiquitin) is responsible for the persistent activation of PKA (which persistently modifies the K^+ and Ca^{2+} channels mentioned in the previous section). The second type is important for structural and morphological changes to the synapse (Kandel et al., 1995), similar to that described for long-term habituation. For example, Bailey and Chen (1983) found that in control animals 41% of sensory neurons had active zones compared to 65% in sensitised animals. Further, in control animals the mean surface area of these active zones was 0.195 μm^2, which was occupied by a mean of 13 vesicles. This compares to a mean area of 0.35 μm^2 occupied by 22 vesicles in sensitised animals. These results indicate that the number, size and vesicle complement have all increased as a result of long-term sensitisation.

Other newly synthesised proteins break down neuronal cell adhesion molecules, thereby allowing reconstruction of the axon and the formation of new sensory connections (Bailey et al., 1992; Bailey and Chen, 1983) via transforming growth factor beta (TGF-beta; Zhang et al., 1997) and Aplysia tolloid/BMP-1-like protein (apTBL-1; Liu et al., 1997). Such details are currently being worked out.

Summary

It is hard to imagine that complicated cognitive processes such as learning and memory could ever be understood in biological terms. Yet, here we have shown that by the careful selection of an appropriate model, the examination of a particular behaviour and the thorough, often painstaking, analysis of neural connections along with their biochemistry, understanding can be achieved. Although both habituation and sensitisation are probably the simplest forms of learning, they are universal behaviours and are essential for survival. Despite the simplicity of the behaviours, we have seen that the molecular mechanisms underlying such behaviours can get quite complex. But if scientists can figure out these mechanisms, then more complicated forms of learning and indeed other behaviours may also be understood.

Questions and Topics Under Current Investigation

- How do the neural mechanisms revealed here compare to those of other invertebrates, including other molluscs and insects?
- Can the processes described in *Aplysia* be directly applied to the mammalian brain?
- What are the equivalent circuits and processes underlying habituation and sensitisation in the mammalian brain?
- What are the exact signalling molecules that lead to structural remodelling during long-term memory?

References

Bailey, C.H., and Chen, M. (1983). Morphological basis of long-term habituation and sensitization in Aplysia. *Science*, 220(4592), 91–93.

Bailey, C.H., and Chen, M. (1988). Morphological basis of short-term habituation in Aplysia. *Journal of Neuroscience*, 8(7), 2452–59.

Bailey, C.H., Chen, M., Keller, F., and Kandel, E.R. (1992). Serotonin-mediated endocytosis of apCAM: an early step of learning-related synaptic growth in Aplysia. *Science*, 256(5057), 645–49.

Bailey, C.H., Hawkins, R.D., Chen, M.C., and Kandel, E.R. (1981). Interneurons involved in mediation and modulation of gill-withdrawal reflex in Aplysia. IV. Morphological basis of presynaptic facilitation. *Journal of Neurophysiology*, 45(2), 340–60.

Beggs, J.M., Bloom, F.E., Byrne, J.H., Crow, T., LeDoux, J.E., LeBar, K., and Thompson, R.F. (1999). Learning and memory: basic mechanisms. In: *Fundamental Neuroscience* (M.J. Zigmond, F.E. Flood, S.C. Landis, J.L. Roberts, and L.R. Squire eds.), 1411–1454. Academic Press: San Diego.

Brunelli, M., Castellucci, V., and Kandel, E.R. (1976). Synaptic facilitation and behavioral sensitization in Aplysia: possible role of serotonin and cyclic AMP. *Science*, 194(4270), 1178–81.

Carew, T., Castellucci, V.F., and Kandel, E.R. (1979). Sensitization in Aplysia: restoration of transmission in synapses inactivated by long-term habituation. *Science*, 205(4404), 417–419.

Castellucci, V.F., Carew, T.J., and Kandel, E.R. (1978). Cellular analysis of long-term habituation of the gill-withdrawal reflex of *Aplysia californica*. *Science*, 202(4374), 1306–8.

Commins, S., Cunningham, L., Harvey, D., and Walsh, D. (2003). Massed but not spaced training impairs spatial memory. *Behavioral Brain Research*. 139(1–2), 215–23.

Hawkins, R.D., Castellucci, V.F., Kandel, E.R. (1981). Interneurons involved in mediation and modulation of gill-withdrawal reflex in Aplysia. I. Identification and characterization. *Journal of Neurophysiology*, 45(2), 304–14.

Kaang, B.K., Kandel, E.R., Grant, S.G. (1993). Activation of cAMP-responsive genes by stimuli that produce long-term facilitation in Aplysia sensory neurons. *Neuron*, 10(3), 427–35

Kandel, E.R., and Hawkins, R.D. (1992). The biological basis of learning and individuality. *Scientific American*, Sept. 53–60.

Kandel, E.R., and Schwartz, J.H. (1982). Molecular biology of learning: modulation of transmitter release. *Science*, 218(4571), 433–43.

Kandel, E.R., Schwartz, J.H., Jessell, T.M. (1995). *Essentials of neural science and behavior*. Appleton & Lange, Connecticut.

Liu, Q.R., Hattar, S., Endo, S., MacPhee, K., Zhang, H., Cleary, L.J., Byrne, J.H., Eskin, A. (1997). A developmental gene (Tolloid/BMP-1) is regulated in Aplysia neurons by treatments that induce long-term sensitization. *Journal of Neuroscience*, 17(2), 755–64.

Zhang, F., Endo, S., Cleary, L.J., Eskin, A., Byrne, J.H. (1997). Role of transforming growth factor-beta in long-term synaptic facilitation in Aplysia. *Science*, 275(5304), 1318–20.

6 Classical Conditioning in the *Aplysia*

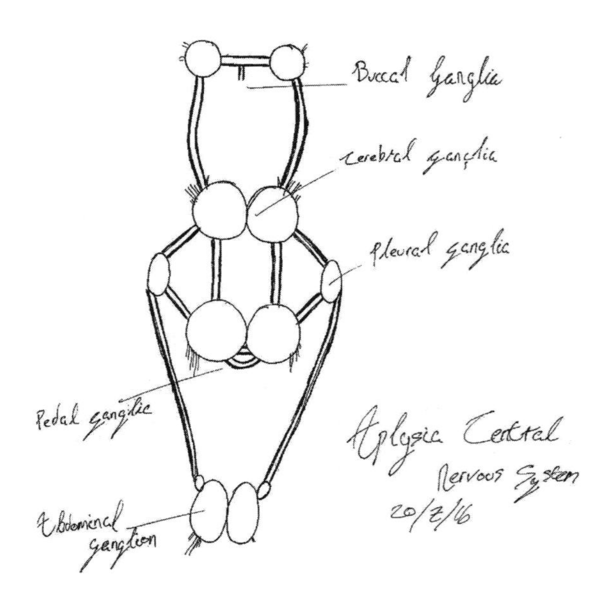

Introduction and Background

Both habituation and sensitisation involve a behavioural change related to a single stimulus and are considered the simplest forms of learning. This type of learning is often referred to as non-associative learning. Other forms of learning are more complicated and involve the association of one thing with another. Two prominent examples of associative learning are instrumental learning and classical conditioning. Instrumental learning, originally described by Thorndike and developed by Skinner, refers to the association of a behaviour with a reward or punishment, which affect the likelihood that the behaviour will recur (see Chapter 4). Classical conditioning refers to learning an association between a stimulus that does not normally produce a behavioural response and a stimulus that does. Classical conditioning was described by Pavlov in 1927 (see Chapter 4), who found that his dogs would not only salivate readily upon the presentation of food but also would salivate when presented with other stimuli that happened to occur around the same time as the food presentation. Anything from the sight of the white lab coat to the ticking of a metronome, the sound of a buzzer to a whistle could all eventually produce a behavioural response. Such stimuli had become associated with the presentation of food. The learning was so sophisticated that even time could become associated with food. For example, if the dog was presented with food every 30 minutes, it would continue to salivate on the half hour, even without the presence of food. The discrimination between stimuli could also be very sensitive. The animal could be trained to respond to and distinguish between a metronome beating at 100 times per minute and one beating at 95 or 105 times. Furthermore, Pavlov showed that animals could respond to music notes that were ascending in tone compared to those that were descending; such abilities form the basis of pattern recognition.

The ability to perform such behaviours should not come as a surprise to us. We associate things with other things all the time. We remember someone's face and associate that with a name, a street with a particular person, a lecturer with a certain topic. However, we have a large brain and a sophisticated nervous system that allows for this to happen. But where does this learning occur and what changes take place at the neural level? As mentioned in the last chapter, our brains are too complex so we are required to turn to simpler organisms to answer these types of questions.

Although we accept that the *Aplysia* may have a very simple learning repertoire, we ask, is it possible, that the *Aplysia*, like Pavlov's dogs, can also be classically conditioned, despite not having a brain? This chapter will describe not only how the *Aplysia* can be conditioned but the site of change within the neural network and the molecular events associated with this form of learning.

Behavioural Conditioning of *Aplysia*

Let us take the classic experiment described by Pavlov as an example before applying it to the *Aplysia*. When presented with food, a dog would normally salivate, i.e. drool. If every time the dog is presented with food a bell is sounded, after a while, the dog will salivate upon presentation of the bell alone. In this case the bell (originally a neutral stimulus) has now become associated with the stimulus (food) that evokes a response (salivation). One of the most important points to note regarding conditioning is that timing is critical. Conditioning only occurs if the food and the bell are presented more or less at the same time. The conditioning effect may disappear if the timing interval between the food and the bell lengthens. Another critical point is that the bell should be presented just before the food presentation; if the bell is presented after the food presentation conditioning will not occur.

Similar steps can be applied to the *Aplysia* using the same experimental procedure as described in the previous chapter. We know from before that squirting water onto the siphon leads to a gill withdrawal response in the small marine creature (Figure 6.1a); it's a reflex, much like the salivation response in the dog. We can continue to do this and keep measuring the amount of withdrawal. We also observed in the previous chapter that if you apply a shock to the tail of the *Aplysia*, the withdrawal response increases. However, in the classical conditioning paradigm, if you shock the tail and *at more or less the same time* you squirt water on the siphon, the gill withdrawal response is a lot stronger, much stronger than even the response during sensitisation (Figure 6.1b and 6.1d). Crucially, if you return to just squirting the water to the siphon, the very strong gill withdrawal response remains and stays like this for a very long time (Figure 6.1c and 6.1d). The squirting of the water has become associated with the shock to the tail. It is as if the shock has been remembered, even though you are now only applying water. So classical conditioning not only produces a stronger response, but the memory for the response remains for longer.

As mentioned earlier, the timing and the sequence of the stimuli presentation are both critical. Hawkins et al. (1986) found a critical time window for conditioning to occur in the *Aplysia*. Strong conditioning occurred when siphon stimulation preceded the tail shock by 0.5 second, but only marginal conditioning occurred when the shock was preceded by siphon stimulation 1.0 second before. Furthermore no significant conditioning occurred when siphon stimulation preceded the shock by 2, 5, or 10 seconds, or when the stimuli were presented simultaneously. No conditioning occurred when the shock preceded the siphon stimulation by 0.5, 1.0 or 1.5 seconds.

Circuits and Behavioural Mechanisms Underlying Classical Conditioning

As described in the previous chapter, information from the *Aplysia*'s siphon travels along sensory neurons that connect with motor neurons which, in turn, causes the gill to contract. Sensory neurons also receive information from the shock to the tail. This information travels along an interneuron (L29) and finally makes contact with the sensory-motor synapse. Similar to habituation and sensitisation, the critical junction for classical conditioning seems to be this sensory neuron–motor neuron synapse (square box in Figure 6.2a). One of the advantages of knowing the exact circuitry in the *Aplysia* and how all the neurons are interconnected is that you can experimentally test something on one neural pathway and compare the results of this to a related pathway (control pathway; see

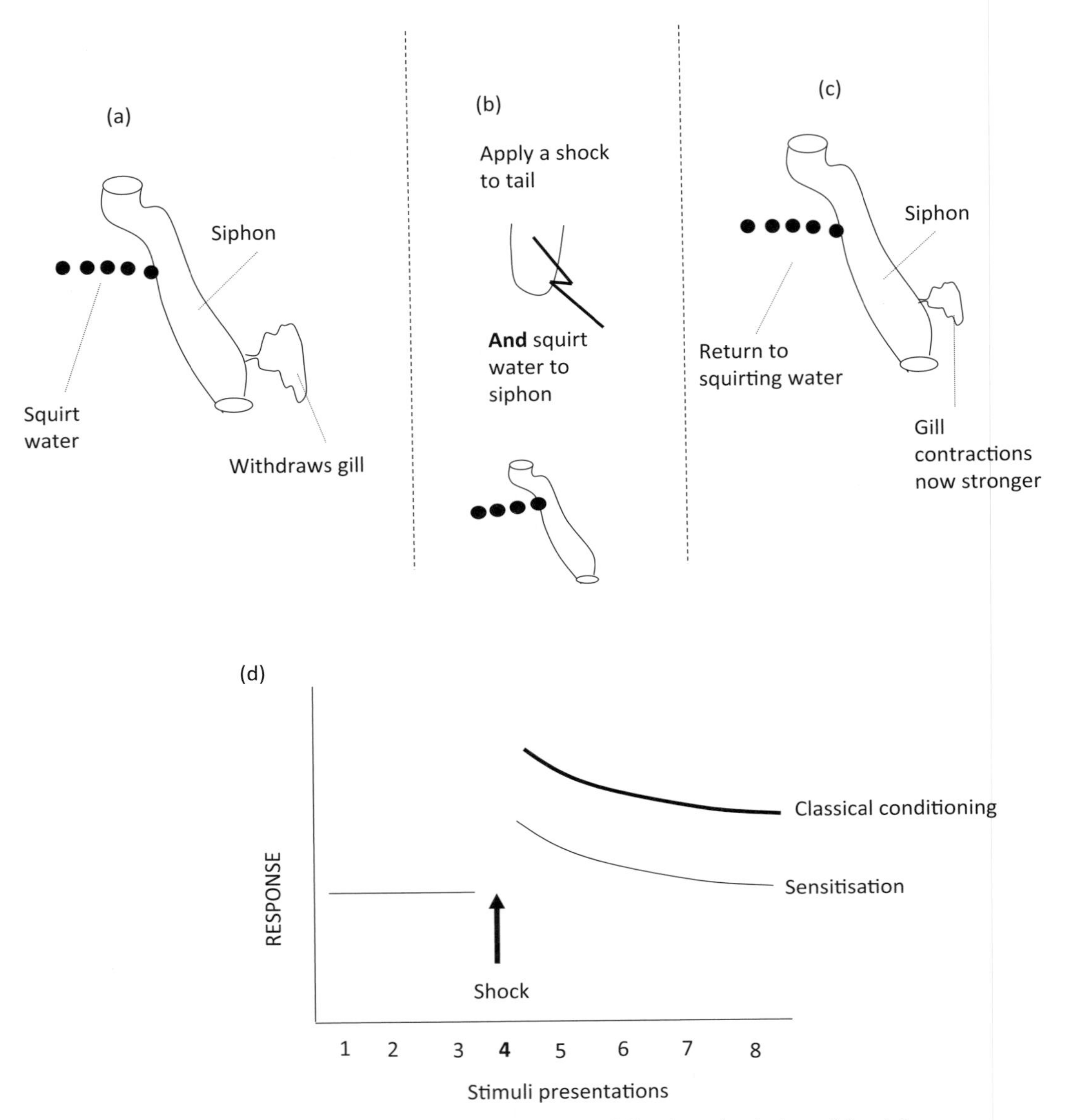

Figure 6.1 Change in the strength of gill withdrawal response following stimulation of the siphon simultaneously with a shock intervention: classical conditioning. See text for details.

Carew et al., 1983; Kandel et al., 1995). So if, for example, you squirt water onto the mantle shelf (protective membrane) of the *Aplysia,* the gill will also withdraw. Similar to that described for the siphon, information from the mantle shelf travels down sensory neurons that synapse onto motor neurons, which in turn innervate the gill. Likewise, a shock to the tail excites interneurons that synapse on the presynaptic terminals of the sensory neurons emanating from both the siphon and mantle shelf (Kandel et al., 1995; see Figure 6.2b). Now we have two separate pathways that converge.

Therefore, each animal can act as its own control, allowing for the determination of the cellular mechanisms that underlie classical conditioning (Carew et al., 1983). It is now possible to compare the behavioural and electrophysiological reactions on both pathways

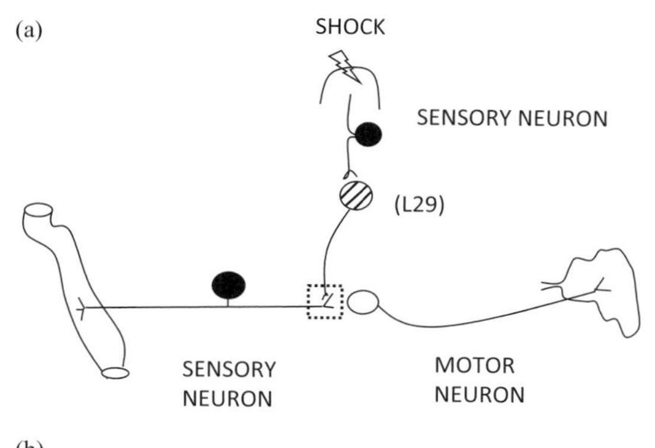

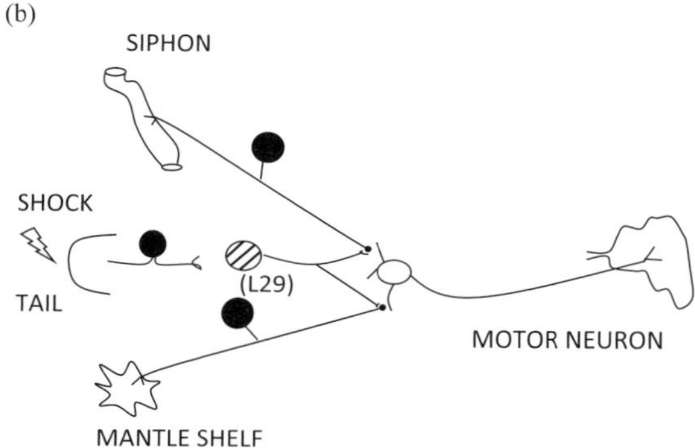

Figure 6.2 The neural connectivity of the *Aplysia* used for classical conditioning in (a) a simple setup or (b) a more complicated one.

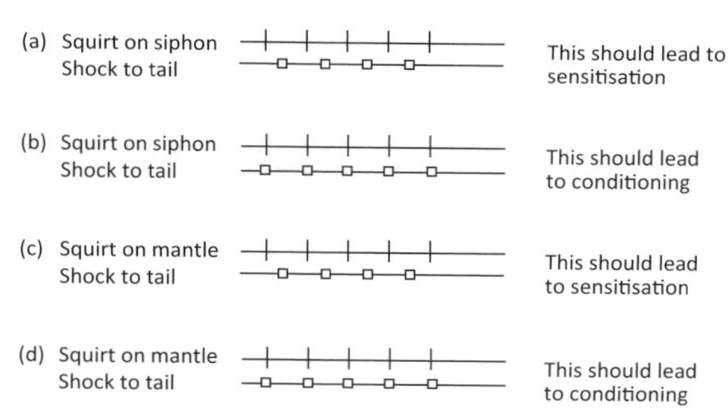

Figure 6.3 Diagrammatical representation of how different circuits can be used to test different paradigms. Vertical lines crossing horizontal lines indicate timing of squirts of water to part of *Aplysia*. Boxes indicate timing of shocks. See text for details.

in response to different sequences of stimuli. As suggested in previous chapter, sensitisation occurs when the shock to the tail comes after (or before) a squirt of water to either the siphon or mantle. However, when the shock to the tail occurs more or less *at the same time* as the squirt of water to either the siphon or mantle, conditioning ensues. Diagrammatically we can represent this as shown in Figure 6.3.

Carew et al. (1983) formally tested this idea and found that 30 minutes after training, animals that had received 15 siphon-paired training trials (Figure 6.3b) showed longer withdrawal response (43 seconds) than 15 mantle-unpaired trials (27 seconds; Figure 6.3c). The reverse was also proven to be true. Animals that had received 15 mantle-paired training trials (Figure 6.3d) showed longer withdrawal response (49.4 seconds) than 15 siphon-unpaired trials (10.2 seconds; Figure 6.3a). Therefore, paired presentations (irrespective of the pathway) produced greater withdrawal responses than the unpaired conditions, demonstrating that the withdrawal response is a lot stronger with conditioning than with sensitisation. Carew et al. (1983) also demonstrated that the number of training trials had an effect on the withdrawal response 24 hours post-training. Following a single paired trial, the mean withdrawal time had increased by 4.8 seconds when compared to the pre-training baseline withdrawal response. With 5 paired trials the mean withdrawal time had increased by 9.8 seconds. With 15 paired trials the time increased by 45.3 seconds (Carew et al., 1983). Increasing the number of unpaired trails (sensitisation) also produces an increasing withdrawal response, although not of the same magnitude as produced in a conditioning response (see Chapter 5 and Carew et al., 1983). Therefore, conditioning produces strong and long-lasting memories.

Circuits and Electrophysiological Mechanisms Underlying Classical Conditioning

Using the same circuitry as described above for studying behaviour, the electrophysiological mechanisms that underlie classical conditioning can be also examined. However, instead of squirting water onto the siphon or the mantle

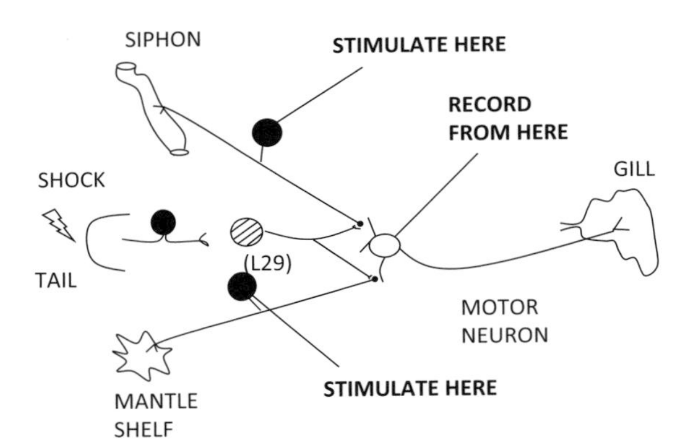

Figure 6.4 Representation of the neural connectivity setup for producing classical conditioning in *Aplysia*. The black circles represent the sensory neurons, the striped circle represents the interneuron and the white circle represents the motor neuron.

of the *Aplysia*, a stimulating electrode can be placed in the sensory neuron coming from either the siphon or mantle (see Figure 6.4). In a sense, an action potential can be created within the sensory neurons, eliminating the need for squirting water. Also, instead of measuring the withdrawal response of the gill, a recording electrode can be placed in the motor neuron (see Figure 6.4), which allows for the measurement of excitatory postsynaptic potentials (EPSPs).

Using this setup and the protocol described for the behavioural experiments (Figure 6.3), Hawkins et al. (1983) examined the effects of paired and unpaired sensory neuron stimulation on EPSP magnitude in the motor neuron. It was predicted that when the shock to the tail was presented at the same time as the sensory neuron stimulation (either of the siphon or mantle), this would lead to conditioning of the evoked motor neuron response (increased magnitude; Figure 6.5a, top), whereas an unpaired shock and stimulation of the sensory neuron would lead to sensitisation (see Figure 6.5a, bottom). The authors found that, similar to the gill withdrawal response, EPSP amplitudes from the paired protocol were much greater than those from the unpaired protocol. At both 5 and 15 minutes, after a series of training trials, the EPSP amplitude from the paired neuron was significantly higher compared to its pre-training amplitude (Figure 6.5b top). In comparison, the authors found in the unpaired protocol that although there was an initial increase in the magnitude of EPSP response, this decreased with time. When measured, the EPSP amplitude from the unpaired neuron did not significantly change from the pre-training magnitudes (Figure 6.5b bottom).

Biochemical and Molecular Mechanisms Underlying Classical Conditioning

Having discovered the circuitry involved with classical conditioning of the gill withdrawal response and the electrophysiological mechanism underlying the behaviour, Kandel and his team went one step further in an attempt to elucidate the biochemistry and molecular underpinnings. Here we will break the process into three steps and attempt to explain what occurs during each stage: pre-pairing (siphon stimulation alone), during pairing of shock and siphon stimulation, and post-pairing (siphon stimulation alone).

Pre-Pairing or Baseline

As described in the previous chapter, under normal circumstances a squirt of water to the siphon or mantle of *Aplysia* leads to the generation of an action potential in the sensory neuron. This action potential reaches the presynaptic terminal, depolarising the membrane, and causing an influx of calcium and subsequent release of glutamate. The neurotransmitter crosses the cleft, attaches to receptors which, in turn, open channels on the postsynaptic membrane, allowing ions to enter the motor neuron. This generates an EPSP resulting in the eventual contraction of the gill (see Figure 6.6).

During Pairing

As previously outlined in Chapter 5, a shock to the tail activates sensory neurons, which in turn, leads to the generation of action potentials

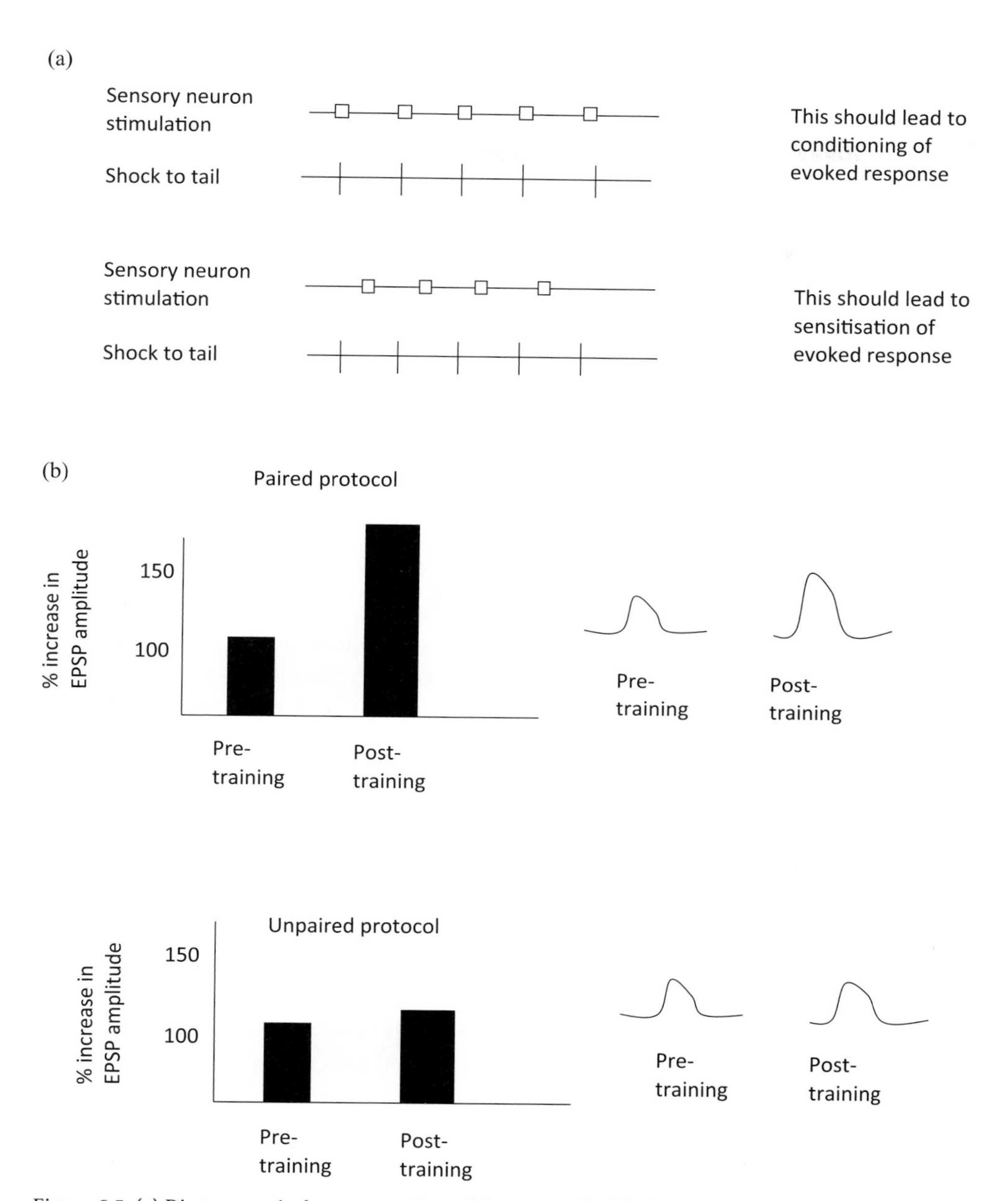

Figure 6.5 (a) Diagrammatical representation of the protocols, (b) changes in the neural response depending on protocol used.

in the interneuron (L29). This action potential reaches the synaptic terminal (Figure 6.7, 1), depolarising the membrane and causing an influx of calcium and subsequent release of serotonin (2). Serotonin crosses the cleft and attaches to a G-protein-coupled receptor on the sensory neuron of the siphon (3). Activation of the receptor leads the conversion of ATP to cAMP via the converting enzyme adenylyl cyclase (4).

However recall (earlier in the chapter) that strong conditioning occurs when siphon stimulation precedes the shock to the tail by 0.5 second.

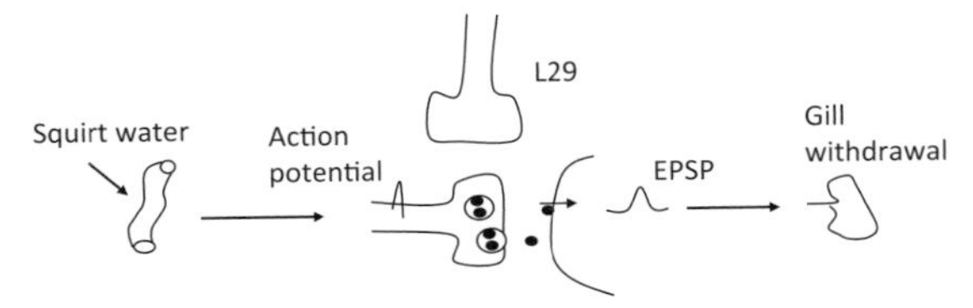

Figure 6.6 Representation of the sequence of events that occur at the pre-pairing (baseline) phase of conditioning.

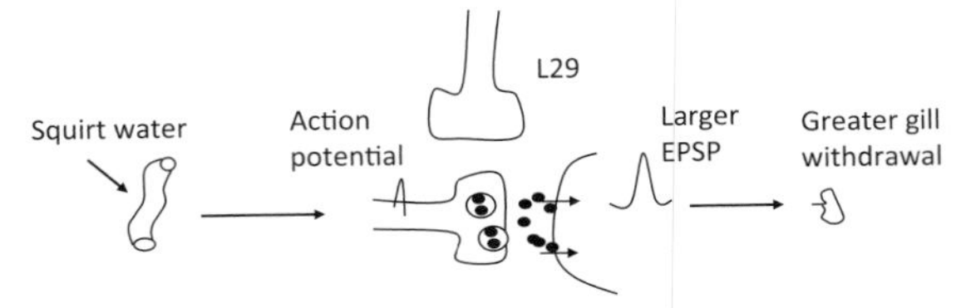

Figure 6.8 Representation of the sequence of events that occur at the post-pairing phase of conditioning. See text for details.

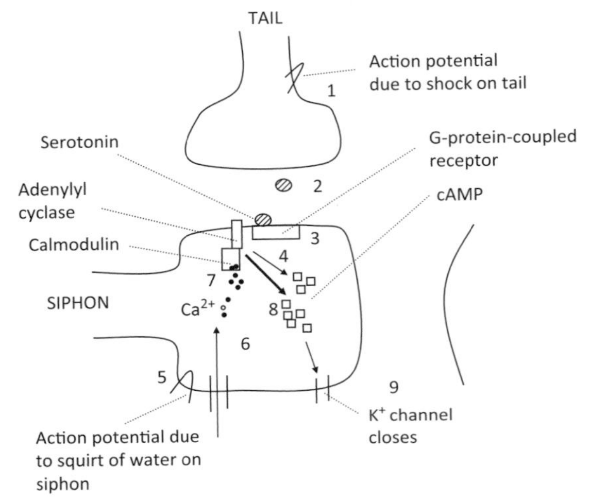

Figure 6.7 Representation of the sequence of events that occur during the pairing phase of classical conditioning. See text for details.

Therefore the action potential from the sensory neuron of the siphon has arrived at the synaptic terminal just before the serotoninergic activation of the G-protein-coupled receptor (5). This arrival depolarises the membrane, causing an influx of calcium into the presynaptic terminal (6). Abrams et al. (1991) discovered that this calcium binds to calmodulin, which in turn activates and causes a conformational change to adenylyl cyclase (7). As a result of the conformational change to adenylyl cyclase, this enzyme can *increase* the production of cAMP from ATP in response to the serotoninergic activation of the tail shock (8). This increased amount of cAMP activates more cAMP-dependent protein kinase A (PKA) from its inactive state (see Chapter 5). This, in turn, phosphorylates many more potassium channels. Similar to what we saw with respect to sensitisation, this has the effect of modifying the potassium channels, essentially closing them down, thereby, prolonging any action potential that should arrive at the presynaptic terminal much more (9), even more than during sensitisation.

Post-pairing

Similar to what happens during sensitisation, a squirt of water to the siphon alone leads to a generation of an action potential in the sensory neuron. This action potential reaches the synaptic terminal, depolarising the membrane. However, due to the fact that a large number of K^+ channels have been modified, the action potential is prolonged, so even a greater amount of calcium enters the presynaptic cell. Therefore, even more synaptic vesicles are mobilised, with more neurotransmitter released from the sensory neuron to the motor neuron. The increased glutamate release crosses the synaptic cleft and attaches to receptors on the motor neuron which, in turn, opens channels on the postsynaptic membrane, allowing more ions to enter. This generates a larger EPSP in the motor neuron, resulting in even greater contractions of the gill (Figure 6.8).

Other Mechanisms?

So far we have described classical conditioning in the *Aplysia* in what has often been referred to as activity-dependent presynaptic facilitation (ADPF; Hawkins et al., 1983). However, there are still a number of questions that have yet to be answered. For example, despite a lot of evidence that suggests ADPF plays a central role in conditioning, other mechanisms may also play a role (Glanzman, 1995) and should not be ignored. One possibility, as highlighted by Lin and Glanzman (1994a,b), is a possible role for the involvement of a postsynaptic mechanism in classical conditioning in *Aplysia*, that is, at the motor neuron section of the sensorimotor junction, rather than on the sensory neuron side. These authors suggest that the synapses undergo a form of long-term potentiation (LTP, see Chapter 7 for further details). To test this idea, Lin and Glanzman (1994a and b) conducted a series of experiments in which they had a number of different groups. In the first group, they stimulated the sensory neuron once every 10 minutes and recorded the magnitude of the EPSPs of the motor neuron (Figure 6.9). They continued this for a total of 100 minutes. In a second group, they also stimulated the sensory neuron and recorded the EPSP from the motor neuron. However, after 10 minutes they stimulated the sensory neuron, but this time with 25 Hz (a strong stimulation). Simultaneously, the authors depolarised the motor neuron (i.e. made it more excitable). Ten minutes later, they returned to normal stimulation and continued for this for 100 minutes. In a third group, they again stimulated the sensory neuron and recorded the EPSP from the motor neuron. Ten minutes later, they stimulated the sensory neuron with 25 Hz. In the third group, however, the motor neuron was not depolarised. Following this, they returned to normal stimulation and continued for 100 minutes.

What the authors discovered was that a single pairing of strong sensory neuron stimulation and motor neuron depolarisation resulted in a prolonged increase in magnitude of the EPSP (Group 2). No other group had a significant increase in the EPSP magnitude. This suggests that if a sensory neuron is stimulated and at the same time the motor neuron is excited, the EPSPs of the motor neuron are increased and remain increased for a period of time. Earlier in the chapter we observed a similar increase in motor neuron EPSPs with the aid of a facilitatory interneuron synapsing onto the sensory neuron. The Lin and Glanzman experiments suggest that a facilitatory neuron may not be actually needed (perhaps, only the sensory and motor neurons are required). How can a motor neuron become excited? There is some evidence to suggest that a tail shock not only activates facilitatory interneurons within the CNS of *Aplysia* (Mackey et al., 1989; Glanzman, 1995) but also strongly depolarises many motor neurons (Frost et al., 1988; Frost and Kandel, 1995). So it may be possible that the classical conditioning withdrawal response can be explained by the two processes, activity-dependent presynaptic facilitation and also by a postsynaptic mechanism that does not require the facilitatory interneurons.

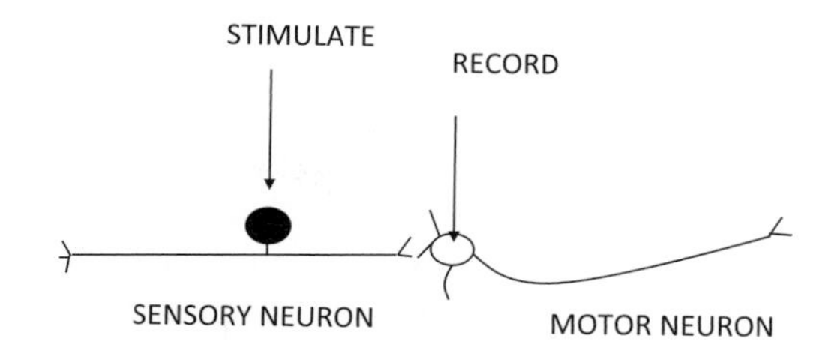

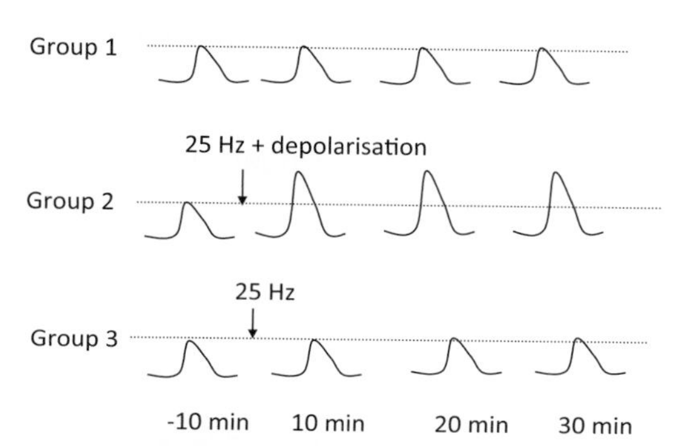

Figure 6.9 Stimulation of a sensory neuron combined with depolarisation of the motor neuron produces a lasting enhanced response (Group 2). This enhancement is brought about without the need for interneurons. See text for details.

Summary

In the previous chapter we demonstrated that very simple forms of learning can be understood in terms of changes that occur at both the neural and molecular levels. In this chapter we have gone one step further and have demonstrated that more complicated forms of learning – that is, associative learning – can also be examined and explained in terms of the underlying cellular mechanics. Classical conditioning is a fundamental form of learning and one of the cornerstones of animal and human behaviour. Understanding the basic processes involved offers the possibility of understanding even more complicated forms of behaviour in animals and humans.

Questions and Topics Under Current Investigation

- Can the processes underlying classical conditioning discovered in *Aplysia* be directly applied to the mammalian brain? If so, what are the mechanisms?
- How does the neural circuit accommodate more complicated associations – for example, when more than two things are paired?
- There have been recent studies examining other forms of associative learning including operant conditioning in *Aplysia* (see Nargeot and Simmers, 2011) but the mechanisms have yet to be fully elucidated. What are these mechanisms and how do they compare to classical conditioning?

References

Abrams, T.W., Karl, K.A., and Kandel, E.R. (1991). Biochemical studies of convergence during classical conditioning in *Aplysia*: dual regulation of adenylate cyclase by Ca+/calmodulin and transmitter. *Journal of Neuroscience*, 11(9), 2655–65.

Carew, T.J., Hawkins, R.D., and Kandel, E.R. (1983). Differential classical conditioning of a defensive withdrawal reflex in *Aplysia californica*. *Science*, 219, 397–400.

Frost, W.N., Clark, G.A., and Kandel, E.R. (1988). Parallel processing of short-term memory for sensitization in *Aplysia*. *Journal of Neurobiology*, 19(4), 297–334.

Frost, W.N., and Kandel, E.R. (1995). Structure of the network mediating siphon-elicited siphon withdrawal in *Aplysia*. *Journal of Neurophysiology*, 73(6), 2413–27.

Glanzman, D.L. (1995). The cellular basis of classical conditioning in *Aplysia californica*- it's less simple then you think. *Trends in Neuroscience*, 18, 30–36.

Hawkins, R.D., Abrams, T.W., Carew, T.J., and Kandel, E.R. (1983). A cellular mechanism of classical conditioning in *Aplysia*: Activity-dependent amplification of presynaptic facilitation. *Science*, 219, 400–405.

Hawkins, R.D., Carew, T.J., and Kandel, E.R. (1986). Effects of interstimulus interval and contingency on classical conditioning of the *Aplysia* siphon withdrawal reflex. *Journal of Neuroscience*, 6(6), 1695–701.

Kandel, E.R., Schwartz, J.H., and Jessell, T.M. (1995). *Essentials of neural science and behavior*. Appleton & Lange, Norwalk, CT.

Lin, X.Y., and Glanzman, D.L. (1994a). Long-term potentiation of *Aplysia* sensorimotor synapses in cell culture: regulation by postsynaptic voltage. *Proceedings of Royal Society of London B Biological Science*, 255(1343), 113–118.

Lin, X.Y., and Glanzman, D.L. (1994b). Hebbian induction of long-term potentiation of *Aplysia* sensorimotor synapses: partial requirement for activation of an NMDA-related receptor. *Proceedings of Royal Society of London B Biological Science*, 255(1344), 215–21.

Mackey, S.L., Kandel, E.R., and Hawkins, R.D. (1989). Identified serotonergic neurons LCB1 and RCB1 in the cerebral ganglia of *Aplysia* produce presynaptic facilitation of siphon sensory neurons. *Journal of Neuroscience*, 9(12), 4227–35.

Nargeot, R., and Simmers, J. (2011). Neural mechanisms of operant conditioning and learning-induced behavioral plasticity in *Aplysia*. *Cell Molecular Life Science*, 68(5):803–16.

Pavlov, I.P. (1927). *Conditioned reflexes*. Oxford: Oxford University Press.

7 Long-Term Synaptic Plasticity in Mammals I: Long-Term Potentiation (LTP)

Introduction and Background

Learning and memory are essential functions for all animals. For us humans, recalling events can help evoke emotions and can also help build a shared experience and add to a knowledge base – all important elements for social interactions, personal development and cultural evolution. More than this, memories allow us to learn from the past and use this knowledge to adapt to new and ever-changing situations. Memories essentially allow us to make future decisions. However, in order for learning to occur and memories to be stored over the long term, the brain must have some mechanism that can facilitate this process. In 1949 the Canadian psychologist Donald Hebb wrote a very influential book entitled *The Organisation of Behaviour*, in which he suggested that changes in neural connectivity underpin behavioural change, particularly changes associated with learning and memory. It is worth quoting his famous postulate here:

> When an axon of cell A is near enough to excite a cell B and repeatedly or persistently takes part in firing it, some growth process or metabolic change takes place in one or both cells such that A's efficiency, as one of the cells firing B, is increased (Hebb, 1949, p. 62).

In what is now referred to as *Hebb's law*, the idea can be summarised as 'cells that fire together, wire together'. The importance of this concept is that it suggests that neurons and their connections are not fixed; rather, they can be modified. The connectivity between neurons can strengthen or weaken; they are malleable or plastic, adjusting their response to various environmental stimuli. This is known as *synaptic plasticity* or *Hebbian plasticity*. In the 1940s and 1950s there was not much evidence for this idea, as many neuroscientists were working on neurons of the peripheral nervous system, which we now know are relatively nonplastic. However, interest in the concept grew when many scientists in the 1960s, including Eric Kandel, Tom Carew, Jimmy Schwartz and others, discovered that neurons in the 'simple' invertebrate nervous systems could be modified and are subject to changes that reflect learning and memory. Such changes involved an increase in presynaptic neurotransmission release, leading to an increase in neural firing and a subsequent change in behaviour. Furthermore, long-term memory effects were based on morphological changes in the circuit, that is, an increase in the number of synaptic contacts between neurons (see Chapters 5 and 6 for more details). These findings provided very strong evidence for the Hebbian concepts of 'growth process', 'metabolic changes' and 'increased firing'. However, could similar findings be observed in the mammalian brain?

One of the major difficulties with the mammalian brain is its sheer complexity. With billions of neurons, how was it possible to identify a particular circuit or site that would show some form of synaptic plasticity? In the late 1960s this was achieved. Terje Lømo and Tim Bliss, working with rabbits in Per Anderson's lab in Oslo, demonstrated a form of synaptic plasticity that fulfilled many of the characteristics described by Hebb, in a region of the brain known as the hippocampus. In 1973 Lømo and Bliss described how synapses could increase their response to certain stimulation; they demonstrated that synapses were plastic and could be modified. More importantly, they were also able to show that this response could remain increased over a long period of time. These are the very properties that you expect of memory, changes that can last over time. They termed this phenomenon *long-term potentiation* (LTP). LTP and its various related forms remain today as the most acceptable biological mechanism thought to underlie mammalian learning and memory.

The Hippocampus as a Site for Memory

Why did Bliss and Lømo investigate the hippocampus as a possible site for plasticity? The hippocampus became synonymous with memory following the publication of a famous case study in 1957 of a patient called Henry Molaison (HM). HM suffered from epilepsy following a seemingly innocuous fall from a bike at the age of 9. By the time he was 27 his seizures were so severe and frequent that he could hardly function on a daily basis. To alleviate the symptoms, HM underwent an experimental surgery that removed his hippocampus in both hemispheres. Although the surgery was successful in reducing the frequency of epileptic attacks, it had much deeper consequences. HM was no longer able to form any new memories. It was as if HM was stuck in time – in 1953. Despite the passing years and right up to his death in 2008, HM was unable to recall any events from either his own personal life or from the world that might have occurred after his surgery. However, HM's short-term memory and long-term memory from his childhood remained intact. Similarly, his personality, language and other abilities were unaffected.

HM's case was remarkable and led to a number of major contributions to the study of memory. First, memory seems to be localised in the brain, to the region called the *hippocampus*. Second, there are different types of memory. HM seemed to forget factual information about his own life and about world events (known as *declarative memories*) but he performed well on tasks that required motor memory and practice, such as riding a bicycle – often referred to as *procedural memories*. Third, short-term memory and long-term memory do not rely on the hippocampus; these types of memories are stored elsewhere. However, the hippocampus is involved in the consolidation process, the conversion of short-term memories to more enduring and stable long-term memories; therefore, if there was any region of the brain that should be plastic it had to be the hippocampus.

Electrophysiological Mechanisms Underlying Long-Term Potentiation

In previous chapters we introduced the idea that it was possible to insert a stimulating electrode that evoked a small current in one set of neurons, and simultaneously record the response from a connecting set of neurons by using a second electrode. This is essentially what Bliss and Lømo did in the hippocampus. Signals typically arrive into the dentate gyrus of the hippocampus from an adjoining structure (termed the *entorhinal cortex*) via a group of fibres called the *perforant path*. By placing a stimulating electrode into the perforant path, Bliss and Lømo were able to stimulate the fibres and record the response of this stimulation in the adjoining dentate gyrus. We represent this diagrammatically by two single neurons (Figure 7.1a). By stimulating a neuron you essentially create an action potential (1) which travels down the axon, releasing the neurotransmitter into the synaptic cleft (2). The transmitter attaches to receptors in the second neuron (e.g. dentate gyrus), opening channels and allowing ions to enter the postsynaptic membrane (3). The entry of ions results in excitatory or inhibitory (depending on the ion type) postsynaptic potentials (4), or PSPs. These PSPs (represented by the wave in Figure 7.1a) then can be measured in a number of different ways; the amplitude or the height of the PSP, which is typically measured in millivolts (mV), and the slope of the upward deflection or latency (the time it takes for the response to occur following stimulation). Latency is typically measured in milliseconds (ms).

Now let's imagine that we stimulate the perforant path (or neuron 1) once every 20 seconds for approximately 10 minutes. This is often referred to as low-frequency stimulation. We can then record the response evoked in neuron 2 and plot the amplitude of each PSP. This would give us approximately 30 PSPs in total (Figure 7.1b).

Once we have recorded a stable number of PSPs and established a baseline response, we can then apply a high-frequency stimulation (HFS) to neuron 1. This may consist of large number of stimuli in a very short time period. So rather than stimulating the neuron once every 20 seconds, up to 100 stimuli can be given each second for a total of 10 seconds. This is a very quick event and is represented by multiple action potentials in Figure 7.2a (top) and by the arrow in Figure 7.2a (bottom). Following this

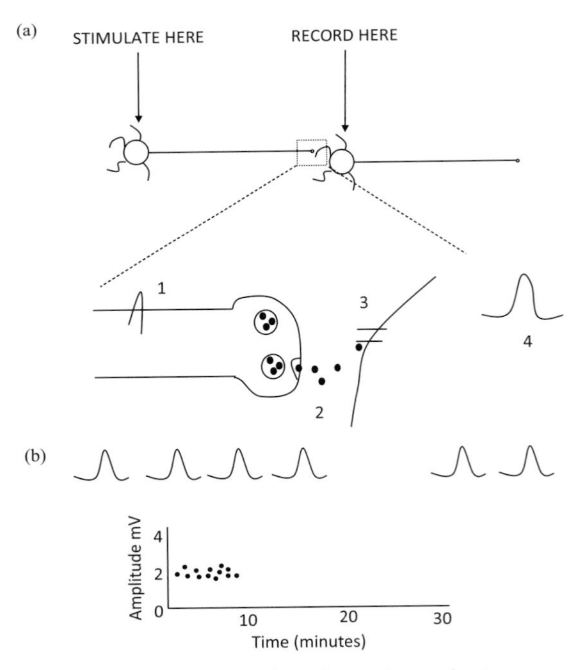

Figure 7.1 (a) Representation of experimental setup for recording LTP (top) and how stimulation of one neuron evokes a potential in a second. (b) Evoked responses (represented by magnitude of the PSP (wave)) can be recorded and measured over time, providing a baseline for the synapse under investigation.

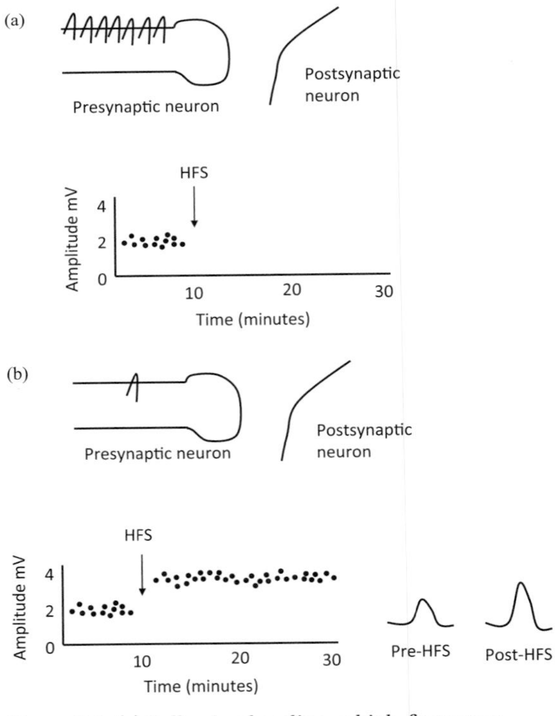

Figure 7.2 (a) Following baseline, a high-frequency stimulation (HFS) is applied. (b) Returning to baseline stimulation produces significantly larger evoked responses that are long-lasting. This is termed long-term potentiation (LTP).

intervention, if we now return to stimulating the neuron once every 20 seconds, we would find (as did Bliss and Lømo) that the amplitudes of the PSPs have increased in size, maybe as much as 50% (Figure 7.2b). This increase in PSP response is not just a temporary response to the high-frequency intervention but can actually last for minutes, hours and weeks. Indeed, some researchers have even reported increases in neural responses that can last over a year. This increase in neural response is called long-term potentiation. We have shown that neurons can change their response as a result of some sort of intervention. Neurons are plastic; their responses can vary and are not fixed. More than this, the change is long-lasting, something that is similar to a memory. Note the similarity between this and the sensitisation effect observed in the *Aplysia* (Chapter 5): a baseline response, an intervention and a long-lasting change in the response as a result of the intervention.

Cellular and Molecular Mechanisms Underlying Long-Term Potentiation

Now that we understand the basic phenomenon of LTP, the next question to address is how it occurs. To explain this I have broken the next section into three different stages: cellular and molecular events that occur prior to the HFS intervention, what happens during HFS and events that occur post-HFS.

Prior to High-Frequency Stimulation

During the initial low-frequency stimulation (once every 20 seconds), an action potential reaches the presynaptic terminal (Figure 7.3, top, 1), depolarising the membrane and causing the opening of voltage-dependent calcium channels. This causes an influx of calcium into the presynaptic terminal (2), which, in turn, triggers the subsequent release of glutamate into the synaptic cleft (3). The neurotransmitter then crosses the cleft and attaches to receptors, which in turn open channels on the postsynaptic membrane, allowing ions to enter the postsynaptic neuron (4). This in turn generates a excitatory postsynaptic potential (5).

This is the basic mechanism of synaptic transmission that has been described many times before. Glutamate is the major excitatory neurotransmiter in the brain and can attach onto a number of receptors on the postsynaptic membrane, including receptors for AMPA (α-amino-3-hydroxy-5-methyl-4-isoxazolepropionate), receptors for NMDA (N-methyl-D-aspartate) and kainate receptors. There is another class of glutamate receptor called the *metabotropic receptor* which is a G-protein-coupled receptor, which we met in the previous chapters. However, the two receptors that we will deal with here are the AMPA and NMDA receptors. When glutamate attaches to AMPA receptors, the receptor opens and allows the entry of sodium (Na^+) ions into the postsynaptic cell (Figure 7.3, bottom). This causes a depolarisation of the postsynaptic membrane, resulting in an EPSP. NMDA receptors differ from AMPA receptors in two important respects. First, they allow both calcium and sodium ions to enter the cell, but second, they are transmitter-gated *and* voltage-dependent. It is a double-lock system. At resting membrane potential and low frequency stimulation, the NMDA receptor is blocked by a magnesium ion. So although glutamate may attach onto NMDA receptors, they will not allow the entry of ions into the postsynaptic cell due to the presence of this magnesium ion.

This magnesium block will only be lifted if the postsynaptic membrane is depolarised – that is, if the inside of the cell becomes more positive than −30 mV. When this occurs, ions such as calcium are then free to enter the cell. In summary, during low-frequency stimulation, glutamate attaches to AMPA and NMDA receptors, but only the AMPA receptors open, causing the influx of sodium into the postsynaptic cell and the generation of an EPSP (Figure 7.3 bottom).

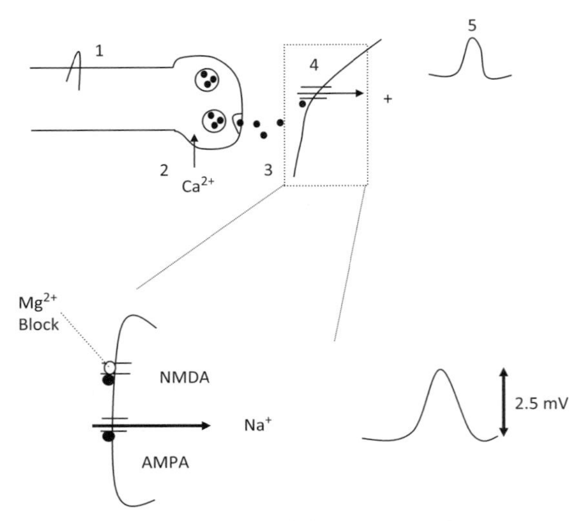

Figure 7.3 Representation of the sequence of events that occur during baseline (before high-frequency stimulation). See text for details.

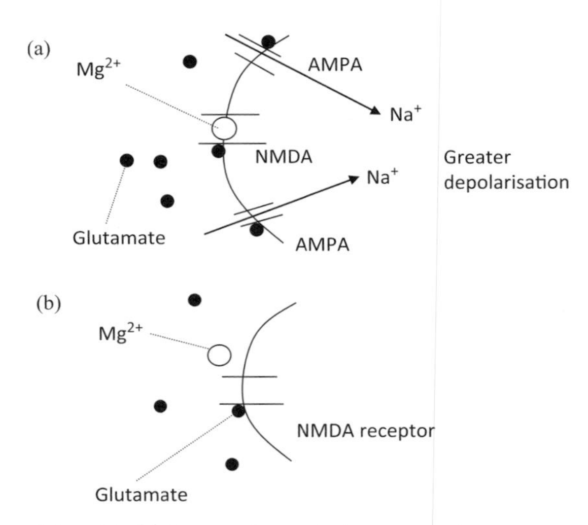

Figure 7.4 (a) Increased stimulation causes greater depolarisation in the postsynaptic membrane; (b) this in, turn removes the Mg^{2+} block on the NMDA receptor.

During High-Frequency Stimulation

During high-frequency stimulation there may be anywhere between 50 and 100 stimuli presented every second for a period of 10 seconds. As a result of this, the presynaptic terminal is bombarded with action potentials. This causes a large amount of glutamate to be released into the synaptic cleft and to attach to a large number of receptors on the postsynaptic membrane. As a result, there is an increased amount of AMPA receptor activation, allowing an increased amount of sodium into the postsynaptic membrane. This increased amount of positive ion influx causes the negatively charged membrane to become more depolarised (Figure 7.4a). With greater depolarisation of the postsynaptic membrane, the magnesium block of the NMDA receptors is now lifted (Figure 7.4b).

After High-Frequency Stimulation

We then return to a low-frequency stimulation, stimulating once every 20 seconds. When each generated action potential reaches the presynaptic terminal, gluatamate is released and attaches to receptors on the presynaptic

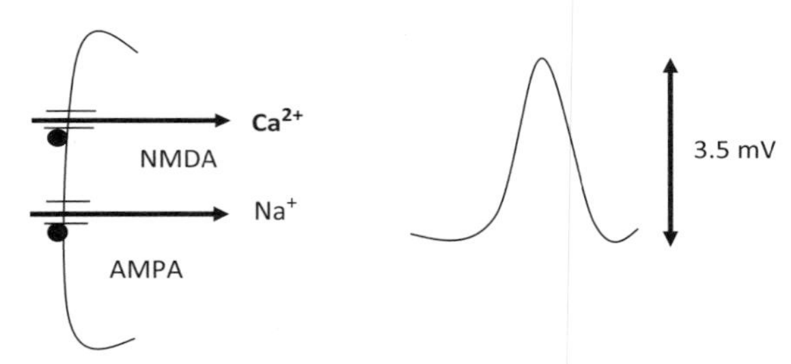

Figure 7.5 Removal of the NMDA block means more ions (specifically Ca^{2+}) can enter the cell when the neuron is next stimulated, producing an increase in response.

membrane. Now both AMPA and NMDA receptors open, allowing the entry of sodium and, more critically, calcium into the cell (Figure 7.5). There is, therefore, an increase in the amount of ions entering the second cell, which causes greater depolarisation and a larger magnitude of EPSP.

Maintenance of Long-Term Potentiation: Early and Late LTP

We know from the case of HM and other work that memory is not a unitary phenomenon but rather it comprises different types and different time frames. As its name suggests, short-term memory allows us to hold information for a short period of time and is thought to have a limited capacity. For example, it is difficult to remember a long string of numbers; we have to continuously rehearse it. Long-term memory has a larger capacity; we seem to be able to store a vast amount of information that can last a whole lifetime. Interestingly, LTP also seems to have two phases; early and late LTP. From our description of LTP earlier, we know that NMDA receptors are critical for LTP (Bliss and Collingridge, 1993; Nicoll and Malenka, 1995; Malenka and Nicoll, 1999), but in addition to this, the increase in calcium concentration in the postsynaptic cell is also of vital importance (Lynch et al., 1983; Malenka et al., 1992). The concentration of calcium in the postsynaptic cell needs to reach a threshold in order for LTP to be maintained over the longer term; if this threshold is not reached, a short-term potentiation may ensue (Malenka and Nicoll, 1999) that lasts anywhere between 5 and 15 minutes. Long-term potentiation too can be divided into two phases: a phase that lasts between 1–3 hours (early LTP) and a phase that lasts longer than this (late LTP).

Early Long-Term Potentiation

The increased concentrations in calcium trigger a cascade of biochemical processes. The interaction between the various kinases and signal molecules is still very much debated. However, evidence implicates the importance of calmodulin and α-calcium/calmodulin-dependent protein kinase II (CaMKII) as key components in LTP maintenance (Nicoll and Malenka, 1995; Malenka and Nicoll, 1999). Inhibitors of or genetic deletion of CaMKII, for example, impairs LTP. It is thought that as calcium concentrations increase they bind to calmodulin (Figure 7.6a, 1) which, in turn, activates CaMKII (2). An interesting property of CaMKII is that it becomes autophosphorylated, that is, it adds a phosphate group to itself (3). A direct consequence of this autophosphorylation is that CaMKII can remain active even without the presence of calcium. Therefore when the calcium concentration returns to its normal levels, CaMKII still remains active. With CaMKII now activated, what does it do? One suggestion is that CaMKII phosphorylates the AMPA receptor directly (Barria et al., 1997), thereby increasing the conductance through the channel (Derkach et al., 1999). The channels allow more ions to enter the postsynaptic membrane, thereby increasing and maintaining the magnitude of the PSP (Figure 7.6b top). A second suggestion is that CaMKII may cause the delivery or clustering of AMPA receptors to the cell membrane. This can occur at synapses already containing AMPA receptors or at synapses that do not (Figure 7.6b middle left; Malenka and Nicoll, 1999; Shi et al., 1999). Evidence suggests that CaMKII binds to the NMDA receptor and forms a tight complex (Gardoni et al., 1998; Bayer et al., 2001). This complex may then become an anchor for new AMPA receptors (Figure 7.6b middle right; Lisman and Zhabotinsky,

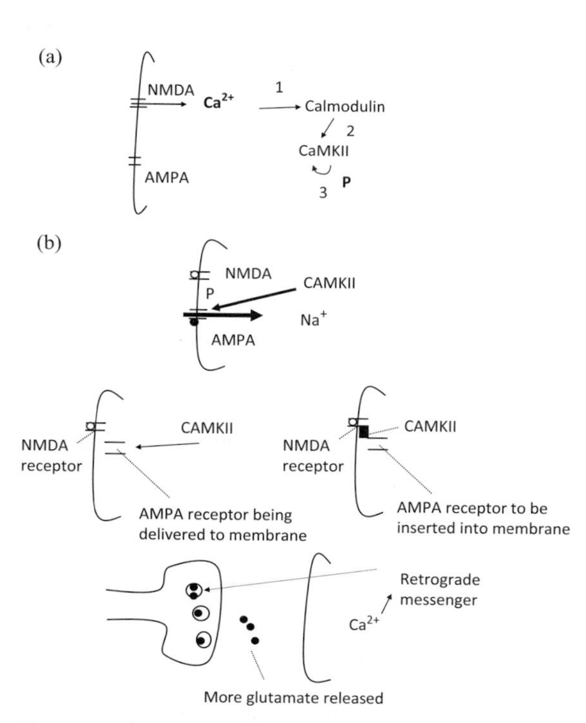

Figure 7.6 (a) Representation of the sequence of events to activate CaMKII. CaMKII may have a number of functions including modifying existing AMPA receptors (b, top), inserting new AMPA receptors (b. middle) and modifying the signal presynaptically (b, bottom).

2001). Now there are more AMPA receptors which, in turn, allow more ions to enter the postsynaptic membrane, thereby increasing and maintaining the magnitude of the postsynaptic potential (PSP).

The above suggestions all involve changes to the postsynaptic membrane. Although the concept is currently out of favour, there has been evidence that changes can also occur at the presynaptic terminal, allowing LTP to be maintained (Bekkers and Stevens, 1990). Such changes include the increase in glutamate release from the presynaptic terminal (Malgaroli and Tsien, 1992). However, in order for this to occur, a retrograde messenger is thought to be released from the postsynaptic cell and travel backwards to change the presynaptic function (Figure 7.6b bottom). A number of molecules have been suggested to serve as a retrograde messenger. These include nitric oxide, carbon monoxide, arachidonic acid and platelet-activating factor (Williams et al., 1993; Hawkins et al., 1998). However the evidence for a presynaptic role remains controversial (Malenka and Nicoll, 1999; Huganir and Nicoll, 2013).

Late Long-Term Potentiation

So far what we have described is a relatively short-term potentiation effect that lasts less than 3 hours. But phosphorylation of a protein is not permanent and eventually the phosphate groups can be eliminated. Indeed, proteins themselves continually undergo renewal and are replaced every couple of weeks. Therefore, if synaptic changes are to be considered as a realistic biological model of learning and memory, more permanent changes at the molecular level must ensue. Some memories last a lifetime! Similar to what we have seen in the *Aplysia* (see Chapters 5 and 6) long-term changes require the synthesis of new proteins and even structural changes to the synapse.

Over the last number of decades, a number of kinases have been identified and seem to be important for the synthesis of proteins and the maintenance of LTP over a long period of time. Such molecules include protein kinase C (PKC), cyclic adenosine 3', 5'-monophosphate (cAMP), protein kinase (PKA) and mitogen-activated protein kinase (MAPK).

Huang et al. (1996) and Alberini et al. (1995) suggested a possible mechanism involving the above molecules that may help in the maintenance of LTP. Figure 7.7a provides a summary of this, describing the various steps thought to be involved: initially, calcium binds to calmodulin (1) which activates adenylyl cyclase, converting ATP to cAMP (2). Prolonged activation of cAMP, in turn, activates PKA from its inactive state (3). PKA then translocates to the nucleus of the cell, where it phosphorylates proteins such as cyclic AMP response element-binding proteins (CREB) (4). This results in the activation of a number of genes, which encode certain proteins that are used for structural functions. Remarkably, this biochemical cascade and process is very similar to what occurs in long-term sensitisation (see Chapter 5). Other proteins and kinases are also thought to be involved, but their exact roles are currently being researched. We have described in the previous chapters, particularly with respect to sensitisation in the *Aplysia*, how the newly synthesised proteins were involved in structural and morphological changes and maintaining the activity of the kinases.

Is this similar for maintaining LTP over a long period of time? While structural and morphological changes are very hard to observe in the mammalian adult brain, changes in spine density on the dendrites (Nikonenko et al., 2002; Toni et al., 2001) have been observed (Figure 7.7b). Such changes include an increase in the number of actual number of spines and also an increase in their length (Buchs and Muller, 1996). With such changes, the surface area of the postsynaptic membrane increases and therefore increases its functionality. Such morphological and structural changes echo Hebb's sentiment of a synaptic 'growth process'.

There are a number of key characteristics of LTP that map very well onto how memories work. These include cooperativity, associativity and input specificity (Bliss and Collingridge, 1993). *Cooperativity* refers to the idea that a threshold exists for the induction of LTP. For example,

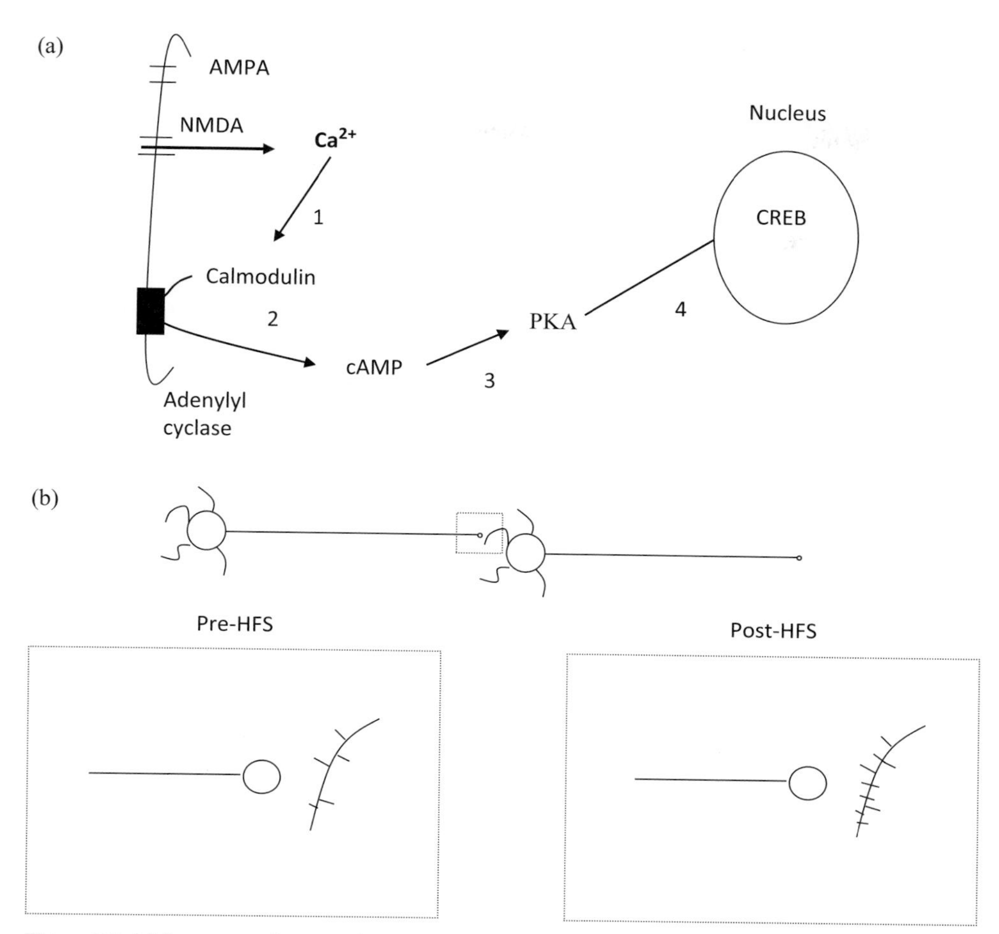

Figure 7.7 (a) Sequence of events thought to underlie the maintenance of LTP. See text for details. (b) Morphological changes at the synapse (increased spine density) thought to be related to long-term modification.

low-frequency stimulation does not induce LTP, an intermediate stimulation protocol tends to induce short-term potentiation, while a strong stimulation and certain patterns of HFS produce robust LTP. This concept is similar to the idea that only certain events are remembered. For example, if we pay strong attention to something, if an event is particularly novel or provokes a certain emotion, it tends to be remembered longer. Other events are completely forgotten. *Associativity* refers to the idea that a weak stimulation can be potentiated if it is paired with a strong stimulation. This allows for the association of different events that are close together in time and is something that we have observed in classical conditioning (see Chapter 6).

Another key characteristic of LTP is that it is *input-specific*; this means that potentiation is confined to the set of synapses that has been stimulated. This allows us to sort one memory from another. So although dendrites receive information from multiple neurons, only the synapse that has been activated undergoes LTP. However, this leaves us with a slight problem. If, as suggested earlier, newly synthesised proteins are thought to travel back to the synapse and help in the maintenance of LTP, how do they know where to go? How do the proteins know which of the multiple synapses has been stimulated?

Research has shown that a type of molecular tagging may exist. Active synapses may be tagged, which act as beacons for the newly synthesised proteins. This idea of molecular tagging has been demonstrated in a number of experiments carried out by Frey and Morris (1997), and it is worth describing some of them here. Imagine HFS is applied to input A (Figure 7.8a top; 1) leading to the molecular changes already described for early LTP. In addition, the molecular changes initiate the synthesis of proteins in the nucleus (2, white diamonds); these proteins then return to the active synapse (3), stabilising the connections, adding extra spines and preventing the decay of LTP – ultimately leading to late LTP (see Figure 7.8a bottom). Let us also imagine that a synaptic tag has been created at the active synapse that serves as a beacon to the newly synthesised proteins (Figure 7.8a top, 4, black star). There seem to be at least two processes that the tag may be dependent on: HFS and protein synthesis. The authors initially tested whether HFS was required by applying HFS to input A and low-frequency stimulation (LFS) to a second set of synapses (input B). If HFS is needed then you would predict that the synapses at input A would display late LTP whereas input B would show only early LTP. This is exactly what was observed (Figure 7.8b). Further, to test whether protein synthesis is necessary, the authors applied HFS to both input A and input B but injected a protein synthesis inhibitor at around the same time. This led only to early LTP in both inputs. Therefore, although synaptic tags were created at both inputs there were no proteins to be drawn to the beacons (Figure 7.8c). Although the idea of a synaptic tag has been generally accepted, much of the current research involves identifying this tag and understanding how it may work.

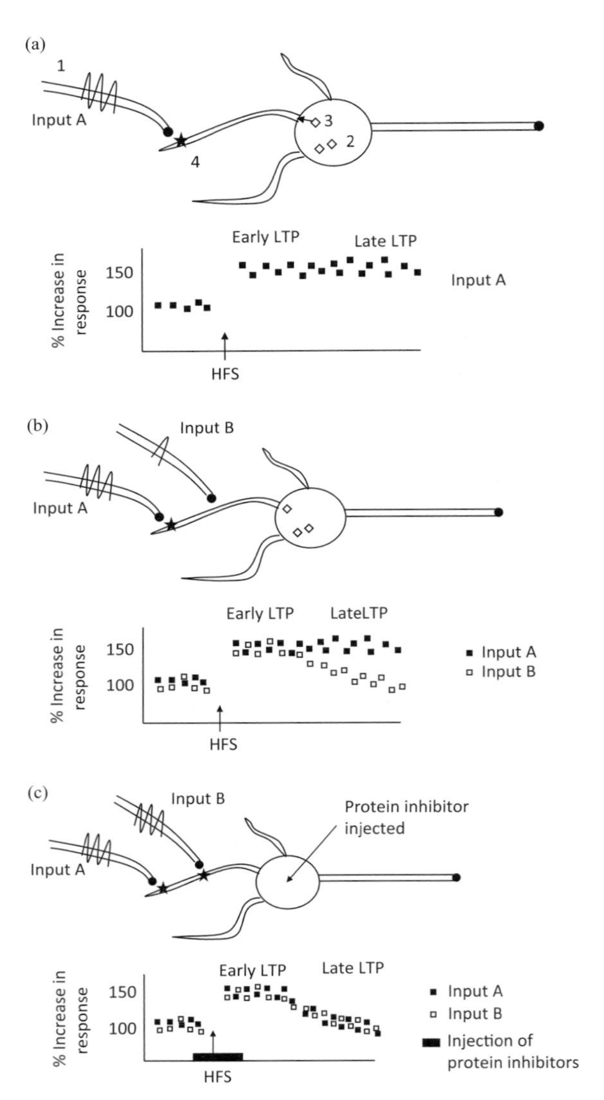

Figure 7.8 (a) Sequence of steps that brings about late- and stable LTP through the synthesis of proteins that then make their way towards tagged synapses. (b) Proteins alone will not produce late LTP; HFS is also required. (c) HFS alone will not produce late LTP; protein synthesis is also required.

LTP as a Mechanism for Learning?

Long-term potentiation does have many of the characteristics that seem to make it an ideal candidate for learning and memory. For example, it fulfills many of the concepts suggested by Hebb, it is rapidly induced, it can last for a long time and it is induced in areas that are known to be involved in learning and memory. But is there any direct evidence that links LTP to learning? Martin et al. (2000) and Martin and Morris (2002) present good reviews examining this very question. Here we will outline some of the major arguments regarding the hypothesis that LTP may be considered as a good model for learning. For a more complete review, refer to Martin et al. (2000) and Martin and Morris (2002).

Detectability

One idea outlined by Martin and Morris is detectability. If an animal learns a specific task, then this learning should be associated with synaptic changes in the area of the brain related to the task. In other words, if we learn a skill that is based on the cortex, then we should get LTP in the cortex. If we acquire a skill that is dependent on another area, for example the hippocampus, we should see LTP in the hippocampus. Is this true?

In one study Rioult-Pedotti et al. (1998) trained animals to reach for food with their preferred paw through a hole in a box. This is a motor skill that would theoretically result in LTP-like changes in the motor cortex. Indeed, following a few days of training, brain slices were taken from the animals and stimulating and recording electrodes were placed into the motor cortex of the brain slice. The motor cortex corresponding to the preferred paw produced postsynaptic potentials (PSPs) significantly larger than did the motor cortex of the untrained paw. Similar findings have been found in other brain regions. For example, Sharp et al. (1985) and Green et al. (1990) recorded EPSPs from the hippocampus of a rat while it explored the environment (spatial exploration and navigation are hippocampal-dependent activities). Would LTP-like changes be found in this region? These authors and others (e.g. Moser et al., 1994; Moser, 1995) demonstrated that rats did in fact show a small and short-lasting increase in hippocampal PSP amplitudes that was related to exploration rather than to any other factors (such as temperature). Likewise, a fear-related task would be expected to evoke plastic changes in the amygdala rather than in any other region: again this does seem to be the case; see Rogan et al. (1997) in Chapter 10 for details regarding this.

Saturation of Long-Term Potentiation

Another angle of approach is saturation of LTP. If you apply high-frequency stimulation (HFS) to a set of synapses and keep applying such stimulations, the responses should increase, but after a number of HFS trains the responses cannot increase any further (Figure 7.9). The set of synapses has reached its saturation point. In theory, as these synapses cannot increase their response any further, they cannot hold any more information; learning as a result can no longer occur. McNaughton et al. (1986) and Castro et al. (1989) found that this was the case. Repeated HFS saturated the inducible LTP, and this led to a severe deficit in a spatial learning task. The results of these studies, however, were found to be difficult to reproduce. Jeffery and Morris (1993), for example, were unable to replicate these findings but did suggest that there was a strong positive correlation between the final level of LTP and subsequent performance of rats in the water maze task. However, using a protocol that was designed to maximally activate perforant path fibres, Moser et al. (1998) were able to demonstrate that saturation did impair water maze learning.

Blocking Long-Term Potentiation

Another powerful method often used is to try to prevent LTP, by targeting various molecules known to be involved in either its induction or maintenance. In theory, if LTP can be prevented, then learning should be impaired.

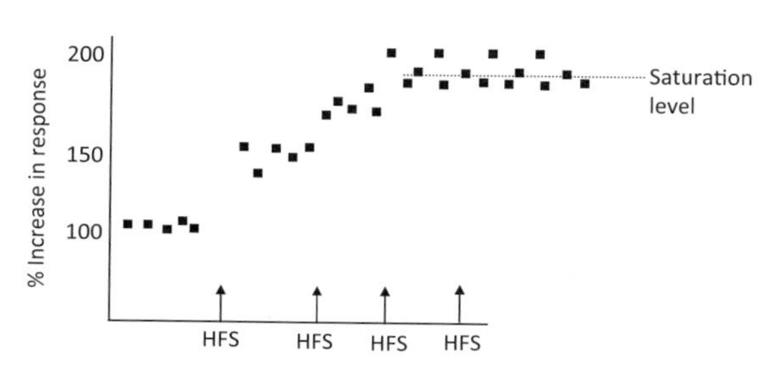

Figure 7.9 Multiple trains of HFS will eventually saturate the synapse so that no further increases in response are observed.

The Role of the NMDA Receptor

We have shown above that the NMDA receptor plays a critical role in the induction of LTP. Blocking the NMDA receptor using antagonists can prevent LTP (Collingridge et al., 1983), but does it also prevent learning? In 1986 Richard Morris and colleagues demonstrated that the NMDA antagonist (AP5) also blocked spatial learning in the water maze. Furthermore, following learning the hippocampi were then removed and LTP was attempted in both control and AP5-treated animals. It was found that LTP was impaired in the AP5-treated hippocampi and not in the control animals, providing good evidence for the role of the NMDA receptor in both LTP and learning. However, it should be noted that learning is a complex skill and consists of encoding, consolidation and retention phases; Morris (1989) and Morris et al. (1990) found that AP5 had no effect on retention of a task previously learnt.

Another approach in the examination of the role of the NMDA receptor in LTP and learning uses genetically altered mice. Tonegawa and his colleagues (Tsien et al., 1996) were able to produce mice that lacked the R1 component of the NMDA receptor. Furthermore, this missing component was restricted to NMDA receptors of area CA1 of the hippocampus. When they were tested on a variety of learning tasks, it was found that these mice performed badly; furthermore, the authors found that LTP was also impaired. If mice with missing components of the NMDA receptor could demonstrate both learning and LTP impairments, would the opposite also hold true? Overexpression of certain components of the NMDA receptor should enhance learning and memory. Tang et al. (1999), in fact, showed that overexpression of NMDA receptor 2B (NR2B) in mice did lead to increased activation of NMDA receptors and showed robust LTP. Furthermore, these mice exhibited superior ability in learning and memory in various behavioural tasks.

Other Proteins in the Biochemical Cascade

We have illustrated that the NMDA receptor is only the initial starting point of a biochemical cascade that results in LTP. Does a blockade of the other proteins along this cascade also impair LTP and learning? One critical component of the process (which we observed earlier) is that CaMKII becomes autophosphorylated so that it can remain active even without the presence of calcium. Mayford et al. (1996) showed that blockade of CaMKII impairs LTP and learning. In another study, Giese et al. (1998) demonstrated that blockade of the autophorylation process of CaMKII impairs LTP and learning. Blocking other molecules also seems to be of critical importance to memory including PKA (Abel et al., 1997) and CREB (Silva et al., 1998), with much evidence suggesting a role for these molecules in the consolidation process. Most of these studies have used genetically altered mice, which are undoubtedly a powerful tool, however see Martin et al. (2000) for further in-depth analysis and critical comments of this approach.

Erasure/Reversal of Long-Term Potentiation

Another method of tackling the question of whether LTP is a good model for learning is to ask, what would happen if LTP was reversed? Would forgetting occur? One method of reversing LTP is by a process called *depotentiation*. Here the researcher applies particular patterns of low-frequency stimulation (e.g. a number of stimuli at 1 to 5 Hz; Staubli and Lynch, 1990) to a set of synapses already potentiated (Figure 7.10). A second method is to apply certain drugs that disrupt the already established LTP (e.g. kinase inhibitors). However, so far no depotentiation protocol has reliably induced forgetting.

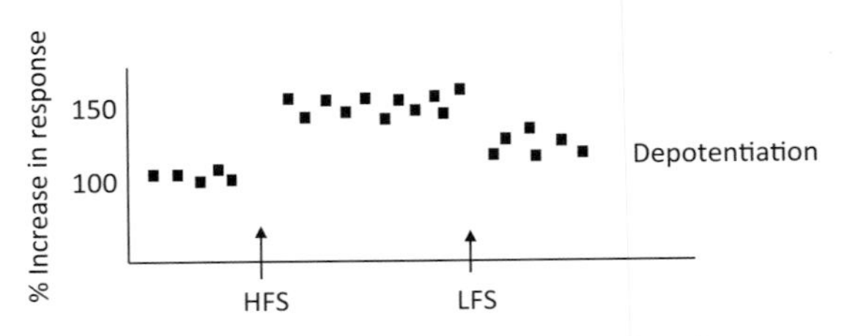

Figure 7.10 Potentiated synapses can be reduced by applying a low-frequency (LFS) stimulation.

Summary

Synaptic plasticity plays a crucial role in information processing, learning and memory. It is increasingly been seen to play a role during brain development as well as in brain dysfunction. Since its discovery over 60 years ago, LTP has sparked an enormous amount of interest as it has been seen as a mechanism by which plasticity in the mammalian brain can occur. More than this, the research that has been built up around the phenomenon has provided huge insights into previously unknown general workings and molecular makeup of the synapse. Some of these basic findings have been described in this chapter, including the role and mechanisms of the AMPA and NMDA receptors, CaMKII and calcium, but so much more has been left out. The importance of the discovery of LTP has in fact led a prominent neuroscientist to remark, 'LTP is the most important scientific discovery that has NOT won the Nobel Prize ... Twice!' (Ted Abel, quoted in Sweatt, 2016).

Questions and Topics Under Current Investigation

- What genes and their products are necessary for the establishment of stable long-term potentiation and memory?
- What is the role of other biochemical cascades not described here?
- How exactly are AMPA receptors trafficked to the membrane and how are they regulated? Are there slots already available and ready to capture the receptors? See Huganir and Nicoll (2013) for a review and possible models.
- How does Hebbian plasticity interact with other forms of "non-Hebbian" plasticity, that is, plasticity not solely confined to one synapse but involving cell-wide changes? How do these cell-wide changes result in behavioural changes?
- Although not discussed here, it has been shown that action potentials not only propagate along the axon, but some 'back-propagate' towards the dendrites. This can lead to a depolarisation of the postsynaptic membrane and 'spike timing-dependent potentiation'. How do the various forms of synaptic plasticity relate to each other?
- What is the connection between changes in synaptic strength and various brain disorders?

References

Abel, T., Nguyen, P.V., Barad, M., Deuel, T.A., Kandel, E.R., and Bourtchouladze, R. (1997). Genetic demonstration of a role for PKA in the late phase of LTP and in hippocampal-based long-term memory. *Cell*, 88, 615–626.

Alberini, C.M., Ghirardi, M., Huang, Y.Y., Nguyen, P.V., and Kandel, E.R. (1995). A molecular switch for the consolidation of long-term memory: cAMP-inducible gene expression. *Annals N Y Academy Science*, 758, 261–286.

Barria, A., Muller, D., Derkach, V., Griffith, L.C., and Soderling, T.R. (1997). Regulatory phosphorylation of AMPA-type glutamate receptors by CaMKII during long-term potentiation. *Science*, 276, 2042–2045.

Bayer, K.U., De Koninck, P., Leonard, A.S., Hell, J.W., and Schulman, H. (2001). Interaction with the NMDA receptor locks CaMKII in an active conformation. *Nature*, 411, 801–805.

Bekkers, J.M., and Stevens, C.F. (1990). Presynaptic mechanism for long-term potentiation in the hippocampus. *Nature*, 346, 724–729.

Bliss, T.V.P., and Collingridge, G.L. (1993). A synaptic model of memory: long-term potentiation in the hippocampus. *Nature*, 361, 31–39.

Bliss, T.V.P., and Lømo, T. (1973). Long-lasting potentiation of synaptic transmission in the dentate area of the anaesthetized rabbit following stimulation of the perforant path. *Journal of Physiology*, 232, 331–356.

Buchs, P.A., and Muller, D. (1996). Induction of long-term potentiation is associated with major ultrastructural changes of activated synapses. *Proceedings of National Academy Science U S A*, 93, 8040–5.

Castro, C.A., Silbert, L.H., McNaughton, B.L., and Barnes, C.A. (1989). Recovery of spatial learning deficits after decay of electrically induced synaptic enhancement in the hippocampus. *Nature*, 342, 545–548.

Collingridge, G.L., Kehl, S.J., and McLennan, H. (1983). The antagonism of amino acid-induced excitations of rat hippocampal CA1 neurones in vitro. *Journal of Physiology*, 334, 19–31.

Derkach, V., Barria, A., and Soderling, T.R. (1999). Ca2+/calmodulin-kinase II enhances channel conductance of alpha-amino-3-hydroxy-5-methyl-4-isoxazolepropionate type glutamate receptors. *Proceedings National Academy Science USA*, 96, 3269–74.

Frey, U., and Morris, R.G.M. (1997). Synaptic tagging and long-term potentiation. *Nature*, 385, 533–536.

Gardoni, F., Caputi, A., Cimino, M., Pastorino, L., Cattabeni, F., and Di Luca, M. (1998). Calcium/calmodulin-dependent protein kinase II is associated with NR2A/B subunits of NMDA receptor in postsynaptic densities. *Journal of Neurochemistry*, 74, 1733–1741.

Giese, K., Fedorov, N.B., Filipkowski, R.K., and Silva, A.J. (1998). Autophosphorylation at Thr286 of the alpha calcium-calmodulin kinase II in LTP and learning. *Science*, 279, 870–873.

Green, E.J., McNaughton, B.L., and Barnes, C.A. (1990). Exploration-dependent modulation of evoked responses in fascia dentata: dissociation of motor, EEG, and sensory factors and evidence for a synaptic efficacy change. *Journal of Neuroscience*, 10, 1455–1471.

Hawkins, R.D., Son, H., and Arancio, O. (1998). Nitric oxide as a retrograde messenger during long-term potentiation in hippocampus. *Progress Brain Research*, 118, 155–172.

Hebb, D. O. (1949). The organization of behavior. Wiley, New York.

Huang, Y.Y., Nguyen, P.V., Abel, T., and Kandel, E.R. (1996). Long-lasting forms of synaptic potentiation in the mammalian hippocampus. *Learning and Memory*, 3, 74–85.

Huganir, R.L., and Nicoll, R.A. (2013). AMPARs and synaptic plasticity: the last 25 years. *Neuron*. 80(3), 704–717.

Jeffery, K.J., and Morris, R.G.M. (1993). Cumulative long-term potentiation in the rat dentate gyrus correlates with, but does not modify, performance in the water maze. *Hippocampus*, 3, 133–140.

Lisman, J.E., and Zhabotinsky, A.M. (2001). A model of synaptic memory: a CaMKII/PP1 switch that potentiates transmission by organizing an AMPA receptor anchoring assembly. *Neuron*, 31, 191–201.

Lynch, G., Larson, J., Kelso, S., Barrionuevo, G., and Schottler, F. (1983). Intracellular injections of EGTA block induction of hippocampal long-term potentiation. *Nature*, 305, 719–721.

Malenka, R.C., Lancaster, B., and Zucker, R.S. (1992). Temporal limits on the rise in postsynaptic calcium required for the induction of long-term potentiation. *Neuron*, 9, 121–128.

Malenka, R.C., and Nicoll, R.A. (1999). Long-term potentiation – a decade of progress? *Science*, 285, 1870–1874.

Malgaroli, A., and Tsien, R.W. (1992). Glutamate-induced long-term potentiation of the frequency of miniature synaptic currents in cultured hippocampal neurons. *Nature*, 357, 134–139.

Martin, S.J., Grimwood, P.D., and Morris, R.G.M. (2000). Synaptic plasticity and memory: an evaluation of the hypothesis. *Annual Review of Neuroscience*, 23, 649–711.

Martin, S.J., and Morris, R.G.M. (2002). New life in an old idea: The synaptic plasticity and memory hypothesis revisited. *Hippocampus*, 12, 609–636.

Mayford, M., Bach, M.E., Huang, Y.Y., Wang, L., Hawkins, R.D., and Kandel, E.R. (1996). Control of memory formation through regulated expression of a CaMKII transgene. *Science*, 274, 1678–1683.

McNaughton, B.L., Barnes, C.A., Rao, G., Baldwin, J., and Rasmussen, M. (1986). Long-term enhancement of hippocampal synaptic transmission and the acquisition of spatial information. *Journal of Neuroscience*, 6, 563–571.

Morris, R.G.M. (1989). Synaptic plasticity and learning: selective impairment of learning in rats and blockade of long-term potentiation in vivo by N-methyl-D-aspartate receptor antagonist AP5. *Journal of Neuroscience*, 9, 3040–3057.

Morris, R.G.M., Anderson, E., Lynch, G.S., and Baudry, M. (1986). Selective impairment of learning and blockade of long-term potentiation by an N-methyl-D-aspartate receptor antagonist, AP5. *Nature*, 319, 774–776.

Morris, R.G.M., Davis, S., and Butcher, S.P. (1990). Hippocampal synaptic plasticity and NMDA receptors: a role in information storage? *Philosophical Transactions Royal Society of London B Biological Sciences*, 329, 187–204.

Moser, E.I. (1995). Learning-related changes in hippocampal field potentials. *Behavioural Brain Research*, 71, 11–18.

Moser, E.I., Krobert, K.A., Moser, M-B., and Morris, R.G.M. (1998). Impaired spatial learning after saturation of long-term potentiation. *Science*, 281, 2038–2042.

Moser, E.I., Moser, M-B., and Anderson, P. (1994). Potentiation of dentate synapses initiated by exploratory learning in rats: dissociation from brain temperature, motor activity, and arousal. *Learning and Memory*, 1, 55–73.

Nicoll, R.A., and Malenka, R.C. (1995). Contrasting properties of two forms of long-term potentiation in the hippocampus. *Nature*, 377, 115–8.

Nikonenko, I., Jourdain, P., Alberi, S., Toni, N., and Muller, D. (2002). Activity-induced changes of spine morphology. *Hippocampus*, 12, 585–591.

Rioult-Pedotti, M-S., Friedman, D., Hess, G., and Donoghue, J.P. (1998). Strengthening of horizontal cortical connections following skill learning. *Nature Neuroscience*, 1, 230–234.

Rogan, M.T., Staubli, U.V., and LeDoux, J.E. (1997). Fear conditioning induces associative long-term potentiation in the amygdala. *Nature*, 390, 604–607.

Sharp, P.E., McNaughton, B.L., and Barnes, C.A. (1985). Enhancement of hippocampal field potentials in rats exposed to a novel, complex environment. *Brain Research*, 339, 361–5.

Shi, S.H., Hayashi, Y., Petralia, R.S., Zaman, S.H., Wenthold, R.J., Svoboda, K., and Malinow, R. (1999). Rapid spine delivery and redistribution of AMPA receptors after synaptic NMDA receptor activation. *Science*, 284, 1811–1816.

Silva, A.J., Kogan, J.H., Frankland, P.W., and Kida, S. (1998). CREB and memory. *Annual Review Neuroscience*, 21: 127–148.

Staubli, U., and Lynch, G. (1990). Stable depression of potentiated synaptic responses in the hippocampus with 1–5 Hz stimulation. *Brain Research*, 513, 113–118.

Sweatt, J.D. (2016). Neural plasticity and behavior – sixty years of conceptual advances. *Journal of Neurochemistry*. 139, 179–199.

Tang, Y.P., Shimizu, E., Dube, G.R., Rampon, C., Kerchner, G.A., Zhuo, M., Liu, G., and Tsien, J.Z. (1999). Genetic enhancement of learning and memory in mice. *Nature*, 401, 63–69.

Toni, N., Buchs, P.A., Nikonenko, I., Povilaitite, P., Parisi, L., and Muller, D. (2001). Remodeling of synaptic membranes after induction of long-term potentiation. *Journal of Neuroscience*, 21, 6245–6251.

Tsien, J.Z., Huerta, P.T., and Tonegawa, S. (1996). The essential role of hippocampal CA1 NMDA receptor-dependent synaptic plasticity in spatial memory. *Cell*, 87, 1327–1338.

Williams, J.H., Li, Y.G., Nayak, A., Errington, M.L., Murphy, K.P., and Bliss, T.V. (1993). The suppression of long-term potentiation in rat hippocampus by inhibitors of nitric oxide synthase is temperature and age dependent. *Neuron*, 11, 877–884.

8 Long-Term Synaptic Plasticity in Mammals II: Long-Term Depression (LTD)

Introduction and Background

The previous chapter introduced the concept of brain plasticity and described how synapses can increase in strength as a result of stimulation, leading to the long-term modification of neurons. This discovery led to great excitement because it provided, for the first time, a mechanism by which learning and memory in the vertebrate brain could be understood from a biological viewpoint. However, as often is the case in science, one discovery leads to many other often unanswered questions. Such questions included whether or not synaptic strength could keep increasing. Is there a limit beyond which synapses could no longer increase? Further, is it possible for synapses to decrease in strength? If so, what could this decrease represent? Intuitively, it makes sense that if synapses can vary in strength, their strength should be able to decrease as well as increase. We know that although certain events are often remembered with great detail and many memories can indeed last a lifetime, most of what we encounter on a daily basis is actually forgotten. The majority of mundane everyday occurrences are not stored; for example, we tend not to remember what we had for lunch three days ago or the song that was playing on the radio while we were driving to work last Monday.

There are, however, a number of very rare cases of people who have fantastic memories. Such people, termed *savants*, do not seem to forget and can recall events and facts in remarkable detail; their memories seem to be without limits. Stephen Wiltshire, a British architectural artist, has an amazing ability to produce drawings of buildings and even cities in exquisite detail following a single exposure. For example, he has drawn London, Tokyo and New York in exacting detail totally from memory following a single 15- to 20-minute helicopter ride. Examples of his work can be seen at www.stephenwiltshire.co.uk/. Another very famous savant is Kim Peek (1951–2009) who was the inspiration behind the 1988 Oscar-winning movie *Rain Man*. Kim Peek was unique not only due to his phenomenal memory for facts, dates and events, but, unlike most savants who tend to have specialised knowledge, his recall spanned across multiple subjects including politics, history, sports and literature. Unfortunately, having such unique memory abilities often comes at a cost; Kim, for example, needed help with general everyday living and was dependent on his father to do many often simple tasks for his entire life.

Although savants are rare they do allow neuroscientists opportunities to further understand the nature of memory. Under normal circumstances some events are remembered and some or not; some events are remembered but are soon forgotten. This suggests that the brain should be very flexible and be able to deal with all of these circumstances. Neural processing should be capable of adjustments that involve both the strengthening and *weakening* of synaptic connections. A biological mechanism that is thought to reflect this synaptic weakening is called long-term depression or LTD.

Long-Term Depression in the Cerebellum

One of the first areas of the brain that was shown to undergo LTD was the cerebellum. The cerebellum takes its name from the Latin, meaning 'little brain'. It is a peach-sized structure resting on the brain stem and is located at the base of the cerebrum (Figure 8.1a). Figure 8.1b shows a section through the beautifully stained black-and-white patterned cerebellum. These black and white regions depict the various layers of the cerebellum and are composed of different cell types: the molecular cell layer, the Purkinje cell layer and the granular cell layer (Figure 8.1c, top). It was in 1837 that a Czech neuroanatomist, Jan Purkinje (1787–1869), first observed and described a number of special cells in the cerebellum that were treelike in construction. The dendrites of these cells had multiple branching that extended into the outer molecular layer, and the cell bodies were organised in a row residing in a layer that is now known as the Purkinje cell layer (black cells in Figure 8.1c). Purkinje cells may be considered the output cells of the cerebellum, as their axons extend deep into the cerebellum, traversing through the granular cell layer, a layer containing many small densely packed granule cells (grey cell in Figure 8.1c), before exiting the structure.

The cerebellum in general and the Purkinje cells in particular receive two major inputs, via two different cell types. The first input to the cerebellum is from the medulla (a structure in the brain stem) via *climbing fibres*. Like ivy, a single fibre climbs up a Purkinje cell and twists around the multiple branches of its dendrites (Figure 8.1c). The second input arises from other regions of the brain stem (for example, the pontine nuclei) via *parallel fibres*. The cell bodies of the parallel fibres are granule cells and reside in the granular cell layer. These cells extend their axons into the molecular layer, where they branch and form a T-junction. These axons then run parallel along the extent of the molecular layer, connecting the dendritic branches of multiple Purkinje cells perpendicularly (Figure 8.1c).

Electrophysiological Mechanisms Underlying Cerebellar LTD

Masao Ito, a Japanese neuroscientist working on the cerebellum, became convinced that the structure could provide an ideal model to test synaptic plasticity. Here in the cerebellum you had a situation whereby two inputs converged (climbing fibres and parallel fibres) onto a single cell (Purkinje cell). According to Hebb (see Chapter 7), early neural computation models (Marr, 1969; Albus, 1971) and the work carried out by Kandel (see Chapter 6) coactivation of the two inputs should lead to some sort of synaptic change between them. Although such predictions were made in the early 1970s, it was not until 1980 that Ito decided to test the idea directly. Taking stimulating and recording electrodes as described in the previous chapters, he began to stimulate the parallel fibres and record EPSPs from the Purkinje cell. He continued this for 15 to 20 minutes to order establish a baseline pattern of response (Figure 8.2a and 8.2b, top). Next, he stimulated the parallel fibres and the climbing fibres *simultaneously* and continued to record from the Purkinje cell (Figure 8.2a and Figure 8.2b, middle). He then returned to stimulating the parallel fibres only and recorded the EPSPs from the Purkinje cell for at least an hour.

What he observed was a *decrease* in the size of the EPSP (Figure 8.2b, bottom). The neural response had been depressed as a result of the coactivation. This depression was observed to last many hours and was subsequently termed *long-term*

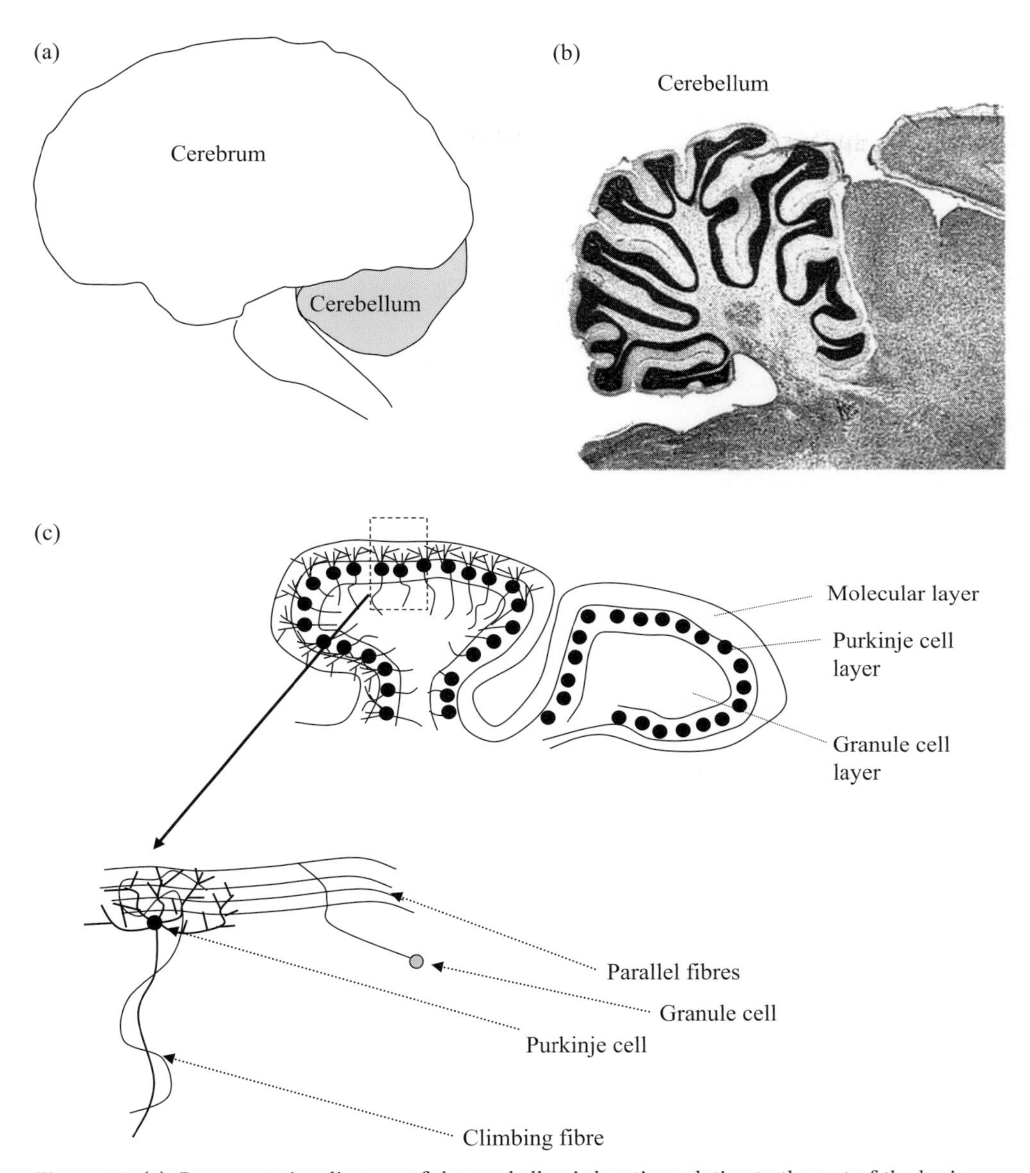

Figure 8.1 (a): Representative diagram of the cerebellum's location relative to the rest of the brain. (b) Section through the cerebellum. Image taken from Pletnikov et al. (2003) Exploring the cerebellum with a new tool: neonatal Borna disease virus (BDV) infection of the rat's brain. *Cerebellum*, 2(1):62–70, with permission from Springer. (c) Representation of the various layers and different cells of the cerebellum.

depression. Despite being confident of his findings, Ito found it hard to convince his colleagues of the results; many felt that no such type of plasticity existed, while others felt potentiation rather than depression should have been observed. Indeed, when he tried to publish his findings in *Nature*, he was turned down but he did eventually get them published later in the *Journal of Physiology*, London (Ito et al., 1982). Over the subsequent years his findings were gradually accepted.

This is only one form of LTD that exists in the cerebellum, and it is termed *conjunctive* or *heterosynaptic LTD* (Ito, 2001) as it requires the joint stimulation of the parallel and the climbing fibres. It is now known that LTD can also occur at the parallel fibre and Purkinje cell synapse, and also at the climbing fibre and Purkinje cell synapse, independently. For example, we can establish a baseline by stimulating the climbing fibres (once every 20 seconds) and recording EPSPs from the Purkinje cell. If we then apply a low-frequency stimulation (of about 5 Hz for 30 seconds) and then return to stimulating once every 20 seconds, depression of the EPSP response should be also observed (Hansel and Linden, 2000). This is termed *homosynaptic LTD* as it only involves one synapse (between the climbing fibres and the Purkinje cell). Such findings show the richness in variety of plasticity that can be observed in neural tissue.

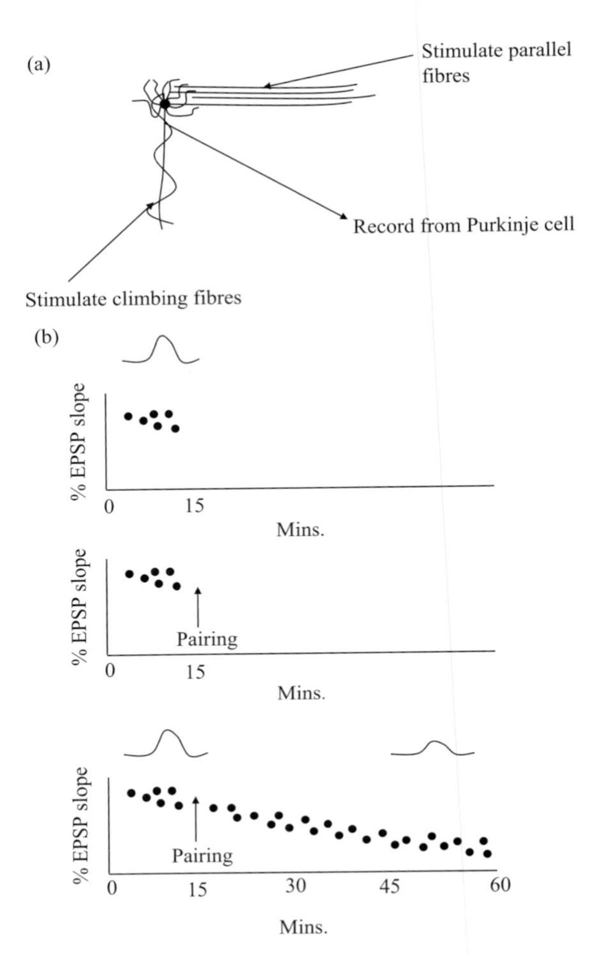

Figure 8.2 (a) Representation of experimental setup for recording LTD in the cerebellum. (b) Stimulation of parallel fibres evokes a response in the Purkinje cell (top). This is followed by the pairing of stimulation of the parallel fibres and climbing fibres (middle). Return to stimulation of parallel fibres now evokes a lower response in the Purkinje cell (bottom). Insets represent the evoked responses.

Molecular Mechanisms Underlying LTD

Following its initial discovery, great strides have been made to try and understand the molecular mechanisms that underpin LTD in the cerebellum. Similar to the approach adopted in the previous chapter on LTP, I will attempt to describe the molecular process step by step.

During Baseline Recordings

During baseline recordings the researcher electrically stimulates the parallel fibres and records the evoked EPSPs from the Purkinje cell. In a sense, an action potential is created along the parallel fibres (Figure 8.3, 1) that reaches the synaptic terminal. At the terminal, glutamate is released into the synaptic cleft (2) between the parallel fibre and the Purkinje cell. Glutamate crosses the cleft and acts on a number of receptors; it can attach to AMPA receptors or to metabotropic glutamate receptors (3). Activation of AMPA receptors allows the entry of mostly Na^+ into the postsynaptic cell (4). This causes rapid depolarisation of the postsynaptic cell (the inside of the cell becomes more positive). Glutamate also binds to a metabotropic receptor that is coupled with guanine nucleotide-binding protein (G protein). Activation of the G protein in turn activates the enzyme phospholipase C (5, PLC). PLC breaks down one lipid molecule (phosphatidylinositol) into two lipid molecules: inositol triphosphate (IP3) and diacylglycerol (DAG) (6). DAG in turn activates protein kinase C (PKC, 7).

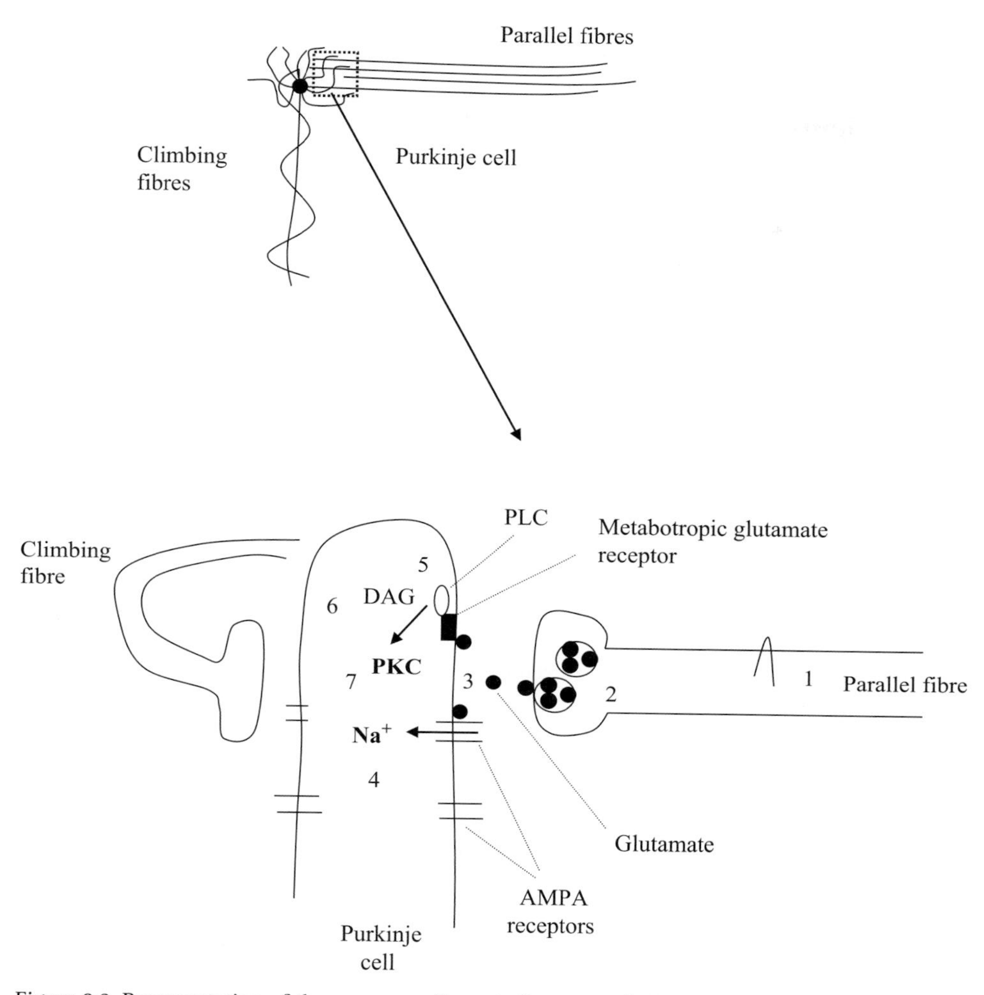

Figure 8.3 Representation of the sequence of events that occur during stimulation of the parallel fibres and recording from the Purkinje cell (baseline). See text for details.

During Pairing

During pairing, the parallel fibres are stimulated and EPSPs are recorded from the Purkinje cell. *At the same time* the climbing fibre is stimulated, again in a sense creating an action potential (Figure 8.4, 1) that travels down the axon and reaches the synaptic terminal. A neurotransmitter (2) is released; the evidence suggests that glutamate is the neurotransmitter (Konnerth et al., 1990), but this remains to be confirmed (Ito, 2001)). The neurotransmitter attaches to the receptor (probably the AMPA receptor) and allows the influx of Na^+ into the Purkinje cell (3). The influx of sodium into the cell leads to greater depolarisation, making the inside of the cell become even more positive. This, in turn, activates calcium channels, which respond to voltage changes (4), thereby opening them and allowing the influx of calcium into the cell (5). Calcium entry is essential for LTD induction (Ito, 2001).

Post-Pairing

Post-pairing, if we return to stimulating the parallel fibres once every 20 seconds and record the EPSPs

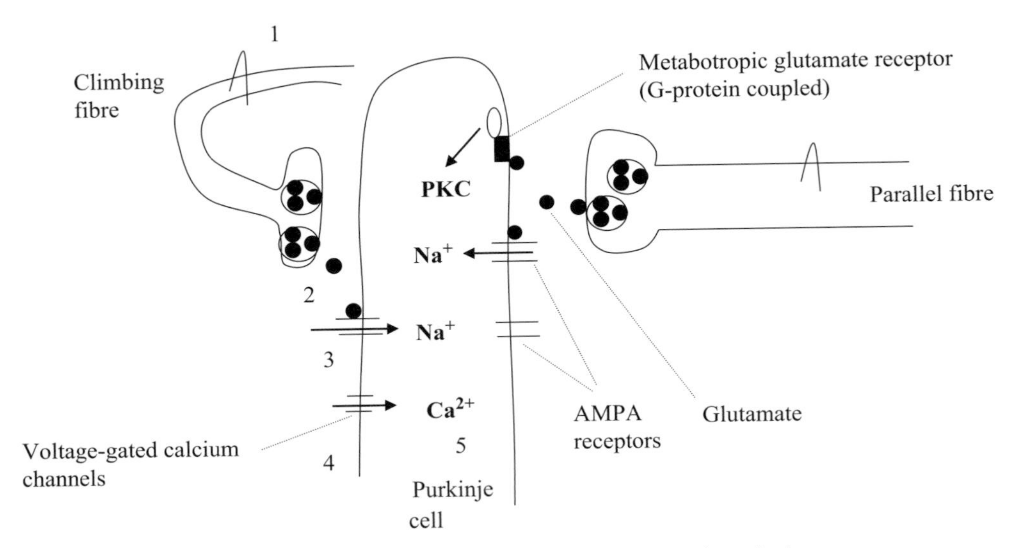

Figure 8.4 Representation of the sequence of events that occur during the paired stimulation of the parallel and climbing fibres. See text for details.

from the Purkinje cell, we would observe a decrease in the size of the EPSP recorded from the Purkinje cell. How does this occur? The Purkinje cell now contains an increased concentration of calcium, sodium and also activation of the PKC cascade. It is thought that the above complex biochemical events leads to the *inactivation* of AMPA receptors involved in the parallel fibre–Purkinje transmission. Currently, there is speculation that 3 possible mechanisms may account for the AMPA receptor inactivation: phosphorylation and internalisation and/or desensitisation.

Phosphorylation and Internalisation

Protein phosphorylation (addition of a phosphate group to proteins) is an important mechanism for many regulatory processes, and we have come across this earlier in our chapter on LTP (Chapter 7). One function of phosphorylation is the regulation of transmitter-gated ion channels, either by increasing or decreasing the conductance through the channel or by interfering with the structure of the receptor. The AMPA receptor is a class that can be subdivided into families based on its subunits, namely GluR1, GluR2, GluR3 and GluR4. The AMPA receptor is known to be particularly susceptible to phosphorylation. For example, Chung et al. (2003) have shown that Purkinje cells from mice that lack the AMPA receptor subunit GluR2 do not show LTD, demonstrating the importance of the AMPA in LTD. These authors further demonstrated that PKC has the ability to phosphorylate the subunit GluR2 AMPA receptor (Chung et al., 2000). Importantly, prevention of phosphorylation of the AMPA receptor prevents LTD (Chung et al., 2003).

Proteins – e.g. glutamate receptor-interacting protein (GRIP), AMPA receptor binding protein (ABP) and protein interacting with C kinase (PICK) – that are attached to the AMPA receptor seem to function as a type of anchor, anchoring the receptor to the membrane. Breaking the affinity of these proteins for the AMPA receptor or removal of these proteins reduces the accumulation of AMPA receptors at the synapse (Osten et al., 2000). There is accumulating evidence that phosphorylation by PKC of the AMPA receptor (especially at the GluR2 subunit) can cause the reduction of the affinity of GRIP to the AMPA receptor (Hirai, 2001). Indeed, disruption of the interaction between the AMPA receptor and these proteins has been shown to block LTD (Xia et al., 2000). If AMPA receptors become loose and are no longer anchored to the membrane, what happens to the AMPA receptors? The receptors now become internalised (Wang and Linden, 2000; Matsuda et al., 2000). Here, a process termed *endocytosis* may occur. The AMPA receptor becomes wrapped in a membrane and removed from the cell surface and brought into the centre of the cell. It is unknown what happens after this.

In summary, if we stimulate the parallel fibres and record the EPSPs from the Purkinje cell, we would observe a decrease in the size of the EPSP recorded. A mechanism for this may run along the lines as follows: The increase in PKC leads to an addition of a phosphate group to the AMPA receptor (Figure 8.5, 1). This addition leads to a breaking of the anchoring process, which normally stabilises the AMPA receptor to a particular area (2). Once this anchor is broken the AMPA receptor becomes free (3). The receptor is then encapsulated in a membrane (4) and is internalised (5). The net effect of this process is the removal of AMPA receptors from the postsynaptic membrane. Therefore, when glutamate is released from the parallel fibre upon each subsequent stimulation (post-pairing), there are no receptors to which it can attach, so there is no influx of sodium into the Purkinje cell, resulting in a reduced EPSP response (6) or LTD.

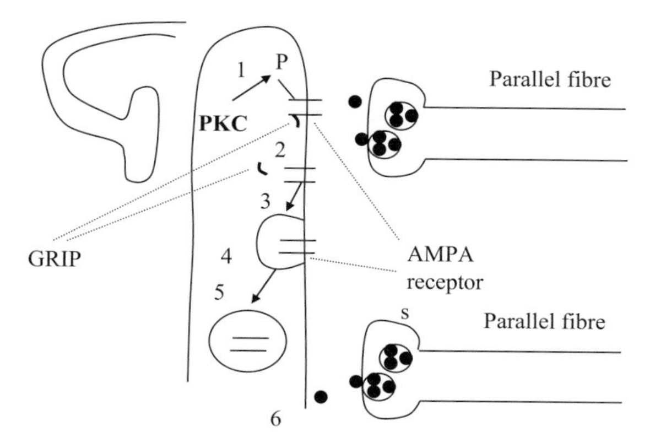

Figure 8.5 Representation of the sequence of events leading to the removal of AMPA receptors, which are thought to underpin LTD. See text for details.

Desensitisation

Another suggestion is that AMPA receptor becomes desensitised, that is, the receptor does not respond as efficiently to the neurotransmitter as it would normally. Again, this desensitisation would arise due to the phosphorylation of the AMPA receptor. However, Linden (2001) argues that LTD is not associated with changes in AMPA receptor kinetics. LTD is not, for example, altered by application of a compound (cyclothiazide) that is known to desensitise AMPA receptors (but see Hemart et al., 1994). In addition, the author suggests that there is no change in glutamate affinity towards the receptors after LTD. Linden (2001) suggests that cerebellum LTD is solely due to the reduction in the number of functional AMPA receptors, most probably due to the internalisation process described earlier.

LTD in the Hippocampus

LTD has been described in many areas of the brain, including the neocortex (Kirkwood et al., 1993), the striatum (Lovinger et al., 1993) and the above-mentioned cerebellum (Linden and Connor, 1992, Ito, 2001). Around the same time as its discovery in the cerebellum, attention soon switched to the examination of LTD in the hippocampus. The hippocampus, as well as being the primary site for learning and memory, is an anatomically well-studied structure in which the properties of long-term potentiation have been elucidated (see Chapter 7). Heterosynaptic depression was initially observed in area CA1 of the hippocampus by Gary Lynch in 1977; it was shown that LTP in one set of synapses seemed to cause a depression in a second set of synapses (Lynch et al., 1977; see also Bear and Malenka, 1994 for further details). Following this, the question arose whether synapses in the hippocampus could also show homosynaptic LTD. The search for this form of LTD in the hippocampus focused particularly on the Schaffer collaterals, the group of neurons that project from area CA3 to area CA1 (Figure 8.6a, see also Bear, 2003).

Over the years, many attempts were made to weaken these synapses; different frequencies, different intensities and different durations were all used in the stimulation protocol to find LTD in this pathway. Then, in 1992 Serena Dudek and Mark Bear, using a prolonged low-frequency stimulation protocol, found that LTD could be reliably obtained (Dudek and Bear, 1992). In a setup similar to that previously described, a stimulating electrode was placed into the Schaffer collaterals and a small electrical pulse was generated once every 20 seconds. At the same time, EPSPs were recorded from area CA1 (Figure 8.6a); recordings were made for 15 minutes to establish a baseline response (Figure 8.6b, top). Next, the authors applied a low-frequency stimulation (LFS) protocol of 900 pulses at a frequency of 1 Hz (1 per second, Figure 8.6b, middle). Then, following a return to normal baseline stimulation, a stable long-lasting decrease in the size of the EPSP recorded from area CA1 was observed (Figure 8.6b, bottom). LTD in the hippocampus for the first time could be reliably demonstrated.

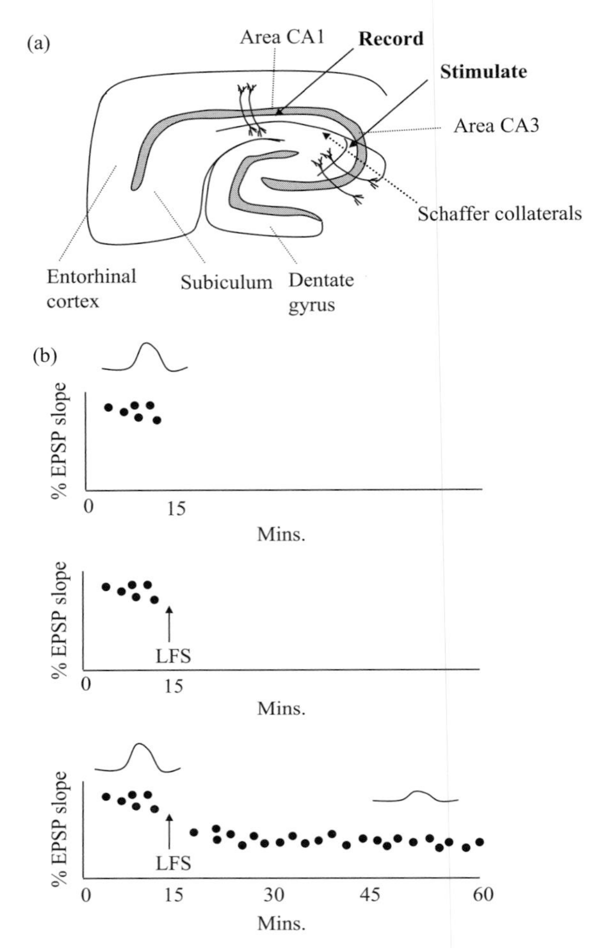

Figure 8.6 (a) Representation of experimental setup for recording LTD in the hippocampus. (b) Stimulation of Schaffer collaterals evokes a response (top) in area CA1. This is followed by low-frequency stimulation (LFS, middle). Return to baseline stimulation of Schaffer collaterals now evokes a lower response (bottom). Insets represent the evoked responses.

Properties of Hippocampal LTD

Over the following years a number of critical properties of LTD emerged. One key finding of hippocampal LTD was its specificity: only the synapses that had received the conditioning stimuli showed depression. Unlike the type of synaptic depression reported by Gary Lynch earlier (Lynch et al., 1977), other synapses using this protocol were unaffected. Therefore, it was concluded that the LTD was homosynaptic in nature, and Dudek and Bear (1992) were confident that a new form of hippocampal plasticity had been discovered. However, the high number of stimuli used was a concern (900 pulses in all); perhaps the observed depression of the neural response was simply a result of using this stimulation protocol (Bear, 2003). To test for this Dudek and Bear varied the frequency while keeping the number of pulses constant. The authors found that this resulted in different responses. Although 900 pulses at a low frequency (e.g. 1 Hz) resulted in LTD, 900 pulses at an intermediate frequency (e.g. 10 Hz) resulted in no change in synaptic response, and 900 pulses at a high frequency (e.g. 50 Hz) resulted in LTP (Figure 8.7 and Bear, 2003). This suggested that it was the frequency applied, rather the number of stimuli, that was important for changes in neural plasticity.

LTD also seems to be age-dependent. For example, Mulkey and Malenka (1992) found

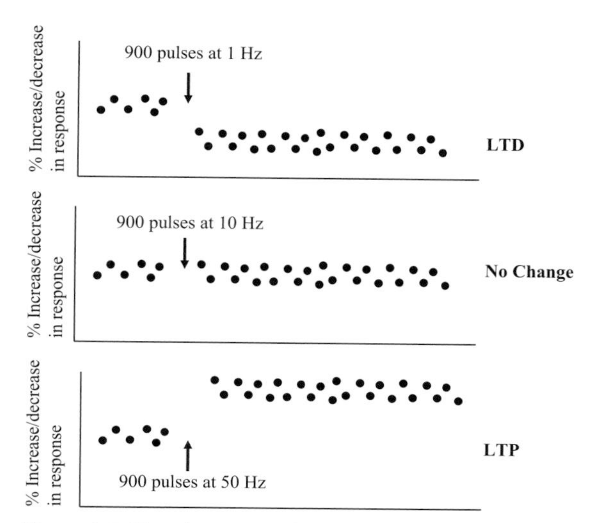

Figure 8.7 Neural response depends on the frequency applied. Low-frequency stimulation produces LTD (top), intermediate levels of frequency stimulation produce no change in neural response (middle), high-frequency stimulation produces LTP (bottom).

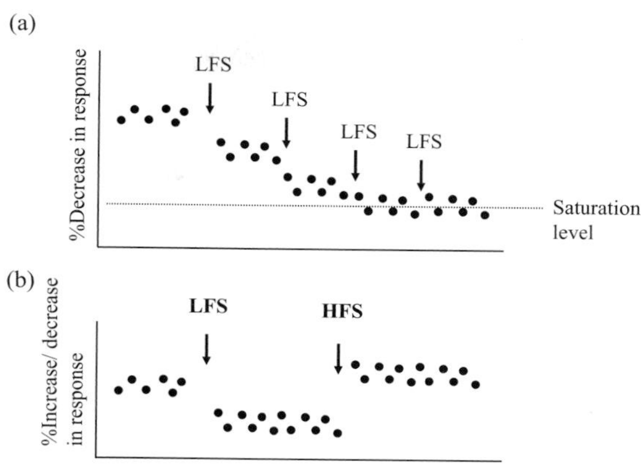

Figure 8.8 (a) Multiple trains of LFS will eventually saturate the synapse so that no further decreases in response are observed. (b) Long-term depression can be reversed by an application of HFS.

that the same low-frequency stimulation in area CA1 from 2-week-old rats produced a relative decrease in response that was double that seen in area CA1 of a 5-week-old rat. Other properties of LTD are similar to those observed with LTP (see Chapter 7). For example, LTD reaches a saturation point following multiple applications of LFS (Figure 8.8a), beyond which the neural response does not seem get any lower. Another important property of LTD is that it is reversible. It has been well-established that a low-frequency stimulation protocol can partially reverse LTP ((Barrionuevo et al., 1980; see Chapter 7), a phenomenon termed *depotentiation*. Likewise, if high-frequency stimulation (HFS) is applied to depressed synapses, a partial reversal can be obtained; this is called *de-depression*. However, limited research has been done examining this (Figure 8.8b).

The Mechanism of Hippocampal LTD

The mechanism underlying how hippocampal LTD is produced is similar to that described previously for both LTP (Chapter 7) and cerebellar LTD, with a few important differences. In brief, during baseline stimulation of the Schaffer collaterals, an action potential is generated, which travels down the axon, releasing glutamate into the synaptic cleft. Glutamate attaches onto AMPA receptors on the postsynaptic membrane of area CA1, opening the channel and allowing Na^+ ions to enter the cell. The entry of ions results in excitatory postsynaptic potentials (EPSPs), which are continuously recorded for approximately 10–15 minutes. Next, a low-frequency stimulation protocol (900 pulses at 1 Hz) is applied to the Schaffer collaterals. This results in a prolonged amount of glutamate released into the synaptic cleft over an extended period (e.g. 900 seconds or 15 minutes). The glutamate attaches to a number of AMPA receptors on the postsynaptic membrane. This prolonged activation leads to a weak depolarisation of the postsynaptic membrane. As a result of this, the Mg^{2+} block on the NMDA receptor is *partially* removed (see Chapter 7 for comparison with LTP). We then return to baseline stimulation. Now when an action potential reaches the presynaptic terminal, glutamate is released (Figure 8.9, 1) and attaches to receptors on the presynaptic membrane (2). The AMPA receptors are activated, allowing sodium to enter the cell, and because the Mg^+ block on the NMDA receptor is only partially removed, a trickle

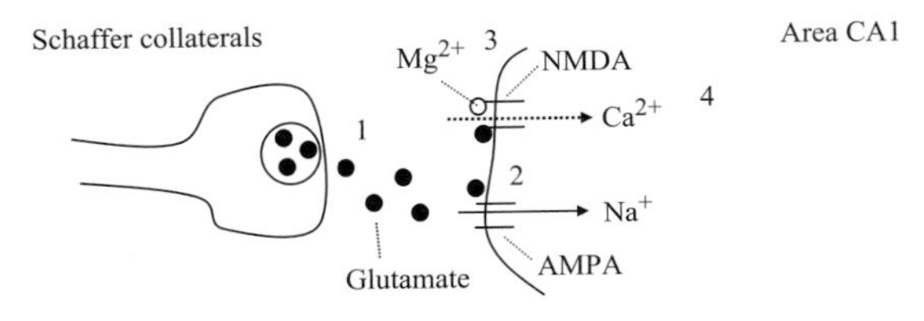

Figure 8.9 Representation of the sequence of events that occur following LFS in hippocampal neurons. See text for details.

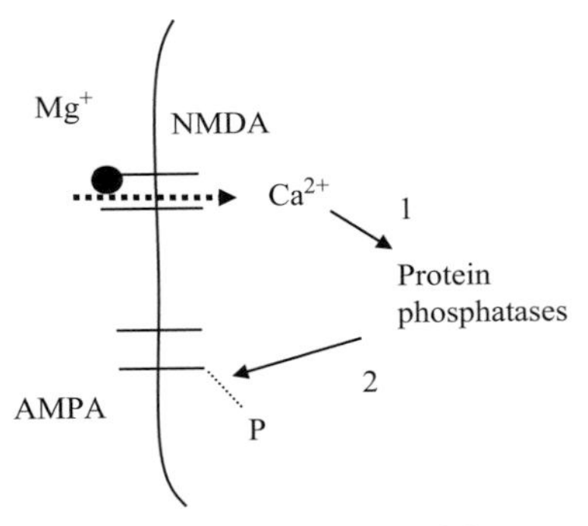

Figure 8.10 Representation of the sequence of events that helps maintain LTD.

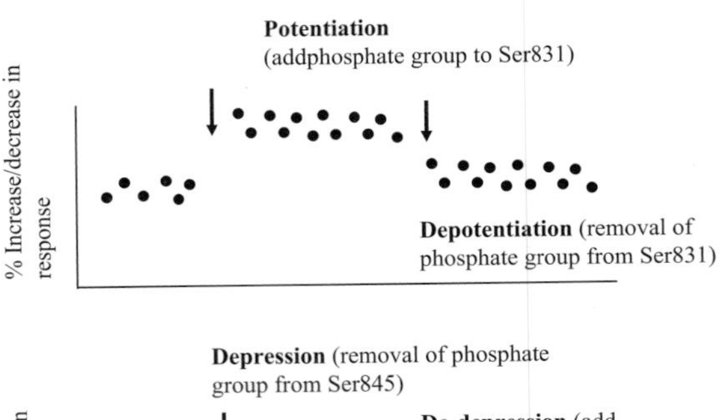

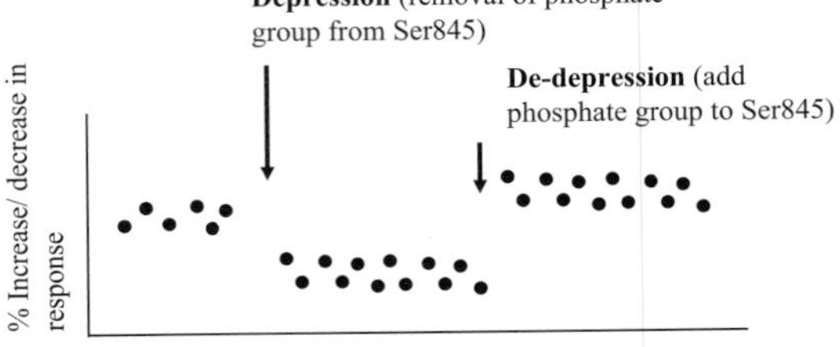

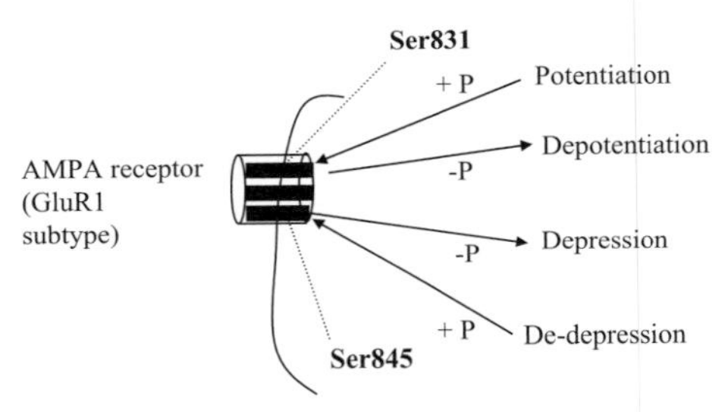

Figure 8.11 Addition or removal of phosphate groups to the Ser831 site of the AMPA receptor is involved in potentiation and depotentiation, respectively (top). Addition and removal of phosphate groups to the Ser845 site of the AMPA receptor are involved in depression and de-depression, respectively (middle). Bottom illustration summarises results.

of calcium ions also enter into the postsynaptic membrane (4). The amount of calcium is the critical factor in determining whether LTD is initiated.

Maintenance of Hippocampal LTD

The discovery that the NMDA receptor and the influx of calcium (Mulkey and Malenka, 1992) were required in LTD came as a major surprise. How was it possible that both LTP and LTD, seemingly opposite phenomena, were dependent on the same mechanisms? The key difference between the two forms of plasticity lies in the amount of calcium that enters the postsynaptic membrane. Different enzymes seem to respond to different amounts of calcium. LTP is believed to occur due to the entry of a large amount of calcium into the postsynaptic membrane which, in turn, activates kinases (see Chapter 7; Bear and Malenka, 1994). On the other hand, prolonged but modest amounts (<1 μM) of calcium activate a different set of enzymes called *protein phosphatases* (Figure 8.10, 1). These protein phosphates act on AMPA receptors and dephosphorylate them (removal of a phosphate group, 2, Mulkey et al., 1994). In particular, the GluR1 subtype of the AMPA receptor seems to be the key element in this (Lee et al., 1998).

Up to now LTD has seemed to act as a mirror image of LTP. Both involve the NMDA receptor, both allow calcium to enter the postsynaptic membrane. During LTP, a large concentration of calcium enters the cell, whereas during LTD a small amount of calcium enters. Both activate a number of enzymes; during LTP, a phosphate group is added to the AMPA receptor via CaMKII, and during LTD a phosphate group is removed from the AMPA receptor via protein phosphatases. However, the phosphate groups that are added or removed do not occur at the same location on the GluR1 receptor, as would be expected if LTP were the exact mirror image of LTD. Rather, during LTP CaMKII seems to activate a specific site on the AMPA receptor called Ser831, but during LTD dephosphorylation occurs at another site on the AMPA receptor called Ser845 (Barria et al., 1997; Lee et al., 1998; Lee et al., 2000). Furthermore, Kameyama et al. (1998) and Lee et al. (2000) found that potentiation and its reversal depotentiation involve the AMPA site Ser831.

Potentiation involves the addition of the phosphate group to this site and depotentiation involves the removal of the phosphate from Ser831 (Figure 8.11 top). On the other hand, depression and its reversal, *de-depression,* involve the AMPA site Ser845. Depression involves the removal of a phosphate group from this site and de-depression involves the addition of a phosphate group to Ser845 (Bear, 2003 and Figure 8.11 (middle); see Figure 8.11 (bottom) for a summary. So the various forms of plasticity within the hippocampus seem to be linked. Finally, the maintenance of hippocampal LTD over the long term, similar to LTD in the cerebellum, is thought to involve a reduction in the number of AMPA receptors in the postsynaptic membrane (Malinow and Malenka, 2002). It is generally thought that this reduction in the number comes about through the internalisation of existing AMPA receptors that have been dephosphorylated (Ehlers, 2000). The net effect of the dephosphorylation process is the removal of AMPA receptors from the postsynaptic membrane. Therefore when glutamate is released from the presynaptic terminal, there are a smaller number of receptors to which the neurotransmitter can attach, therefore there is less influx of sodium into the postsynaptic cell, reducing the EPSP response.

Summary

We have seen that synaptic plasticity plays a crucial role in learning and memory. However, what happens if our information needs to be updated and added to? What happens if we need to remove incorrect material from our memories? How are mistakes corrected as we learn? Neural circuits need to be flexible; they should be able to vary in strength, both upwards and downwards. The discovery of LTD shows us that neurons are indeed capable of being flexible; they can readily adjust the gain, allowing learning itself to be flexible. In this chapter we have described LTD in two very different structures (cerebellum and hippocampus), and although the phenomenon and molecular mechanisms are similar in many ways, LTD in each structure may play a different role.

Questions and Topics Under Current Investigation

- LTD involves the decrease in synaptic strength and has often been associated with forgetting or the suppression of memory; however, LTD (as we shall see in the next chapter) may be important in learning process. Its role in updating and unlearning irrelevant material is currently being investigated.
- We have seen that potentiation can become saturated, leading to no further increases in synaptic strength. How do savants with seemingly unlimited memories overcome such

saturation levels? Is it that saturation levels do not exist for such people, or is there a readjustment of neural strength that allows more material to be constantly added?

- As we age and develop age-related diseases, our neural circuitry becomes less flexible. How does this relate to changes in both LTP and LTD? Could LTD be used as a method to re-establish neural equilibrium in synapses that are overstimulated (e.g. in epilepsy and addiction) and show impaired behavioural flexibility?
- Recent evidence (Coultrap et al., 2014) suggests that CaMKII can mediate LTD (as well as LTP), so much more research is needed at the molecular level.

References

Albus, J.S. (1971). A theory of cerebellar function. *Mathematical Biosciences*, 10, 25–61.

Barria, A., Muller, D., Derkach, V., Griffith, L.C., and Soderling, T.R. (1997). Regulatory phosphorylation of AMPA-type glutamate receptors by CaM-KII during long-term potentiation. *Science*, 276, 2042–2045.

Barrionuevo, G., Schottler, F., and Lynch, G. (1980). The effects of low frequency stimulation on control and 'potentiated' synaptic responses in the hippocampus. *Life Science*, 27, 2385–2391.

Bear, M.F. (2003). Bidirectional synaptic plasticity; from theory to reality. *Philosophical Transactions of the Royal Society of London B*, 358, 649–655.

Bear, M.F., and Malenka, R.C. (1994). Synaptic plasticity: LTP and LTD. *Current Opinion in Neurobiology*, 4, 389–399.

Chung, H.J., Steinberg, J.P., Huganir, R.L., and Linden, D.J. (2003). Requirement of AMPA receptor GluR2 phosphorylation for cerebellar long-term depression. *Science*, 300, 1751–1755.

Chung, H.J., Xia, J., Dcannevin, R.H., Zhang, X., and Huganir, R.L. (2000). Phosphorylation of the AMPA receptor subunit GluR2 differentially regulates its interaction with PDZ domain-containing proteins. *Journal of Neuroscience*, 20, 7258–7267.

Coultrap, S.J., Freund, R.K., O'Leary, H., Sanderson, J.L., Roche, K.W., Dell'Acqua, M.L., and Bayer, K.U. (2014). Autonomous CaMKII mediates both LTP and LTD using a mechanism for differential substrate site selection. *Cell Report*, 6(3):431–437.

Dudek, S.M., and Bear, M.F. (1992). Homosynaptic long-term depression in area CA1 of hippocampus and effects of N-methyl-D-aspartate receptor blockade. *Proceedings of the National Academy of Sciences USA*, 89, 4363–4367.

Ehlers, M.D. (2000). Reinsertion or degradation of AMPA receptors determined by activity-dependent endocytic sorting. *Neuron*, 28, 511–525.

Hansel, C., and Linden, D.J. (2000). Long-term depression of the cerebellar climbing fibre-Purkinje neuron synapse. *Neuron*, 26, 473–482.

Hemart, N., Daniel, H., Jaillard, D., and Crepel, F. (1994). Properties of glutamate receptors are modified during long-term depression in rat cerebellar Purkinje cells. *Neuroscience Research*, 19, 213–21.

Hirai, H. (2001). Modification of AMPA receptor clustering regulates cerebellar synaptic plasticity. *Neuroscience Research*, 39, 261–267.

Ito, M. (2001). Cerebellar long-term depression; Characterization, signal transduction, and functional roles. *Physiological Reviews*, 81, 1143–1195.

Ito, M., Sakurai, M., and Tongroach P. (1982). Climbing fibre induced depression of both mossy fibre responsiveness and glutamate sensitivity of cerebellar Purkinje cells. *Journal of Physiology*, 324, 113–134.

Kameyama, K., Lee, H.K., Bear, M.F., and Huganir, R.L. (1998). Involvement of a postsynaptic protein kinase A substrate in the expression of homosynaptic long-term depression. *Neuron*, 21, 1163–1175.

Kirkwood, A., Dudek, S.M., Gold, J.T., Aizenman, C.D., and Bear, M.F. (1993). Common forms of synaptic plasticity in the hippocampus and neocortex *in vitro*. *Science*, 260, 1518–1521.

Konnerth, A., Llano, I., and Armstrong, C.M. (1990). Synaptic currents in cerebellar Purkinje cells. *Proceedings of the National Academy of Sciences USA*, 87, 2662–2665.

Lee, H.K., Kameyama, K., Huganir, R.L., and Bear, M.F. (1998). NMDA induces long-term synaptic depression and dephosphorylation of GluR1 subunit of AMPA receptors in the hippocampus. *Neuron*, 21, 1151–1162.

Lee, H.K., Barbarosie, M., Kameyama, K., Bear, M.F., and Huganir, R.L. (2000). Regulation of distinct AMPA receptor phosphorylation sites during bidirectional synaptic plasticity. *Nature*, 405, 955–959.

Linden, D.J. (2001). The expression of cerebellar LTD in culture is not associated with changes in AMPA-receptor kinetics, agonist affinity or unitary conductance. *Proceedings of the National Academy of Sciences USA*, 98, 14066–14071.

Linden, D.J., and Connor, J.A. (1992). Long term depression of glutamate currents in cultured cerebellular Purkinje cells does not require nitric oxide signaling. *European Journal of Neuroscience*, 4, 10–15.

Lovinger, D.M., Tyler, E.C., and Merritt, A. (1993). Short- and long-term synaptic depression in rat neostriatum. *Journal of Neurophysiology*, 70, 1937–1949.

Lynch, G.S., Dunwiddie, T., and Gribkoff, V. (1977). Heterosynaptic depression: A postsynaptic correlate of long term potentiation. *Nature*, 266, 737–739.

Malinow, R., and Malenka, R.C. (2002). AMPA receptor trafficking and synaptic plasticity. *Annual Review of Neuroscience*, 25, 103–126.

Marr, D. A. (1969). A theory of the cerebellar cortex. *Journal of Physiology*, 202, 437–470.

Matsuda, S., Launey, T., Mikawa, S., and Hirai, H. (2000). Disruption of AMPA receptor GluR2 clusters following long-term depression induction in cerebellar Purkinje neurons. *EMBO J.* 19, 2765–74.

Mulkey, R.M., Endo, S., Shenolikar, S., and Malenka, R.C. (1994). Calcineurin and inhibitor-1 are components of a protein-phosphatase cascade mediating hippocampal LTD. *Nature*, 369, 486–488.

Mulkey, R.M., and Malenka, R.C. (1992). Mechanisms underlying induction of homosynaptic long-term depression in area CA1 of the hippocampus. *Neuron*, 9, 967–975.

Osten, P., Khatri, L., Perez, J.L., Kohr, G., Giese, G., Daly, C., Schulz, T.W., Wensky, A., Lee, L.M., and Ziff, E.B. (2000). Mutagenesis reveals a role for ABP/GRIP binding to GluR2 in synaptic surface accumulation of the AMPA receptor. *Neuron*, 27, 313–325.

Wang, Y.T., and Linden, D.J. (2000). Expression of cerebellar long-term depression requires postsynaptic clathrin-mediated endocytosis. *Neuron*, 25, 635–647.

Xia, J., Chung, H.J., Wihler, C., Huganir, R.L., and Linden, D.J. (2000). Cerebellar long-term depression requires PKC-regulated interactions between GluR2/3 and PDZ domain-containing proteins. *Neuron*, 28, 499–510.

9 Eye-Blink Conditioning

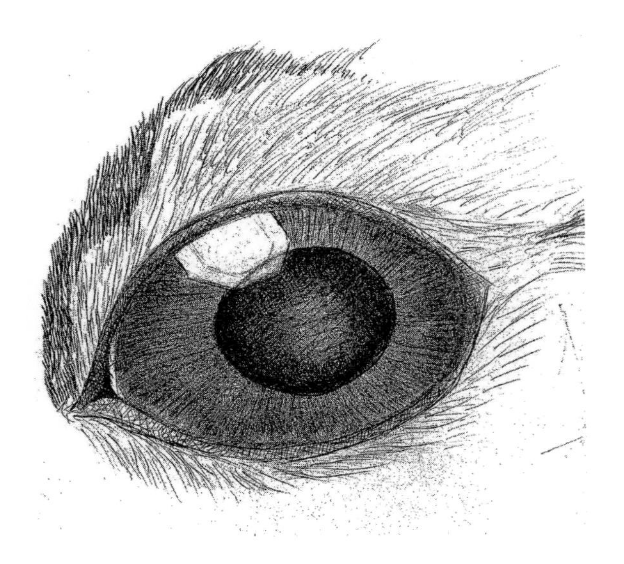

Introduction and Background

Reflexes are involuntary movements that occur almost instantaneously in response to certain stimuli. They are unplanned, uncontrollable motor movements and are considered to involve the most basic and fundamental neural mechanism. As an example, imagine you were to strongly tap a person's bent leg just below the kneecap; however, be careful, because once you do this the person's lower leg immediately strikes up and you may receive an unpleasant kick. This reflex is known as the patellar or knee-jerk reflex (see Figure 9.1a). It is very dramatic to observe, but is also very useful medically to test for cerebellar diseases and some cases of hyperthyroidism and anxiety, whereby exaggerated reflexes are often observed.

Our bodies have many more reflexes; some are even observed in newborn babies that are not observed in adults. One such reflex is the sucking reflex, which causes the newborn to immediately suck as soon as something comes near the roof of its mouth. Another reflex, known as the palmar grasp reflex, causes a baby to tightly grasp its fingers around an object when it is placed in the baby's palm.

The study of reflexes has a long history dating back to the French philosopher and mathematician René Descartes (1596–1650). Descartes viewed behaviour as being akin to the motion of a machine, after being inspired by the moving statutes in the royal garden of St. Germain in Paris. For example, when someone approached the statue of Diana, the goddess would move and hide behind a rosebush. If she was further pursued, a statute of Neptune would appear, pointing his trident down at the curious visitor. All of the statutes operated through a complex hydraulic system; by stepping on a hidden pad, the operator would let water flow through pipes into the statues, causing them to move. Descartes suggested that our bodies also operated in a similar fashion: animal spirits, rather than water, flowed into muscles through nerves, causing them to move. He further suggested that all behaviour could be divided into voluntary and involuntary actions. Indeed, all animal behaviour and involuntary human behaviour could be accounted for by reflexes.

You get a sense of Decartes' philosophy from the following description of the eye-blink reflex described in his work *Les Passions de l'âme* (*Passions of the Soul*, number 13, 1641):

> If someone quickly thrust his hand against your eyes ... even though we know him to

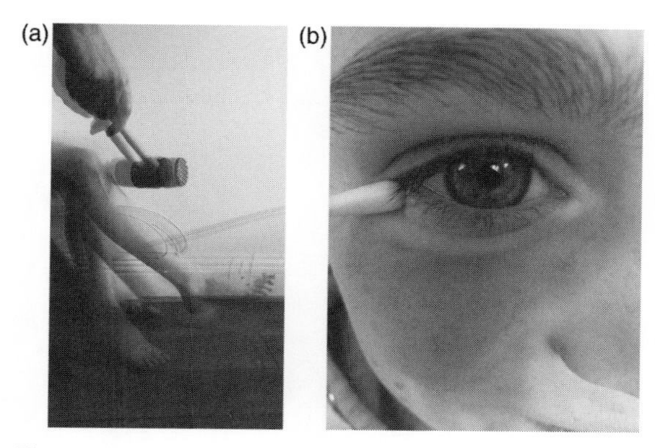

Figure 9.1 Examples of two reflexes: (a) knee-jerk (b) eye-blink.

> be our friend, that he only does it by fun, and will take great care not to hurt us, we have all the same trouble in preventing ourselves from closing them ... it is because the machine of our body is so formed that the movement of this hand towards our eyes excites another movement in our brain, which conducts the animal spirits into the muscles which causes the eye lids to close.
> (quoted in Fearing, 1970, p. 27)

Reflexes perform very important functions. Many serve as part of our body's defence system. Such reflexes include the eye-blink response and the constriction of our pupils in response to light as well as the simple acts of coughing and sneezing. Other reflexes and involuntary moments are critical for maintaining homeostasis and life itself, including the involuntary actions of breathing, blood pressure and our continuous heartbeat. In previous chapters we described the reflex defence mechanisms of the *Aplysia* and how it curled up its gill in order to protect itself. We observed how by associating this reflex with another action (the squirting of water) the animal could learn (become classically conditioned) and form long-term memories. Because the invertebrate had such a simple nervous system, it was possible to identify the neural circuit underlying this behaviour and describe its cellular and molecular machinery. However, the question arose whether it was ever possible to find a similar circuit in the mammalian brain that could also learn a particular response, even with the simplest of behaviours like reflexes.

In this chapter, we examine the eye-blink reflex and show that a circuit underlying this behaviour has indeed been identified. Furthermore, we describe how when a stimulus that does not normally produce an eye-blink reflex is paired with stimulus that does, learning or conditioning can occur. We examine the changes that occur within the reflex circuit, including the molecular mechanisms, and suggest that a mechanism like long-term depression (LTD), rather than one like LTP (see Chapters 7 and 8) is an important element in this learning process.

Behavioural Conditioning of the Eye-Blink Reflex

In a laboratory situation, eye-blink conditioning can be shown easily. In a typical experiment (Figure 9.2a), a tone is sounded in the presence of an animal, usually a rabbit; this has little effect on the animal's behaviour. When, however, a puff of air is delivered to an eyelid of the animal, it immediately blinks, thereby showing a strong defensive reflex (Figure 9.2b). If, every time the tone is sounded the animal also receives a puff of air to the eyelid (Figure 9.2c), eventually the animal will close its eye in response to the tone alone in the absence of the puff of air (Figure 9.2d). Such conditioned responses can persist for a long period of time (Schreurs, 1993), demonstrating a strong memory effect. An alternative method to examine eye-blink conditioning is depicted in Figure 9.3. The focus in this situation is on the pairing trials. Under this experimental paradigm, a tone is sounded for 300 ms; the interval between the tone onset and puff of air onset can range between 100 and 1000 ms, and the puff of air is then delivered 250 ms after the tone onset (Figure 9.3). During the early stages of pairing, the eyelid responds only to the air puff, but over an extended period of time, and with the presentation of multiple paired trials, the eyelid will respond to the tone (even before the air puff is delivered).

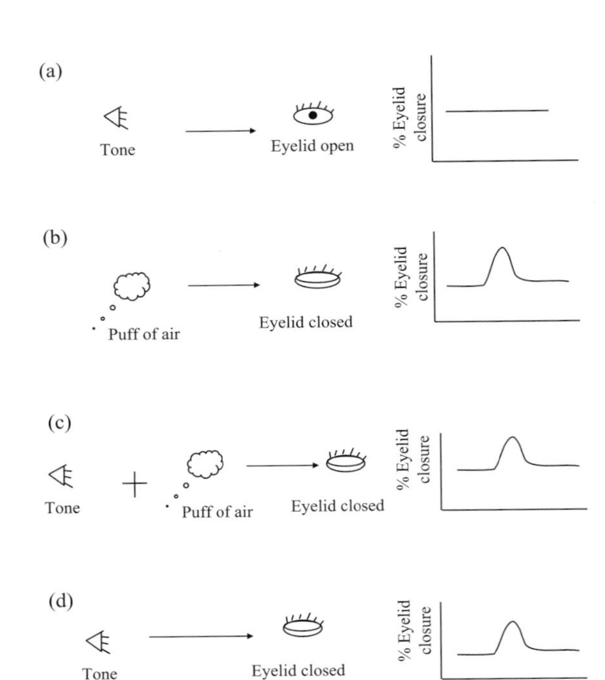

Figure 9.2 Experimental setup for eye-blink conditioning. (a) A tone does not normally produce an eye-blink response, whereas (b) a puff of air to the eye will evoke a strong response. (c) Following multiple pairing of the tone and air puff, the tone alone will now produce a strong eye-blink response (d).

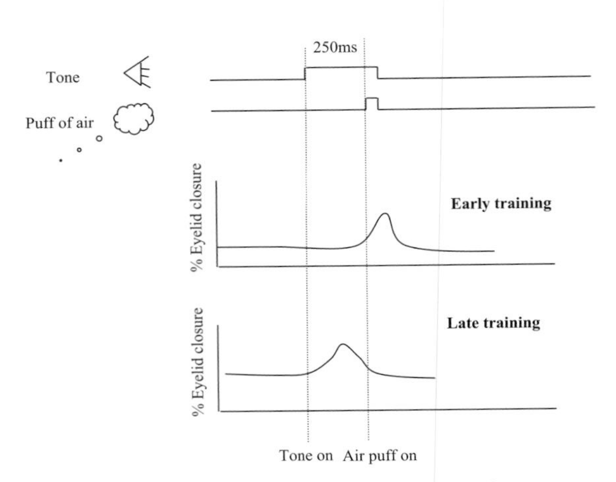

Figure 9.3 An alternative experimental setup that sees a shift in the eye-blink response towards the onset of the tone, rather than immediately at the air puff, with increased training.

Anatomy of the Cerebellum

Considerable research has been done over many decades examining the neural pathways that underlie eye-blink conditioning. Evidence suggests that two brain areas are critically involved in this type of learning. The first area is the hippocampus, particularly areas CA1 and CA3, and the second area is the cerebellum. Although we introduced the cerebellum in the last chapter with the discussion of LTD, it may be useful to highlight some of the important anatomical features of this structure again. Figure 9.4a shows the location of the cerebellum relative to the rest of the brain in the rabbit. It lies to the back of the cerebral cortex and is directly connected to the brain stem. Figure 9.4b shows a close-up representation of the cerebellum. It is divided into two hemispheres (left and right), which are well recognisable due to their compact folding. The cerebellum also contains a narrow midline region called the vermis. If we take a cross-section of the cerebellum (Figure 9.4c), we can see a number of important nuclei, including the dentate, fastigial and globose nuclei. These nuclei are the main output structures of the cerebellum and project to a variety of targets in the cerebral cortex. The majority of neurons within these nuclei use glutamate as their neurotransmitter.

The cerebellum is thought to contain more cells than the rest of the brain, despite its small size. Indeed, it has been suggested that the cerebellum has over 3 times the number of neurons compared to the cerebral cortex. Despite the large number of cells, the cerebellum only takes up a fraction of the brain's overall volume, because the majority of cells within the cerebellum are granule cells, which, as their name suggests, are very small. In addition to granule cells, the cerebellum contains other types of neurons, which we met in the previous chapter: Purkinje cells, Golgi cells and stellate or basket cells. Granule cells are glutamatergic in nature, and Golgi cells and stellate types are inhibitory interneurons. Purkinje cells are large GABAergic neurons and function as the main output source of the cerebellum. As described in Chapter 8, Purkinje cells receive two major inputs; one from *climbing fibres* that arise from the medulla region of the brain stem and another from *parallel*

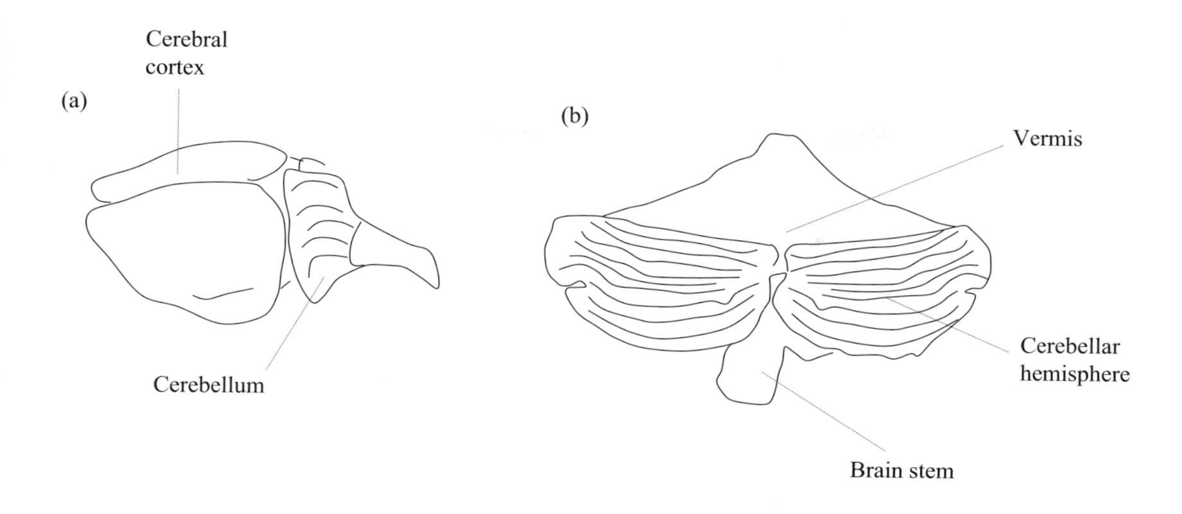

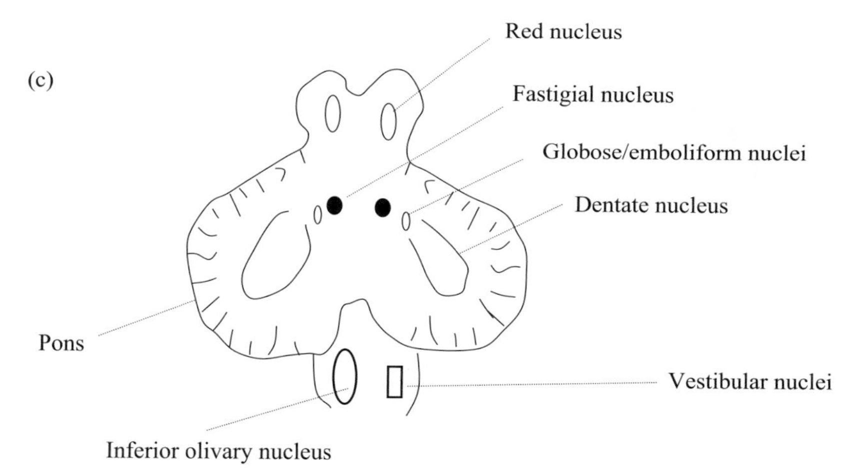

Figure 9.4 (a) Location of cerebellum relative to the rest of the brain. (b) A close-up representation of the cerebellum and its various structures. (c) A section through the cerebellum, highlighting a number of internal structures and nuclei.

fibres that arise from the pontine nuclei of the brain stem. The parallel fibres are from granule cells that extend their axons into the molecular layer, which then run parallel to the brain surface, making contact with multiple Purkinje cells (Figure 9.5).

Cerebellar Circuits Underlying Eye-Blink Conditioning

When a tone (the conditioned stimulus, CS) is presented to the ear of the rabbit, it activates the vestibulocochlear nucleus, a structure located in the brain stem. The information is then sent to the pontine nuclei, also located in the brain stem. From these nuclei, mossy fibres transmit tone information directly to the cerebellum (Eccles et al., 1967; Voogd and Glickstein, 1998). However, within the cerebellum, the mossy fibres divide, with one set innervating the granule cells, while the other set of fibres innervates the deep cerebellar nuclei, particularly the interpositus nucleus. The interpositus nucleus is composed of the globose nucleus and the emboliform nucleus

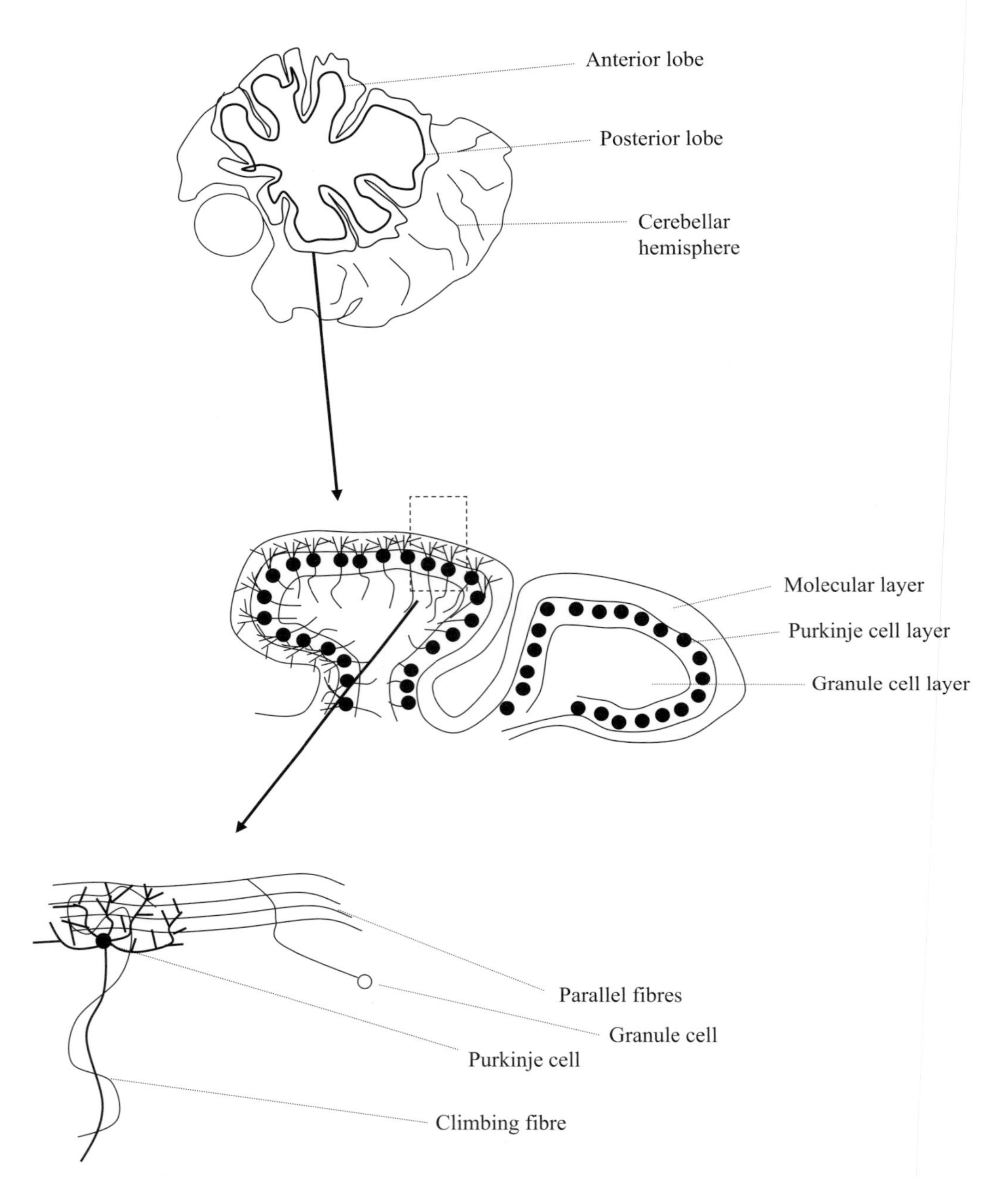

Figure 9.5 (top) A cross section through the rabbit cerebellum. (middle) A close-up view of the section showing the various cell layers of the cerebellum. (bottom) A close-up view of the various cells contained within each layer.

(see Figure 9.4). The granule cells, as noted earlier, are excitatory and form parallel fibres that innervate the Purkinje cells (see Figure 9.6).

A puff of air to the eye (the unconditioned stimulus, US) activates a different set of fibres. The air puff initially activates the trigeminal nerve and sends the information to the inferior olive (see Figure 9.4). Similar to the mossy fibres, the inferior olive sends information directly to the cerebellum via climbing fibres that wrap themselves around

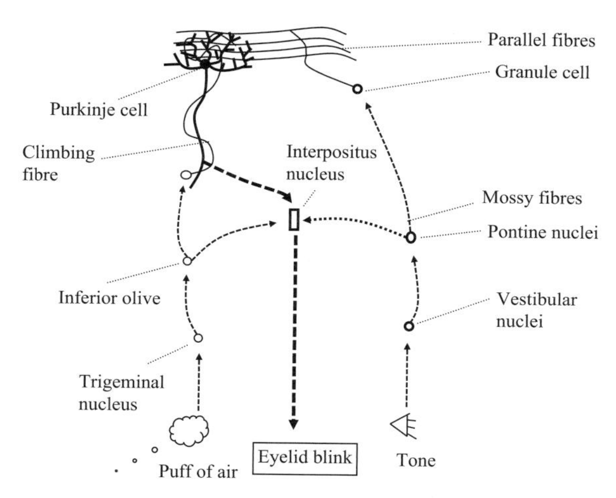

Figure 9.6 Neural circuit in the cerebellum underlying eye-blink conditioning. The cerebellum receives separate inputs from the puff of air and the tone (light dashed lines). The Purkinje cell provides output to cause the eye-blink response (thick dashed lines).

and innervate the Purkinje cells. The inferior olive also sends information to the interpositus nucleus. Therefore, there seem to be two main areas within the cerebellum where information from the tone can integrate with information from the air puff. The first is the interpositus nucleus and the other is at the level of the Purkinje cells (see Figure 9.6). The tone and the air puff are two inputs (represented by the light dashed lines in Figure 9.6), but the motor output is the eye-blink. What circuit is involved with this motor response? And does it involve either of the above-mentioned structures? Following much research, it is now known that the deep cerebellar nuclei send output directly and indirectly to various motor areas (Medina et al., 2002). More specifically, it has been shown that direct stimulation of the anterior interpositus nucleus causes eyelid movement (McCormick and Thompson, 1984a, b; Figure 9.6, thick dashed lines).

We have described the circuit thought to be involved in the different aspects of the eye-blink conditioned response. What is the evidence to suggest that the above-mentioned structures are involved in the various aspects of the task?

General Lesion and Electrophysiological Studies

An early experiment that lesioned one hemisphere of the cerebellum was found to abolish the conditioned response only ipsilateral (on the same side) to the lesion (McCormick et al., 1982). The conditioned response developed normally in the opposite hemisphere to the lesion; that is, one eye developed a blink response to the tone only, but the other did not. A number of other studies refined this initial result and found that lesions of the fastigial, the posterior interpositus or the dentate nucleus did not affect the occurrence of the conditioned response. One area, however, that seemed to be critically involved in the learning and memory of the conditioned eye-blink response was the anterior interpositus nucleus (McCormick and Thompson, 1984a; Clarke et al., 1984). Lesions of this area produced severe and sustained losses of the conditioned response, irrespective of the sensory modality (tone, light or touch) that was paired with the air puff. Interestingly, similar to the motor cortex (Chapter 2) the interpositus nucleus contains a representation of the whole body; stimulation of the medial part causes movements in the lower body and hind legs of the rabbit, whereas stimulation of the lateral region evokes movements in the head region, including eye-blink responses.

Lesions and Stimulation/Recordings of Specific Structures Along the Tone Pathway

As observed in Figure 9.6, critical structures identified in the circuit for projecting tone information to the cerebellum are the pontine nuclei and their mossy fibre outputs. Lesions of the cerebellar region that encompasses the mossy fibres abolish acquisition of the eye-blink response when a tone is paired with the air puff (see Steinmetz et al., 1987; Solomon et al., 1986). The authors demonstrated that this area was critical for tone and not for other sensory modalities. For example, lesions of the pontine nuclei prevented the eye-blink response after the air puff was trained with a tone, but if the puff of air was paired with a light instead, the eye-blink response remained normal. Furthermore, electrophysiological recordings from different areas of the pontine nuclei evoked potentials when sound was presented to the ear (Steinmetz et al., 1987), but not when other sensory stimuli (visual, olfactory etc.) were presented to the animal. It has also been shown that if the mossy fibres are electrically stimulated using an electrode (instead of presenting a tone to the ear) and if this stimulation is paired with the air puff, eyelid conditioning occurs. In addition, if the pontine nuclei themselves are electrically stimulated (rather than the mossy fibres) and this stimulation is again paired with the air puff, eyelid conditioning occurs. All of this points to the critical role of the pontine nuclei and their mossy fibres in processing tone information.

Lesions and Stimulation/Recordings of Specific Structures Along the Air-Puff Pathway

Similar evidence can be gathered with respect to the air-puff pathway. A critical structure in the air-puff pathway, as highlighted in Figure 9.6, is the inferior olive nucleus and its output (the climbing fibres). It has been demonstrated that lesions of the inferior olive abolish conditioning (McCormick et al., 1985; Yeo et al., 1986). In addition to this, Sears and Steinmetz (1991) recorded neurons in the inferior olive and found no response following presentation of a tone but did find an increase in cell activity following presentation of a puff of air. This demonstrates that the inferior olive mediates air-puff information and not tone information. Similar to that described earlier, electrical stimulation (instead of an air puff) of the inferior olive, when paired with a tone, can produce a strong conditioned eyelid response (Mauk et al., 1986).

Electrophysiological Mechanisms Underlying Eye-Blink Conditioning in the Cerebellum

The circuit and lesion studies described earlier suggest that tone information and air-puff information converge in two areas; namely, the Purkinje cells and the interpositus nucleus. What is the contribution of these two areas to learning? Is one of these areas more critical compared to the other or do both contribute equally to learning and memory of the conditioned response? Initial lesion studies demonstrated that damage of the cerebellar cortex (containing the Purkinje cells) had no effect on retention of a well-learned eyelid response, but lesions of the interpositus nucleus completely abolished the response (McCormick and Thompson, 1984a; see Figure 9.7, bottom). This suggested a role for the interpositus nucleus and not the cortex in the recall of the information. However, it should be noted that other studies have shown that lesions of the cerebellar cortex can abolish a previously learned response (Yeo et al., 1985). How can these two opposing views be reconciled? More recently, it has been suggested that the two areas are involved but that they contribute to different aspects of learning (Medina et al., 2002).

If these two areas are critically involved in establishing an association between tone information and air-puff information, then neurons in these areas should display some type of electrophysiological change as training progresses and associations are being formed. In addition, these changes must be persistent and must

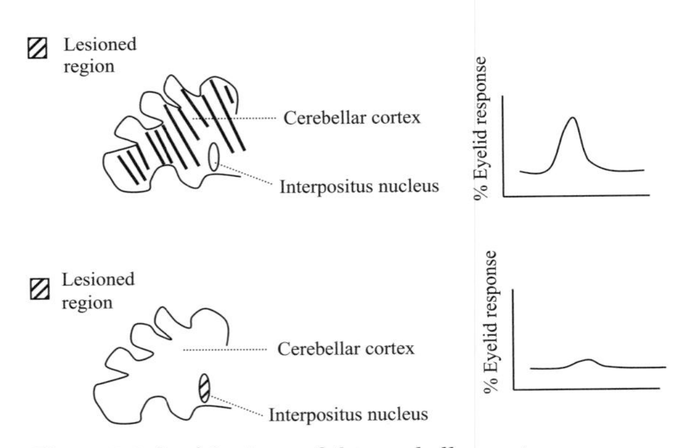

Figure 9.7 (top) Lesions of the cerebellar cortex (and sparing of the interpositus nucleus) has no effect on the eye-blink response. (bottom) Lesions of the interpositus nucleus (and sparing of the cortex) eliminates the response.

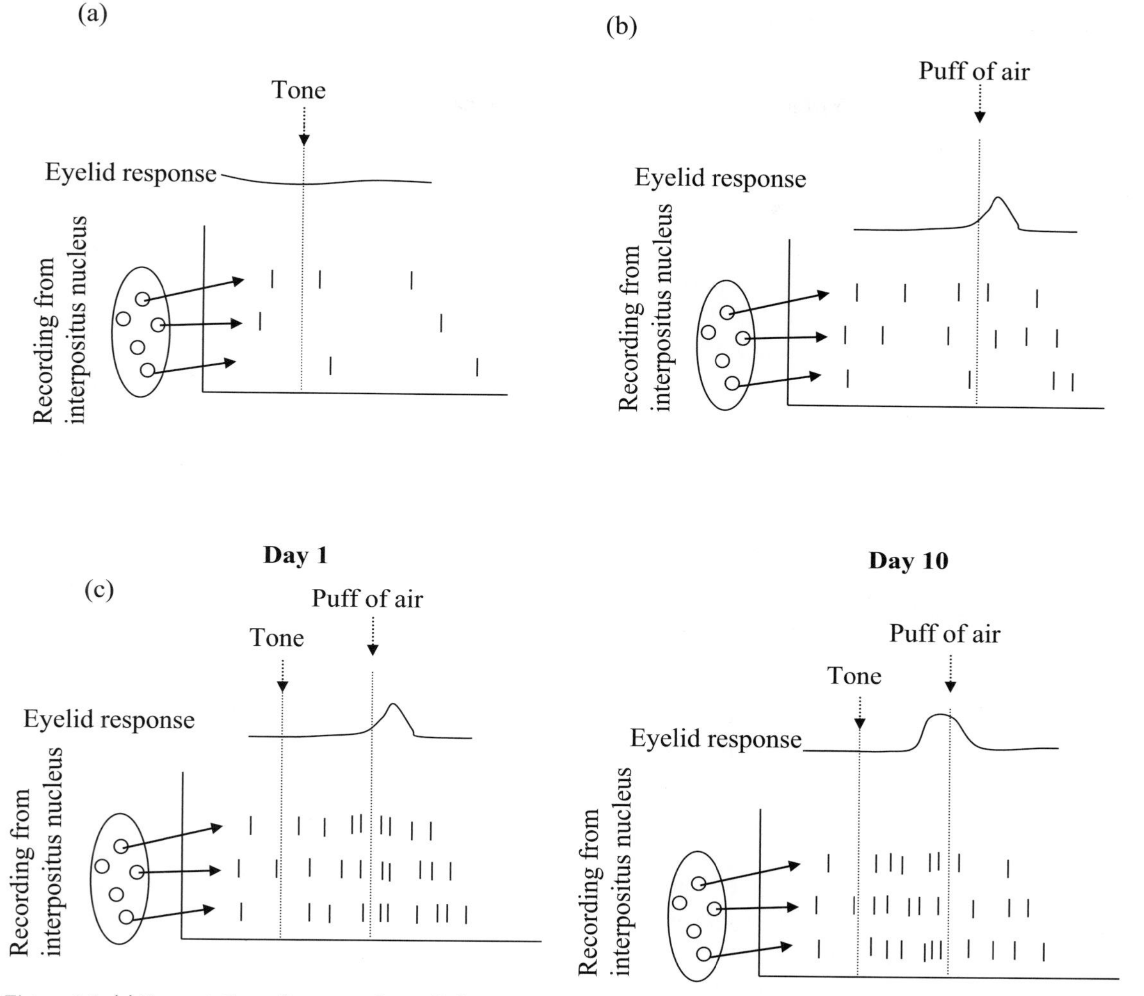

Figure 9.8 (a) Presentation of tone produces little eye-blink response and little neural activity in the interpositus nucleus. (b) Presentation of air puff produces a large eye-blink response, but little neural activity in the interpositus nucleus. (c) Presentation of both tone and air puff produces a strong eye-blink and neural response; however, both behavioural and neural responses shift towards when the tone is presented (rather than when the air puff is presented) with increased training.

remain for a period of time. Electrophysiological recordings of neurons from both the Purkinje cells in the cerebellar cortex and the cells of the interpositus nucleus indicate that they do change their activity during the training period. McCormick and Thompson (1984a), while recording from the interpositus nucleus, showed that there was little or no cellular activity in response to a tone (Figure 9.8a). In addition, there was only a small amount of cellular activity in the interpositus nucleus in response to an air puff (Figure 9.8b). However, on the first day of training when the puff of air was presented in conjunction with the tone, the authors observed an increase in cellular firing (Figure 9.8c, left). As training continued there was a large increase in neural firing, but the firing pattern shifted. This firing occurred after the tone was presented and just before the puff of air was given (as if to predict the onset of the air puff), correlating well with the eyelid response (see Figure 9.8c, right, and Figure 9.3).

There is some controversy about the role of the cerebellar cortex in learning. Early studies demonstrated that lesions of the cerebellar cortex had no effect on responses already acquired (McCormick and Thompson, 1984a, see Figure 9.7 above) but later studies indicated that lesions of the cerebellar cortex (Perrett et al., 1993; Perrett and Mauk, 1995) or the blockade of the cerebral cortical output (Garcia and Mauk, 1998) disrupted the timing of the eye-blink response. This is illustrated in Figure 9.9a. Notice that the eye-blink response in the lesioned animals (Figure 9.9a, lower panel) is not only narrower (shorter latency) compared to the normal response (Figure 9.9a, upper panel) but also occurs after the presentation of the tone – much earlier than even the normal learned response! See Figure 9.3 for comparison. Despite the disruption in timing, animals with cortical damage and especially damage centred on the Purkinje cells still show a learning effect. For example, when an air puff and the tone are paired, a short, mistimed response develops slowly over time, similar to what is observed in non-lesioned animals (Figure 9.9b). Furthermore, Medina et al. (2001) demonstrated that following 60–70 presentations of the tone by itself, the learned conditioned response tends to disappear (extinction). However, in animals that have a disconnection of the cerebellar cortex, the short-term latency eye-blink response remains intact well beyond the 70 tone-only presentations. In fact, even after 3,000 tone-only presentations, there is still a small but noticeable eye-blink response. This suggests that if the interpositus nucleus no longer receives cortical input, memories do not fade away as quickly (or are not overwritten), implying a *strong memory storage* role for the nucleus. Alternatively, the cortex may play a role in removing memories or overwriting memories from the interpositus nucleus. More research is needed on this topic.

Cellular and Molecular Mechanisms Underlying Eye-Blink Conditioning

So far the evidence would suggest that both the cerebellar cortex and the interpositus nucleus are involved in learning and memory of a conditioned response, with the interpositus nucleus necessary for the expression of the behaviour and perhaps for the memory of the response, while the cerebellar cortex is involved in learning the timing of the response. How might this happen? One suggestion is that learning occurs via plastic changes involving both long-term depression (LTD) and long-term potentiation (LTP) within the cerebellar circuitry. We have dealt with both these processes in detail in previous chapters (see Chapters 7 and 8). In brief, LTD involves the persistent weakening of synaptic strength in response to some kind of stimulation, while LTP involves the strengthening of synapses. Much focus over recent years has been on LTD, particularly with the discovery of this form of plasticity and its underlying molecular mechanism at the parallel fibres–Purkinje cell synapses (see Chapter 8 for details).

The question therefore arises whether LTD can be somewhat linked to eye-blink conditioning. There is some evidence that it can (see Fanselow and Poulos, 2005 for details). In brief, it has been demonstrated that AMPA receptors play a key role in LTD; Wang and Linden (2000) have shown that blocking the AMPA receptor prevents LTD and impairs eye-blink conditioning. Similarly, mice that have a deficit in glial fibrillary acidic protein show normal excitatory transmission but impairments in both LTD and eye-blink conditioning (Shibuki et al., 1996). Despite such evidence, we have seen earlier that learning can still occur without Purkinje cells and the cortex. Furthermore, it is the interpositus nucleus that seems to be critical for learning, and this structure has been shown to undergo LTP. How can this be accounted for? A very nice model (Medina et al, 2000) has attempted to include both types of plasticity (both LTP and LTD) to explain how the cerebellum

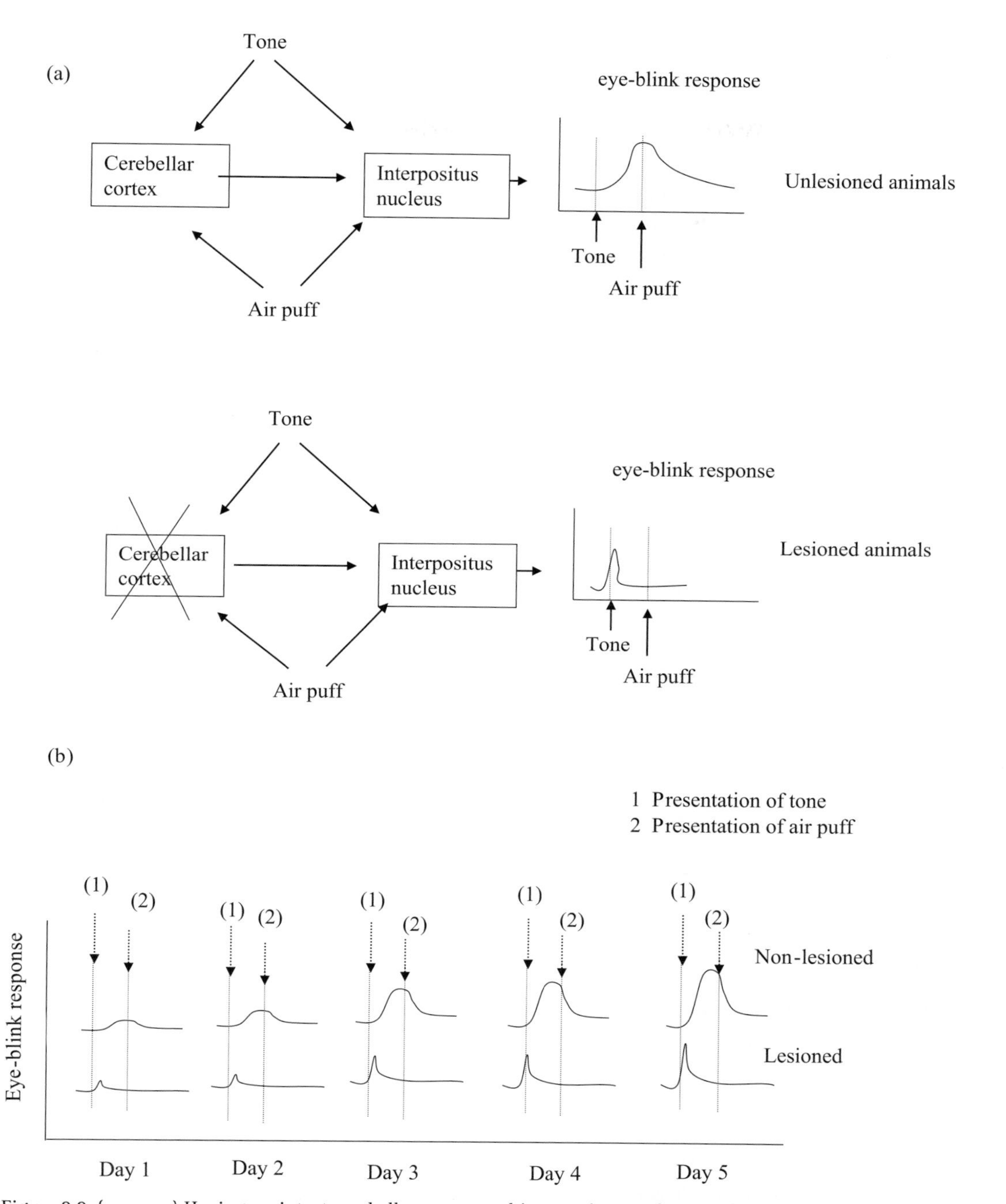

Figure 9.9 (a, upper) Having an intact cerebellar cortex and interpositus nucleus produces a conditioned response (control study). (a, lower) Lesion (crossed-out box) of the cortex (containing Purkinje cells) produces a short-latency and mistimed eye-blink response. (b) Such short-latency eye-blink responses develop 'normally' with training.

might learn the timing of the conditioned response. It is worth reporting the details here.

When a tone is sounded, it activates the mossy fibres (Figure 9.10, 1). The activation of the mossy fibers, in turn, activates different granule cells at different times during the presentation of the tone (2, thick arrows). Meanwhile, the puff of air activates climbing fibres (3, dashed arrows). Those synapses of the granule cells that are active at the exact time upon arrival of the puff of air undergo LTD (4). Those synapses of the granule cells that had already been active at the time of arrival of the air puff undergo LTP (5). In short, synapses of the granule cells that have synchronised firing with cells of the climbing fibres cause depression of response, at that precise time, in the Purkinje cell. In a sense the induction of these changes during learning allows the Purkinje cell to pause/inhibit/depress its activity at the right time, represented by the gap at 6 in Figure 9.10 (Medina et al., 2000). However, the main neurotransmitter of the Purkinje cell is GABA (see Chapter 8); GABA, therefore, has the effect of inhibiting most of the activity in the interpositus nucleus (7), except for the 'paused' region. The climbing fibres, as well as activating the Purkinje cell (3), directly excite the interpositus nucleus (8). Therefore, the pause in activation of the Purkinje cells, combined with the activation of the interpositus nucleus air puff, causes the eyelid response to occur (9) at the appropriate time.

It is possible that the learning of timed responses occurs in the cortex according to this model. However, we have seen earlier that cortical lesions or lesions involving cortical input into the interpositus nucleus show a short-term latency response and, importantly, learning. How does this short-latency response increase during acquisition? Does the interpositus nucleus also undergo some form of plasticity? The answers to these questions are not as clear. Learning-induced changes at the mossy fibre-to-nucleus synapses (8) or changes in the excitability of the cells of the interpositus nucleus themselves may produce this short-latency response (see Aizenman et al., 1998; Aizenman and Linden, 2000).

The Hippocampus and Eye-Blink Conditioning

We have examined the role of the cerebellum in the eye-blink conditioning in some detail. However, the cerebellum is not the only brain structure involved in this type of reflex learning. Another area that seems to be important for eye-blink conditioning is the hippocampus, a structure that we have already encountered (Chapter 8) and will describe in more detail later on (Chapter 14). An hippocampus is situated deep within each of the temporal lobes. The structure is present in all mammals and seems to have been well-conserved across species.

Figure 9.11a shows the location of the hippocampus in the rat brain; it is a banana-like structure that encompasses a large area. The hippocampus is readily recognisable when dissected out due to its prominent layering and shape. Indeed, to the early anatomists, its shape resembled a seahorse and thus they named it 'hippo-campus'. Figure 9.11b shows a cross section through the structure, revealing its different subregions and the inner circuitry. The subregions of the hippocampus include the dentate gyrus and areas CA1 and CA3 (CA stands for *Cornu ammonis*; named after the horns of the Greek god Ammon). Surrounding the hippocampus are the entorhinal and perirhinal cortices (not shown).

Sensory information gains access to the hippocampus through a series of direct and indirect projections via these surrounding cortices. In brief, the entorhinal cortex projects directly to the dentate gyrus which, in turn, projects to area CA3 via mossy fibres. A series of fibres termed the Schaffer collaterals then send information from area CA3 onto area CA1 before exiting back to the entorhinal cortex and perirhinal cortices directly or indirectly.

Electrophysiological Mechanisms Underlying Eye-Blink Conditioning in the Hippocampus

The conditioning procedure in the hippocampus is identical to that used in the cerebellum, in

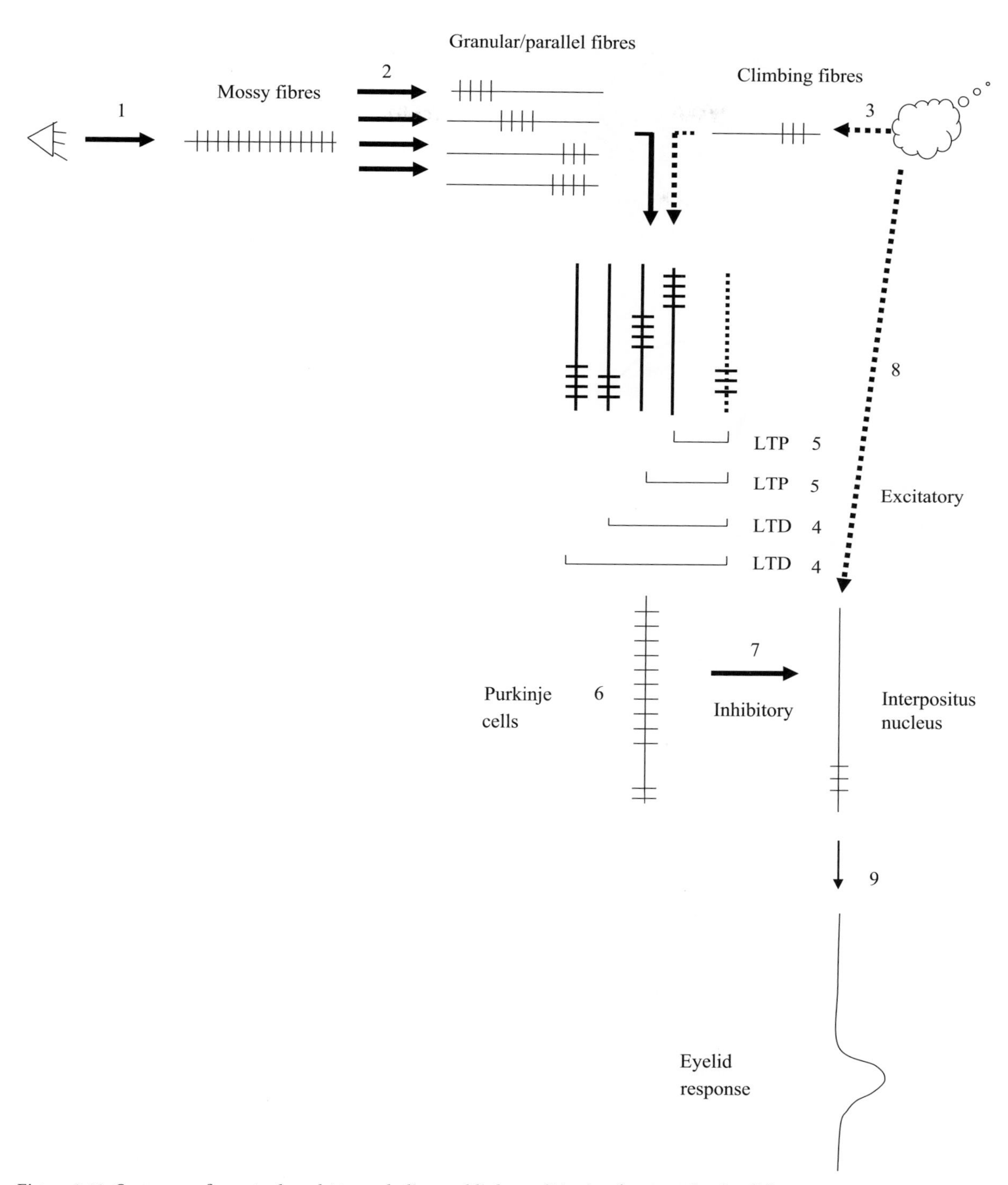

Figure 9.10 Sequence of events thought to underlie eye-blink conditioning (see text for details).

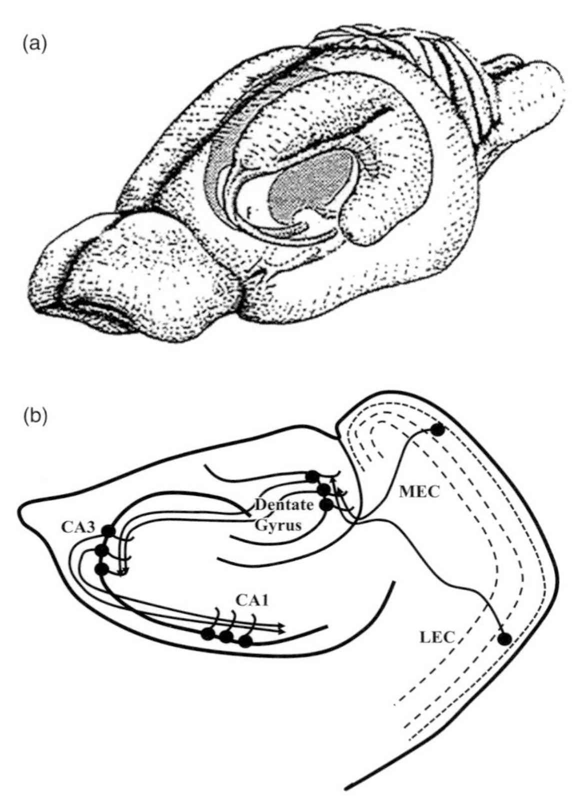

Figure 9.11 (a) Location of hippocampus in the rat brain. Image taken with permission from Amaral and Witter (1995), Hippocampal formation, in G. Paxinos (ed.), *The Rat Nervous System*, San Diego, Academic Press, (b) Section through the hippocampus showing its subregions and connectivity. Image redrawn from F. Torrealba and J. L. Valdes (2008), The parietal association cortex of the rat, *Biological Research* 41: 369–377 (http://creativecommons.org/licenses/by/4.0/).

which a puff of air to the eye is presented shortly after a tone. After repeated pairings, animals learn to associate these stimuli with each other and then display the eye-blink response to the tone alone. Lesion studies have demonstrated that the hippocampus is important for learning this response, especially when the tone and air puff are presented separately (e.g. with an interval of 500 ms between the stimuli). If the tone ends at the same time or overlaps with the air-puff presentation, conditioning does not occur (Solomon et al., 1986; Moyer et al., 1990). Therefore it is suggested that the hippocampus plays an important role in time. Animals with hippocampal lesions can learn associations that have no delay without difficulty; however, introducing a delay period requires an intact hippocampus (Figure 9.12). It seems that the hippocampus is important for making an association between stimuli that are separated in time (Sweatt, 2003). This important property is the key difference between the hippocampus and the cerebellum in relation to the eyelid response task.

Finally, it has been demonstrated that during the training period many neurons, particularly in areas CA3 and CA1, show an increase in firing when the tone is paired with the puff of air (Berger et al., 1986), compared to when the tone or air puff is presented alone. More recent studies suggest that during training the dynamics of CA1 neurons change. McEchron and Disterhoft (1997) demonstrated that during the initial stages of pairing there are large increases in hippocampal activity after both the tone and air-puff presentations, particularly as the eye-blink response starts to increase in size. Subsequent training sees more moderate increases in cellular activity. However, once the conditioned response has been strongly established, an actual decrease in hippocampal activity is observed, particularly following the air-puff presentation (Figure 9.13). These results would suggest that the large increases in cellular activity represent a strong role for the hippocampus in the initial encoding of events but not when learning has been established. Indeed, other studies have similarly reported decreases in hippocampal activity late in training (see, for example, Sears and Steinmetz, 1990). Furthermore, lesion studies involving the hippocampus after a certain period of time have no effect on the learned response, suggesting that the hippocampus has a time-limited role in eyelid conditioning. Kim et al. (1995), for example, demonstrated that hippocampal lesions one day but not 30 days after the eye-blink response had been firmly established disrupted the learned response.

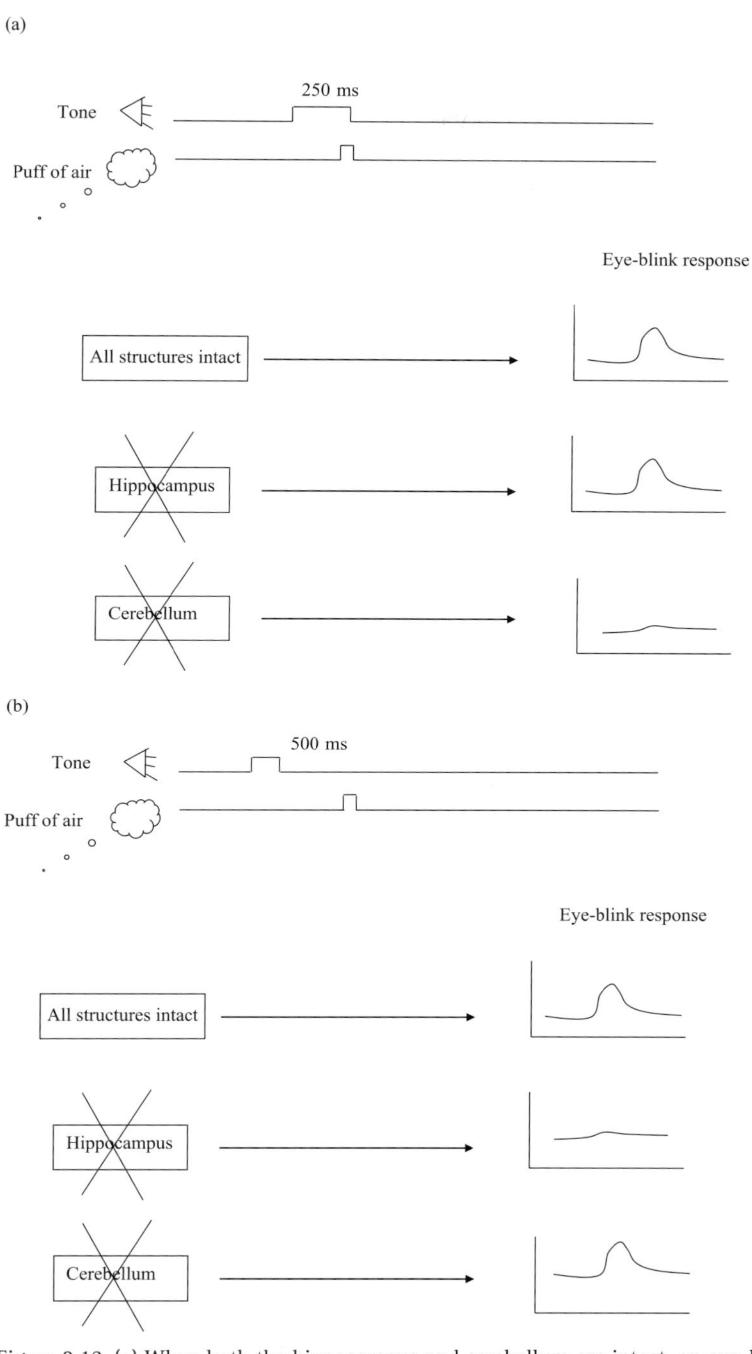

Figure 9.12 (a) When both the hippocampus and cerebellum are intact, an eye-blink response occurs normally following pairing of the tone and air puff *together*. When the hippocampus is lesioned (hippocampus crossed-out box) the eye-blink response remains strong. However, when the cerebellum is lesioned (cerebellum crossed-out box) the eye-blink response does not occur. (b) When both the hippocampus and cerebellum are intact, an eye-blink response occurs normally following the *separate* pairing of the tone and air puff. When the hippocampus is lesioned (hippocampus crossed-out box) the eye-blink response does not occur. However, when the cerebellum is lesioned (cerebellum crossed-out box) the eye-blink response remains strong.

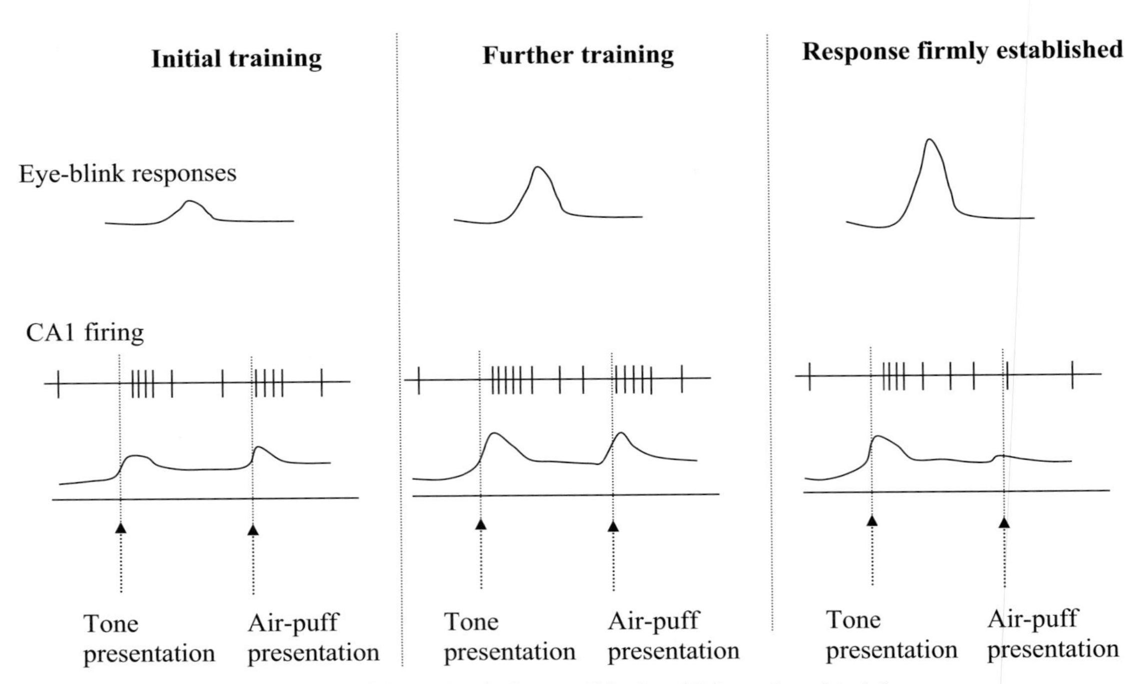

Figure 9.13 Changes in hippocampal dynamics during conditioning. With continued training, eye-blink responses increase and remain strong (left panel and centre panel); however, once conditioning has been firmly established, CA1 firing diminishes, despite the initial increases (right panel).

Summary

Reflexes are considered to be some of the most basic behaviours we possess. The eye-blink is a strong defensive response that helps protect our eyes and prevents foreign entities from gaining internal access. Similar to the defensive reaction of the *Aplysia,* this reflex can become associated with other stimuli, making the eye-blink response a powerful model for studying learning and memory in the mammalian brain. Although both the cerebellum and the hippocampus have been identified as playing important roles, most research has focused on cerebellar circuits. In particular, this chapter has shown a mechanism by which tone and eye-blink information can become associated with each other to form a learnt response.

Questions and Topics Under Current Investigation

- What are the molecular mechanisms that underlie eye-blink conditioning in both the cerebellum and hippocampus?
- What is the exact role played by cerebellar LTD in eye-blink learning?
- Is there a role for other forms of plasticity, e.g. LTP, in both the cerebellum and hippocampus regarding eye-blink conditioning?
- Can similar circuits be found for other motor reflexes?
- The timing of when the tone and air puff are presented can bring about different learning responses in the hippocampus and cerebellum. Are there other differences between the structures?
- Do other brain structures play a role in eye-blink conditioning? For example, if the hippocampus is involved in early learning, are other structures more involved during late learning?

References

Aizenman, C.D., and Linden, D.J. (2000). Rapid, synaptically driven increases in the intrinsic excitability of cerebellar deep nuclear neurons. *Nature Neuroscience*, 3(2), 109–11.

Aizenman, C.D., Manis, P.B., and Linden, D.J. (1998). Polarity of long-term synaptic gain change is related to postsynaptic spike firing at a cerebellar inhibitory synapse. *Neuron*, 21, 827–35.

Amaral, D.G., and Witter, M.P. (1995). Hippocampal formation. In G. Paxinos (ed.), *The rat nervous system*. San Diego, Acadamic Press, 443–493.

Berger, T.W., Berry, S.D., and Thompson, R.F. (1986). Role of the hippocampus in classical conditioning of aversive and appetitive behaviours. In: *The Hippocampus* (R.L Isaacson and K.H. Pribraum, eds.), 4, 203–239. Plenum, New York.

Clarke, G.A., McCormick, D.A., Lavond, D.G., and Thompson, R.F. (1984). Effects of lesions of cerebellar nuclei on conditioned behavioral and hippocampal neuronal responses. *Brain Research*, 291, 125–36.

Eccles, J.C., Ito, M., and Szentagothai, J. (1967). *The cerebellum as a neuronal machine*. Springer, New York.

Fanselow, M.S., and Poulos, A.M. (2005). The neuroscience of mammalian associative learning. *Annual Review of Psychology*, 56, 207–34.

Fearing, F. (1970). *Reflex action. A study in the history of physiological psychology*. 2nd Edition. MIT Press, Cambridge, MA.

Garcia, K.S., and Mauk, M.D. (1998). Pharmacological analysis of cerebellar contributions to the timing and expression of conditioned eyelid responses. *Neuropharmacology*, 37, 471–80.

Kim, J.J., Clark, R.E., and Thompson, R.F. (1995). Hippocampectomy impairs the memory of recently, but not remotely, acquired trace eyeblink conditioned responses. *Behavioral Neuroscience*, 109, 195–203.

Mauk, M.D., Steinmetz, J.E., and Thompson, R.F. (1986). Classical conditioning using stimulation of the inferior olive as the unconditioned stimulus. *Proceedings of National Academy of Sciences USA*, 83, 5349–53.

Medina, J.F., Garcia, K.S., and Mauk, M.D. (2001). A mechanism for savings in the cerebellum. *Journal of Neuroscience*, 21, 4081–4089.

Medina, J.F., Nores, W.L., Ohyama, T., and Mauk, M.D. (2000). Mechanisms of cerebellar learning suggested by eye-lid conditioning. *Current Opinion in Neurobiology*, 10, 717–724.

Medina, J.F., Repa, C.J., Mauk, M.D., and LeDoux, J.E. (2002). Parallels between cerebellum- and amygdala-dependent conditioning. *Nature Reviews Neuroscience*, 3, 122–131.

McCormick, D.A., Clark, G.A., Lavond, D.G., and Thompson, R.F. (1982). Initial localization of the memory trace for a basic form of learning. *Proceedings of National Academy of Sciences USA*, 79, 2731–2735.

McCormick, D.A., Steinmetz, J.E., and Thompson, R.F. (1985). Lesions of the inferior olivary complex cause extinction of the classically conditioned eyeblink response. *Brain Research*, 359, 120–130.

McCormick, D.A., and Thompson, R.F. (1984a). Cerebellum: essential involvement in the classically conditioned eyelid response. *Science*, 223, 296–299.

McCormick, D.A., and Thompson, R.F. (1984b). Neuronal responses of rabbit cerebellum during acquisition and performance of classically conditioned nictitating membrane-eyelid response. *Journal of Neuroscience*, 4, 2811–2822.

McEchron, M.D., and Disterhoft, J.F. (1997). Sequence of single neuron changes in CA1 hippocampus of rabbits during acquisition of trace eyeblink conditioned responses. *Journal of Neurophysiology*, 78, 1030–1044.

Moyer, J.R. Jr., Deyo, R.A., and Disterhoft, J.F. (1990). Hippocampectomy disrupts trace eyeblink conditioning in rabbits. *Behavioral Neuroscience*, 104, 243–252.

Perrett, S.P., and Mauk, M.D. (1995). Extinction of conditioned eyelid responses requires the anterior lobe of cerebellar cortex. *Journal of Neuroscience*, 15, 2074–80.

Perrett, S.P., Ruiz, B.P., and Mauk, M.D. (1993). Cerebellar cortex lesions disrupt learning-dependent timing of conditioned eyelid responses. *Journal of Neuroscience*, 13, 1708–1718.

Schreurs, B.G. (1993). Long-term memory and extinction of the classically conditioned rabbit nictitating membrane response. *Learning & Motivation*, 24, 293–302.

Sears, L.L., and Steinmetz, J.E. (1990). Acquisition of classically conditioned-related activity in the hippocampus is affected by lesions of the cerebellar interpositus nucleus. *Behavioral Neuroscience*, 104, 681–692.

Sears, L.L., and Steinmetz, J.E. (1991). Dorsal accessory inferior olive activity diminishes during acquisition of the rabbit classically conditioned eyelid response. *Brain Research*, 545, 114–122.

Shibuki, K., Gomi, H., Chen, L., Bao, S., Kim, J.J., Wakatsuki, H., Fujisaki, T., Fujimoto, K., Katoh, A., Ikeda, T., Chen, C., Thompson, R.F., and Itohara, S. (1996). Deficient cerebellar long-term depression, impaired eyeblink conditioning, and normal motor coordination in GFAP mutant mice. *Neuron*, 16(3), 587–599.

Solomon, P.R., Lewis, J.L., LoTurco, J.J., Steinmetz, J.E. and Thompson, R.F. (1986). The role of the middle cerebellar peduncle in acquisition and retention of the rabbit's classically conditioned nictitating membrane response. *Bulletin of the Psychonomic Society*, 24, 74–78.

Steinmetz, J.E., Logan, C.G., Rosen, D.J., Thompson, J.K., Lavond, D.G., and Thompson, R.F. (1987). Initial localization of the acoustic conditioned stimulus projection system to the cerebellum essential for classical eyelid conditioning. *Proceedings of National Academy of Sciences USA*, 84, 3531–5.

Sweatt, J.D. (2003). *Mechanisms of memory*, Academic Press, San Diego, CA.

Torrealba, F., and Valdes, J.L. (2008). The parietal association cortex of the rat. *Biological Research*, 41, 369–377.

Voogd, J., and Glickstein, M. (1998). The anatomy of the cerebellum. *Trends in the Neurosciences*, 21, 370–375.

Wang, Y.T., and Linden, D.J. (2000). Expression of cerebellar long-term depression requires postsynaptic clathrin-mediated endocytosis. *Neuron*, 25(3), 635–47.

Yeo, C.H., Hardiman, M.J., and Glickstein, M. (1985). Classical conditioning of the nictitating membrane response of the rabbit. II. Lesions of the cerebellar cortex. *Experimental Brain Research*, 60, 99–113.

Yeo, C.H., Hardiman, M.J., and Glickstein, M. (1986). Classical conditioning of the nictitating membrane response of the rabbit. IV. Lesions of the inferior olive. *Experimental Brain Research*, 63, 81–92.

10 Fear Conditioning

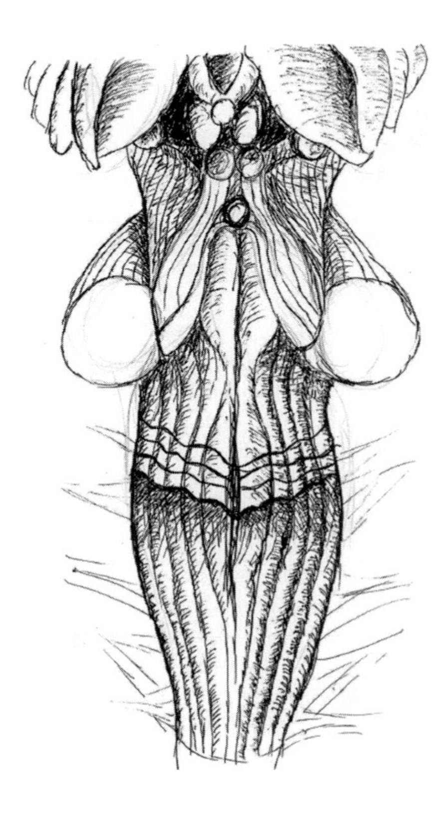

Introduction and Background

Humans are thought to be able to express and recognise at least six different types of emotions. These include happiness, sadness, disgust, surprise, fear and anger. Over the last 100 years and more, psychologists and philosophers have attempted to define and formulate theories of emotion. Despite this, there is no real consensus on what exactly constitutes an emotion, though most would agree that it comprises both a physiological and behavioural response to a stimulus. One emotion that has received much attention and research is fear. To be able to feel and express fear is critical for an animal's survival. Without fear, animals may expose themselves to the risk of injury and death. It is therefore good, at times, to be afraid. However, there is also a risk that fear can become crippling and pathological, and many humans suffer from a range of phobias, including the fear of public speaking, fear of heights and fear of enclosed spaces. In other cases, people may suffer post-traumatic stress disorder and constantly recall and relive major fearful situations. Such conditions often lead to anxiety, stress and, in severe cases, hospitalisation.

Being unable to express fear is an extremely rare condition, but it has been described in the case study of a woman known as SM. Although SM scores in the normal range on standardised tests of intelligence, and has normal perception, language and motor function, she does suffer from a rare genetic disorder known as Urbach-Wiethe disease. With this disorder, SM has particular difficulty in recognising and expressing fear. This difficulty can be illustrated more clearly in an experiment conducted with SM in 2011, at age 44. The experiment attempted to examine the extent to which SM could demonstrate fear when exposed to a number of different situations (see Figure 10.1; Feinstein et al., 2011). In the first situation, SM was brought to a reptile store and, despite stating that she hated snakes and spiders, SM seemed to be immediately drawn towards them. She asked the store manager if she could hold them; when given one, SM stroked the snake as it moved over her. Following this, she then asked to touch one of the more dangerous snakes

Figure 10.1 Different fearful situations that SM had to face, including a snake, spiders and a 'haunted house'. Reprinted from J.S. Feinstein et al. (2011), The human amygdala and the induction and experience of fear, *Curr Biol.* 21(1):34–38, with permission from Elsevier.

in the store. Despite being refused permission, SM felt a strong compulsion to reach towards it and asked 15 times if she could touch it. The store manager also had to stop SM from reaching towards and touching a dangerous tarantula.

In a second situation SM was brought to a 'haunted house' at night, which she and a group of other people were allowed to freely explore. The house was decorated with eerie scenes and played scary music. Throughout the house, actors dressed in ghoulish costumes lay hidden and were ready to jump out at the passing visitors. SM seemed to delight in the situation, often running down dark corridors, leading the way and berating those lagging behind. She never once felt afraid, despite the many attempts of the various actors to scare her. In fact, she scared one of the actors when she poked him in the head, simply out of curiosity!

Exposing SM to such situations may feel a bit contrived, but SM does in fact live in a poor, deprived neighbourhood where exposure to real danger exists on a daily basis. SM has been attacked many times throughout her life. She has been held up both at gunpoint and knifepoint. She has suffered domestic violence and has been physically attacked on more than one occasion. Despite this, her routine behaviour does not display any sense of desperation or fear, and she repeatedly exposes herself to various dangers by walking through dark laneways and alleys. SM has not learnt to be afraid. SM can, however, express other emotions. She laughs at funny movie clips and is appropriately sad at other occasions; it is only with fear that she has a problem. She is also unable to recognise fear in the facial expressions of others as well as having difficulty expressing fear herself. What makes SM so unique is that she has damage to a particular region of the brain known as the amygdala, in both hemispheres.

Behaviour of Fear Conditioning

Research into fear has focused on a particular behavioural paradigm known as *Pavlovian fear conditioning*. This procedure has been very successful as it gives researchers great control over the various stimuli, allows neural circuits to be identified and can mimic real-life situations. For example, a mouse that has been attacked by a cat that has suddenly emerged from a bush may become fearful of rustling leaves. The mouse has not only learned to be afraid and has a memory of the particular incident, but has also generalised this fear to other stimuli present at the time. Likewise, if attacked in a certain part of town, you would tend to avoid this area, and if this was not possible, you would go only with extreme caution. Your senses would be heightened, your heart pumping and your behaviour radically changed.

In a laboratory situation, fear conditioning can be easily shown. When an animal is placed in a cage and a tone is sounded, this generally has little effect on the animal's behaviour (Figure 10.2). However, if every time the tone is sounded the animal receives a small shock to its feet, this results in a change in the animal's behaviour (it freezes), as well as a change in its physiological responses (increase in blood pressure and heart rate) and hormonal responses (release of stress hormones, adrenaline, etc.). This pairing of the tone and shock can be continued for a number of trials. Following this, if the tone is presented alone, it will elicit the same behavioural and hormonal response as if the shock were still present. The rat is said to have been fear-conditioned.

Brain Structures and Circuits Underlying Fear Conditioning

Let us consider the neural circuits that might be involved with both the presentation of a tone and the presentation of the shock; then we can determine where changes during learning may occur. Taking the presentation of the tone first, sounds enter the ear, where they are converted to neural responses. Information is then transmitted via the auditory nerve to the inferior colliculus of

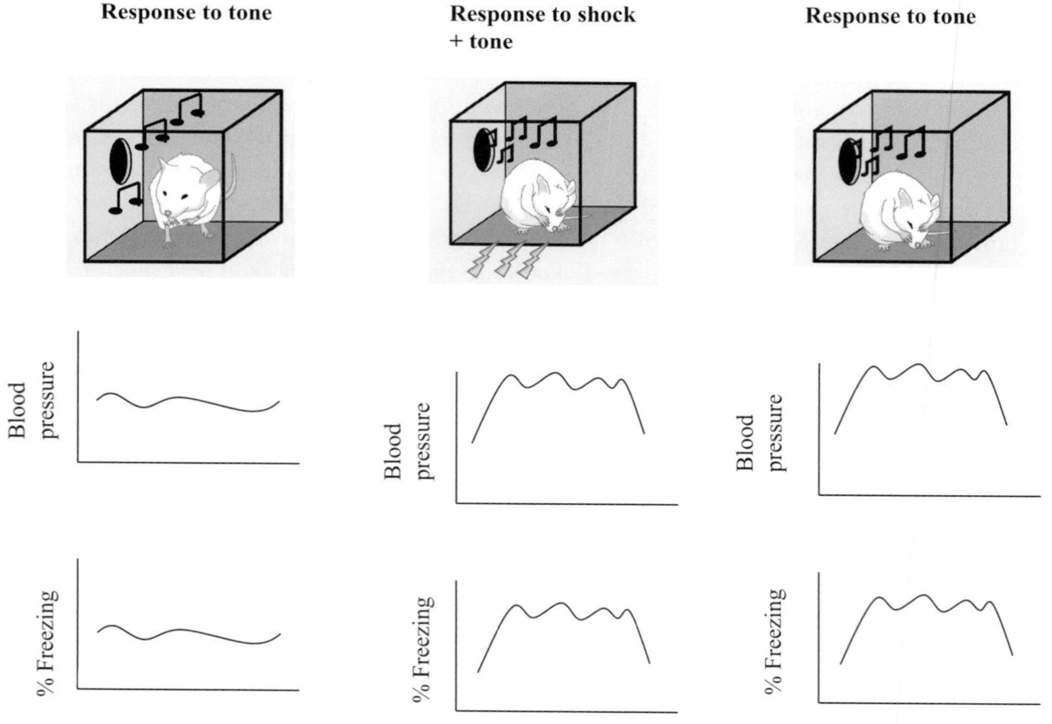

Figure 10.2 Presentation of tone does not elicit a major response from the rat (left panel). However, if a shock is applied at the same time as the tone, the animal will freeze and will show an increase in physiological response, e.g. increased blood pressure (middle panel). Following multiple shock–tone pairings, presentation of the tone alone will produce a behavioural and physiological response (right panel). Reprinted from J.P. Johansen et al. (2011), Molecular mechanisms of fear learning and memory. *Cell* 147(3):509–24, with permission from Elsevier.

the midbrain. From here, signals are sent to the medial geniculate nucleus (MGN) of the thalamus (auditory thalamus). Finally, the thalamus projects to the auditory cortex, where the neural activity is interpreted as a tone (Figure 10.3a). On the other hand, when a shock is presented, it is converted to neural activity and relayed via the spinal cord to the ventral posterior nucleus of the thalamus (somatosensory thalamus), which, in turn, sends signals to the somatosensory cortex/parietal cortex. This neural activity is then interpreted as pain, and the appropriate response ensues. When the tone and the shock are paired, both systems are activated.

Careful examination of these two pathways poses an immediate difficulty. The two neural circuits seem to be completely independent of each other. How is it then possible that, following multiple pairings of the tone with the shock, the systems seem to overlap? How is it that by presenting a tone alone (following multiple pairings) a reaction occurs similar to that of a shock? Is there a structure or group of structures that is common

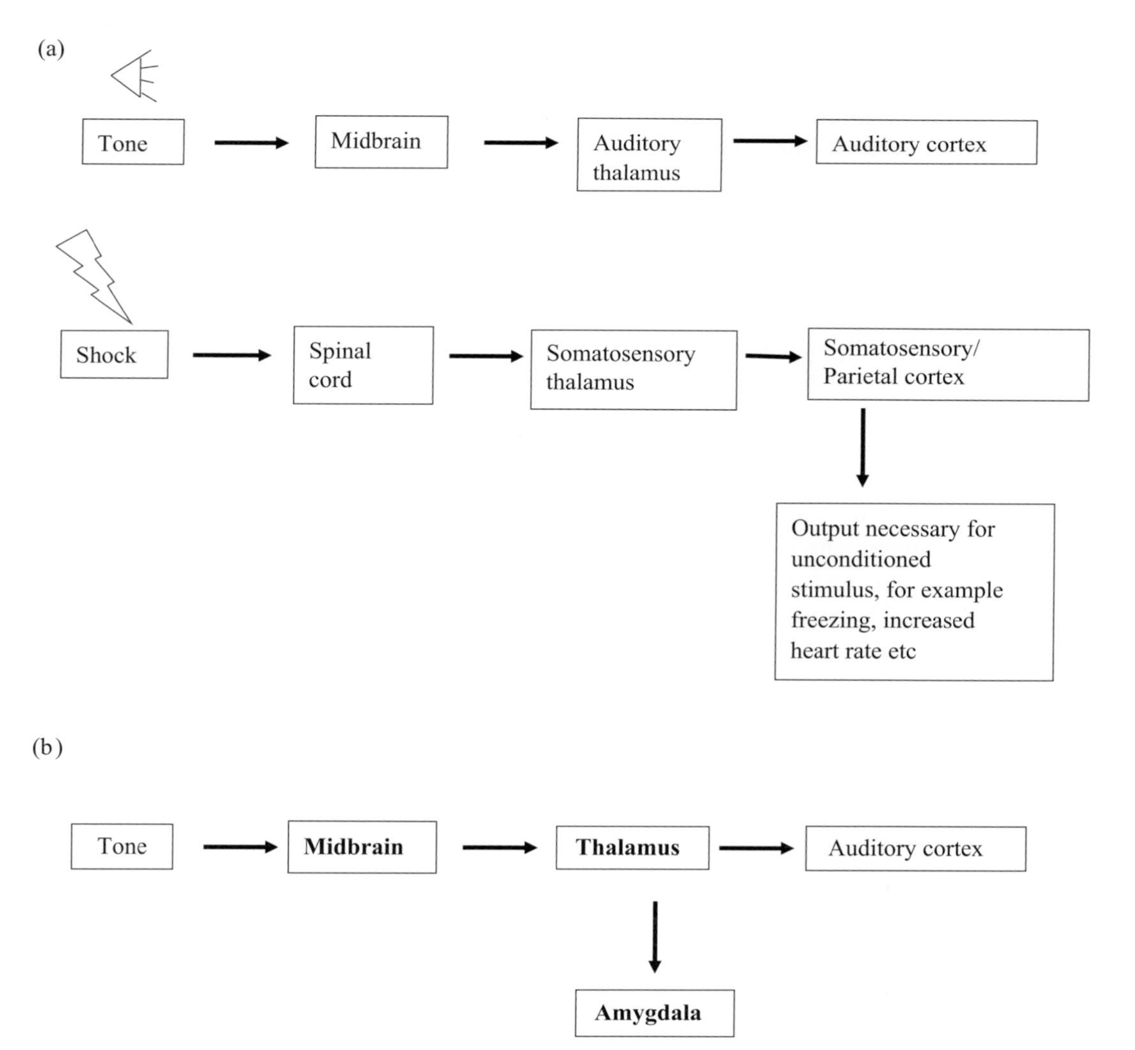

Figure 10.3 (a) Neural circuit involved in processing sound (top) and shock (bottom) information. (b) Structures along the circuit that are necessary for a fear-conditioned response are highlighted in bold.

to both circuits where this interaction may take place? Joseph LeDoux in the early 1990s began to investigate this by examining what structures along the tone pathway are involved in the auditory fear-conditioning response. He found that lesions to both the auditory thalamus and the midbrain prevented the fear-conditioned response, whereas lesions to the auditory cortex had no effect.

If the auditory cortex was not involved in the fear-conditioned response, where did the thalamic output go? Following a number of careful experiments, it was discovered that, as well as innervating the auditory cortex, axons from the thalamus reach a number of different subcortical structures. Lesions to each of these revealed that only one structure, the amygdala, had an effect on the fear-conditioned response (Figure 10.3b).

The amygdala was first recognised as a distinct brain region in the early part of the 1800s; due to the structure's close resemblance in shape to an almond nut it was given the name *amygdala* (the Greek word for 'almond'). The amygdala is located in the temporal lobes, close to the anterior tip. Like many structures, the amygdala can be subdivided and is actually composed of a dozen or more nuclei. The three major nuclei of the amygdala include the basal, the lateral and central nuclei. Although these nuclei can be further subdivided – for example, the basal nucleus can be subdivided into the basolateral and basomedial nuclei (Maren and Fanselow, 1996), and the lateral amygdala consists of three subregions (dorsal, ventrolateral and ventromedial, Pitkanen et al., 1997) – we shall consider only the three main nuclei here (Figure 10.4).

Despite the large number of nuclei within the amygdala, only a few of these seem to be involved in fear conditioning. Lesions of both the lateral and the central nuclei disrupt the fear-conditioned response, demonstrating that these two areas are critically involved (Kapp et al., 1992; LeDoux, 1992). Indeed, neuroanatomical studies have revealed that the lateral nucleus serves as an input zone, receiving sensory information from various structures, including both the auditory thalamus and the auditory cortex (LeDoux et al.,

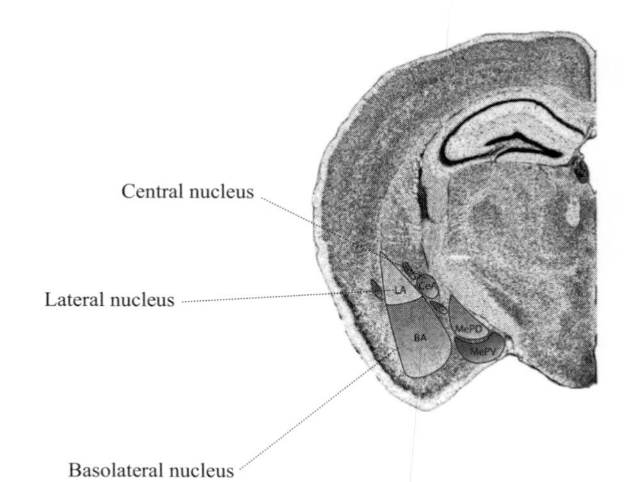

Figure 10.4 The anatomy of the mouse amygdala, highlighting various subregions, including basolateral (BA), lateral (LA) and central (CeA) nuclei. Image from Rob Hurt 2016, reproduced under Creative Commons Attribution License (CCA). For colour version, please refer to the plate section.

1990; Romanski and LeDoux, 1993; McDonald, 1998). The lateral amygdala, in turn, sends projections to the central nucleus (Pare and Smith, 1998), which acts as an output structure projecting to various brain stem regions. Damage to the tiny central nucleus completely abolishes the expression of the fear-conditioned response, including the behavioural freezing response, as well as the various hormonal changes associated with fear. In turn, lesions to the various regions that receive projections from the central nucleus *selectively* interfere with different components of that fear response. For example, lesions to the lateral hypothalamus affect blood pressure but not freezing behaviour, whereas lesions to the periaqueductal grey area affect freezing behaviour only (Figure 10.5, LeDoux et al., 1988).

We suggested earlier that there has to be an overlap between the conditioned stimulus (CS) tone system and the unconditioned stimulus (US) shock system. Given that both the lateral amygdala and the central nucleus seem to be the critical structures in the fear-conditioned response, does shock information actually reach these areas?

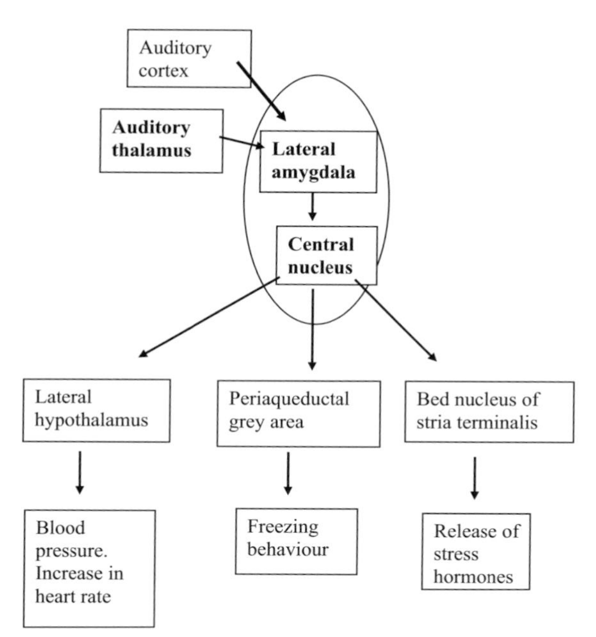

Figure 10.5 Diagram showing how the amygdala can directly affect different aspects of the fear-conditioning response through the activation of specific neural targets.

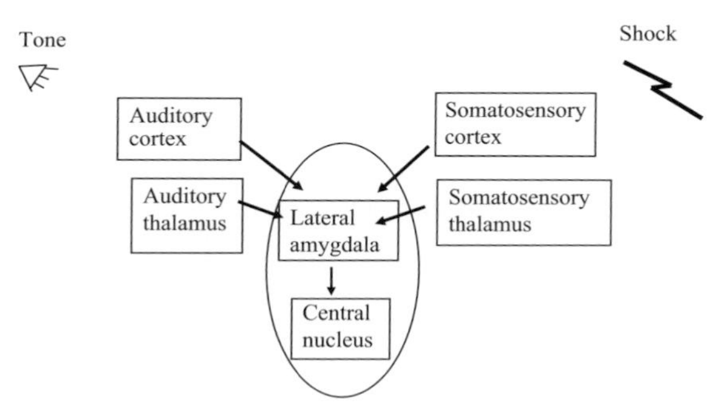

Figure 10.6 The lateral amygdala seems to be the site of convergence of tone and shock information.

Although the central nucleus receives nociceptive (harmful) information, it does not seem to receive direct auditory information (Medina et al., 2002). On the other hand, there is evidence that thalamic areas that receive information from the spinal tract (shock) do project to the lateral amygdala (LeDoux et al., 1987). The lateral amygdala, therefore, seems to be the site of convergence (Figure 10.6). However, as mentioned, the lateral amygdala consists of three subregions, the dorsal, ventrolateral and ventromedial nuclei, but the convergence seems only to occur in the dorsal part of the lateral amygdala (Romanski et al., 1993).

Electrophysiological Mechanisms Underlying Fear Behaviour

As we have identified some of the major structures involved in fear conditioning, the next questions are: How does it occur? What are the electrophysiological changes that take place in the amygdala? There is certainly evidence that neurons in the lateral amygdala become more responsive following pairing of the tone and the shock (Medina et al., 2002). An early study conducted by Rogan and LeDoux (1995) suggested that the auditory thalamic to lateral amygdala pathway is modifiable, and is capable of sustaining long-term potentiation (LTP; see Chapter 7); following high-frequency stimulation of the auditory thalamus, a persistent increase in the amygdalar response would ensue. Rogan and LeDoux (1995) further demonstrated that if a tone was applied, rather than stimulation of the auditory thalamic nucleus, a neural response could be generated in the lateral amygdala. Then, if high-frequency stimulation was applied to the auditory thalamus, and a tone was subsequently presented, the neural response in the amygdala to the tone would be enhanced and remain potentiated over a long period. In another set of experiments, the authors showed that pairing a tone with a shock also led to an enhanced neural response in the lateral amygdala, as well as producing a freezing response in the animal. Following multiple pairings, presentation of the tone alone continued to produce an enhanced neural response in the amygdala, along with a continued freezing behavioural response (see Figure 10.7 and Rogan et al., 1997). These experiments reveal that changes in the neural responsiveness of the lateral nucleus bring about a behavioural change.

In another set of experiments examining plasticity more directly, Quirk et al. (1995) demonstrated that the magnitude of neural responses to a tone increased during the pairing

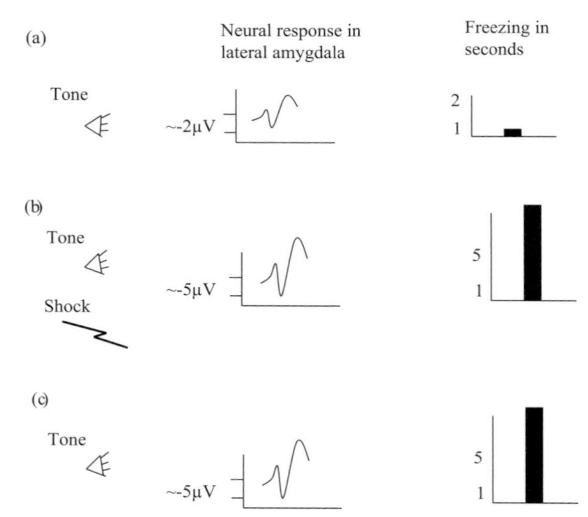

Figure 10.7 (a) Presentation of the tone alone leads to a small field potential in the lateral amygdala and minimum freezing behaviour. (b) Presentation of tone and shock at the same time leads to an increase in freezing behaviour and an increase in the amplitude of field potentials in the lateral amygdala. (c) With presentation of tone alone, the freezing behaviour remains high, as do the field potentials' slope and amplitude.

with a shock. Furthermore, some cells that were initially unresponsive to the tone became responsive when paired with the shock. In a follow-up study worth describing in detail here, Repa et al. (2001) also showed how the firing pattern of amygdalar neurons change during fear conditioning. The authors recorded action potentials (spikes) from single neurons in the dorsal region of the lateral amygdala. Action potentials were recorded during three sessions: exposure to a tone, pairing of the tone and shock, and presentation of the tone alone. The authors discovered two types of neurons with different electrophysiological properties within this region. One population was recorded from the dorsal tip of the dorsal lateral amygdala; the second was recorded from the ventral tip of the dorsal lateral amygdala (Figure 10.8a). Cells in the dorsal tip showed very robust firing in response to the tone. These cells also responded with shorter latencies; for example, in Figure 10.8b top, most of the firing for three representative cells occurs between 10–30 ms after tone presentation. In comparison, cells in the ventral tip did not fire as strongly, and when they did fire their responses were much slower, with latencies of between 50 and 60 ms after the tone (Figure 10.8b bottom).

Then, when the tone was paired with the shock, the two populations of neurons continued to display their different response patterns. There was an enhancement of neural responses in the dorsal tip, with most cells firing at short latencies. In the ventral tip, an increase in firing rate was also observed, with most cells responding at even longer latencies than before – some cells fired at 100 ms post-presentation of the tone and shock. In the experiment, Repa et al. (2001) gave 16 trials where both the tone and shock were presented together. Animals were then returned to their home cage for an hour. Following this, animals were presented with the tone alone for a further 20 trials. As described above, both populations of cells increased their neural firing when the shock was paired with the tone. Did this increase in firing continue throughout the 16 trials of pairing, and what happened when the tone was presented alone one hour later? Again, it seems that the two populations of cells reacted differently. The cells in the dorsal tip increased their firing when the tone was paired with the shock, but this increased firing only lasted for about 8 of the 16 paired trials, and then the level of firing returned to the pre-paired rates. One hour post-pairing, when the tone was presented alone, there was also an initial increase in neural firing, but this quickly returned to baseline levels for the remainder of the 20 trials (Figure 10.8c top).

In contrast, when the tone was paired with the shock, cells in the ventral tip increased their firing gradually and only reached their maximum response level close to the last trial of the paired trials. When the tone was presented alone one hour later, there was also an increase in firing initially, and this increased neural firing remained consistent for the full 20 trials (Figure 10.8c bottom right). What exactly these changes represent is difficult to know. However, this experiment reveals the dynamic nature of different populations of neurons within the amygdala and how they

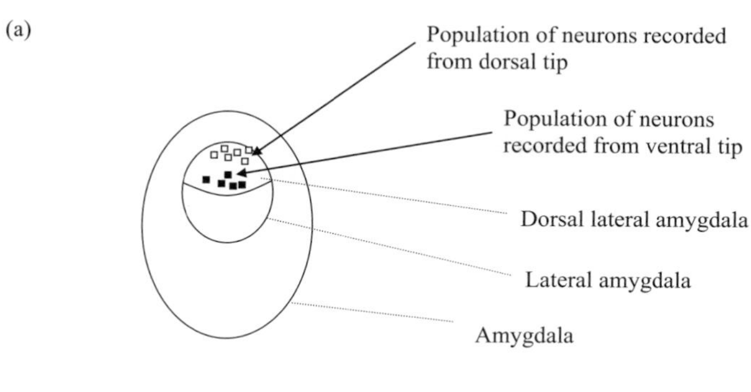

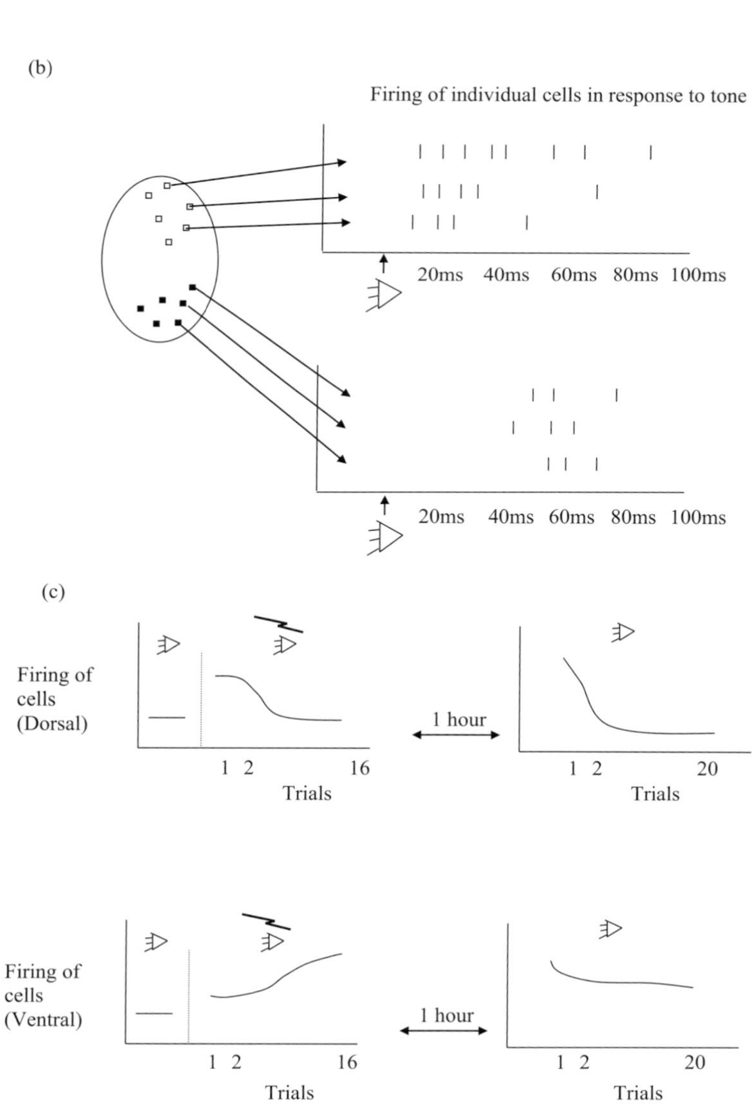

Figure 10.8 (a) Location of recorded neurons within the lateral amygdala. (b) Electrical firing properties of cells recorded from the dorsal (top) and ventral tip (bottom) in response to a tone. (c) The firing pattern changes for both dorsal (top) and ventral (bottom) neurons across multiple tone-shock trials (left) and also when presented with the tone alone 1 hour later (right).

change during fear conditioning. A final point to note about these experiments is that most of the changes in the firing properties of the neurons were observed before the behavioural changes.

Biochemical and Molecular Mechanisms Underlying Fear Behaviour

As we can see from the previous section, neurons change their electrophysiological properties during conditioning. In some cases, the strength of firing is increased; in others, the latencies at which the responses occurred have changed. For some neurons this plasticity is short-term and only lasts for a couple of trials during conditioning. For other neurons, the changes are more gradual and indeed can be stored for at least an hour. Are there molecular and biochemical changes that reflect these changes? For example, what molecular changes are associated with the increased responsiveness of amygdalar neurons, and what biochemical changes are involved in the long-term storage of fear information? Fear conditioning increases neural activity in the lateral amygdala in a fashion similar to that described for LTP (see Chapter 7). Indeed, most research has focused on attempts to correlate the biochemical changes related to LTP to the biochemical changes that occur with fear conditioning (Schafe et al., 2001). For example, pairing of the tone and the shock pathways in the lateral amygdala (during training) leads to an influx of calcium ions through the NMDA receptors. This in turn sets off a cascade of intracellular events, which activates different kinases, depending on whether the neural response is short-term or long-term.

For short-term memories, calcium enters through the NMDA receptor, activating protein kinase C (PKC) which, in turn, activates Ca2+/calmodulin-dependent protein kinase II (alphaCaMKII). When these two kinases are activated, they can remain activated for a short period, even without the presence of calcium. These kinases act by adding a phosphorus atom to other proteins, including the AMPA receptor. If the AMPA receptor is phosphorylated, it can modify its shape (over the short term) to allow more positive ions into the postsynaptic cell, thereby increasing the neural activity of the cell (Figure 10.9a). It is known that blocking the NMDA receptor blocks fear conditioning (Fanselow and Kim, 1994), and mice deficient in either PKC or alphaCaMKII also show impaired fear conditioning (Silva et al., 1996; Weeber et al., 2000).

For long-term fear memories to occur, the synthesis of new proteins is required. Calcium enters through the NMDA receptor, which activates protein kinase A (PKA) and the extracellular-signal-regulated kinase/mitogen-activated protein kinase pathway (ERK/MAPK). Both PKA and ERK/MAPK translocate to the cell nucleus where they activate CREB (cAMP response element-binding protein), which, in turn, can bind to the DNA machinery and synthesise new proteins, resulting in more permanent structural changes (Figure 10.9b). Evidence for the role of these kinases initially came from the LTP literature, whereby inhibitors of both PKA and the ERK/MAPK machinery led to an impairment of LTP (Impey et al., 1996; 1998). However, more recent experiments have shown that inhibitors of PKA and of the ERK/MAPK pathway can directly impair long-term memory of fear conditioning (Schafe and LeDoux, 2000; Schafe et al., 2000). Furthermore, evidence is mounting that the pathway described above is essential for long-term rather than short-term fear conditioning. Infusion of these inhibitors during training, for example, had no effect on freezing 1–4 hours post-training, but markedly decreased freezing around 24 hours later (Schafe et al., 2001).

Although it is only possible here to provide a brief outline of the molecular processes underlying fear conditioning, much recent research has focused on the role of other pathways and molecules within the amygdalar complex, as well as on other neural circuits. For example, learning in highly emotional situations is generally thought to involve noradrenaline (norepinepherine) and dopamine; there is much evidence demonstrating that these

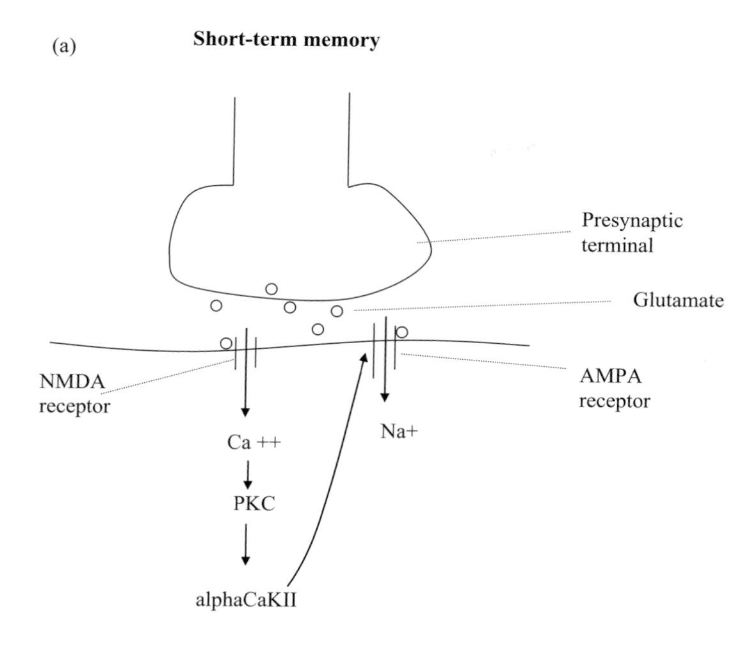

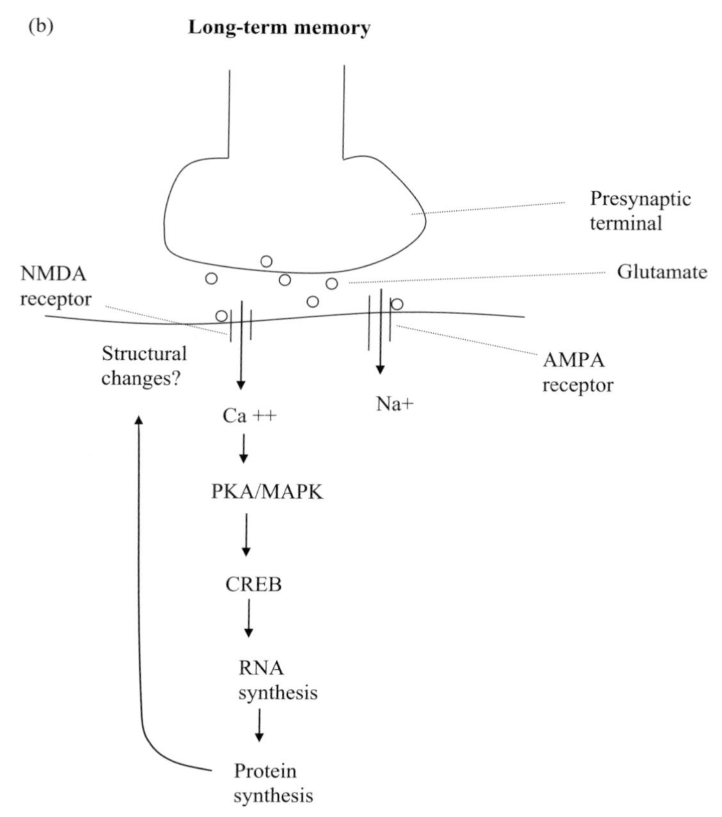

Figure 10.9 Molecular mechanisms thought to underlie (a) short-term and (b) long-term fear-conditioning memory.

neurotransmitters can modify the plastic changes in the lateral amygdala during fear conditioning (Johansen et al., 2011). Further research has focused on the inhibitory neurotransmitter GABA within the central nucleus of the amygdala, as well as neural circuits involving the prefrontal cortex (see Herry and Johansen, 2014). Unfortunately, this work will not be described here.

Summary

Fear is one of the most basic emotions that we can express. Being able to learn fear is critical for an animal's well-being and, at times, its very survival. We saw how SM, being unable to fear, constantly places herself in danger. In this chapter we have been able to show, by using a simple behavioural conditioning paradigm, how fear can be learnt. If a neutral stimulus is associated with something that provokes a strong reaction, both behaviourally and physiologically, the neutral stimulus becomes something to be feared. Through careful investigations of various neural pathways, one critical structure – the amygdala – has been identified as being the key to our understanding of fear. Neurons within this structure have been shown to change their electrophysiological and cellular properties to allow us to both express and recall fearful events.

Questions and Topics Under Current Investigation

- How does understanding the behaviour and the neural basis of fear translate into providing help for those with various phobias, anxiety disorders and post-traumatic stress disorders that can be so debilitating?
- Much research, including this chapter, has focused on the amygdala, but what role do other structures, including the prefrontal cortex, play in fear?
- Can circuits that underlie some of the other emotions, including happiness, sadness, disgust, surprise and anger also be identified and investigated?
- What other neuromodulators (e.g. acetylcholine, serotonin), molecular pathways and specific genes are involved in learning, consolidation and recall of a fearful situation?
- Is *fear conditioning* the correct term to be used? Current debate suggests that this term should be abandoned and replaced with *threat conditioning*, because mechanisms that relate to threats (freezing in response to a shock) are not the same as those that produce conscious fear (e.g, fear of spiders or of heights). See LeDoux (2014) for further discussions.

References

Fanselow, M.S., and Kim, J.J. (1994). Acquisition of contextual Pavlovian fear conditioning is blocked by application of an NMDA receptor antagonist D,L-2-amino-5-phosphonovaleric acid to the basolateral amygdala. *Behavioral Neuroscience*, 108, 210–212.

Feinstein, J.S., Adolphs, R., Damasio, A., and Tranel, D. (2011). The human amygdala and the induction and experience of fear. *Current Biology*, 21(1), 34–38.

Herry, C., and Johansen, J.P. (2014). Encoding of fear learning and memory in distributed neuronal circuits. *Nature Neuroscience*, 17(12), 1644–1654.

Impey, S., Mark, M., Villacres, E.C., Poser, S., Chavkin, C., and Storm, D.R. (1996). Induction of CRE-mediated gene expression by stimuli that generate long-lasting LTP in area CA1 of the hippocampus. *Neuron*, 16(5), 973–982.

Impey, S., Obrietan, K., Wong, S.T., Poser, S., Yano, S., Wayman, G., Deloulme, J.C., Chan, G., and Storm, D.R. (1998). Cross talk between ERK and PKA is required for Ca2+ stimulation of CREB-dependent transcription and ERK nuclear translocation. *Neuron*, 21(4), 869–883.

Johansen, J.P., Cain, C.K., Ostroff, L.E., and LeDoux, J.E. (2011). Molecular mechanisms of fear learning and memory. *Cell*, 147(3):509–524.

Kapp, B.S., Whalen, P.J., Supple, W.F., and Pascoe, J.P. (1992). Amygdaloid contributions to conditioned arousal and sensory information processing. In: *The amygdala: neurobiological aspects of emotion, memory, and mental dysfunction* (ed. J.P. Aggleton,), 229–254. Wiley-Liss, New York.

LeDoux, J.E. (1992). Emotion and the amygdala. In: *The amygdala: neurobiological aspects of emotion, memory, and mental dysfunction* (ed. J.P. Aggleton,), 339–351. Wiley-Liss, New York.

LeDoux, J.E. (2014). Coming to terms with fear. *Proceedings of National Academy of Sciences USA*, 111(8), 2871–2878.

LeDoux, J.E., Farb, C., and Ruggiero, D.A. (1990). Topographic organization of neurons in the acoustic thalamus that project to the amygdala. *Journal of Neuroscience*, 10, 1043–1054.

LeDoux, J.E., Iwata, J., Cicchetti, P., and Reis, D.J. (1988). Different projections of the central amygdaloid nucleus mediate autonomic and behavioural correlates of conditioned fear. *Journal of Neuroscience*, 8, 2517–2529.

LeDoux, J.E., Ruggiero, D.A., Forest, R., Stornetta, R., and Reis, D.J. (1987). Topographic organization of convergent projections to the thalamus from the inferior colliculus and spinal cord in the rat. *Journal of Comparative Neurology*, 264, 123–146.

Maren, S. and Fanselow, M.S. (1996). The amygdala and fear conditioning: has the nut been cracked? *Neuron*, 16, 237–240.

McDonald, A.J. (1998). Cortical pathways to the mammalian amygdala. *Progress in Neurobiology*, 55, 257–332.

Medina, J.F., Repa, J.C., Mauk, M.D., and LeDoux, J.E. (2002). Parallels between cerebellum- and amygdala-dependent conditioning. *Nature Reviews Neuroscience*, 3, 122–131.

Pare, D., and Smith, Y. (1998). Intrinsic circuitry of the amygdaloid complex: common principles of organization in rats and cats. *Trends in Neurosciences*, 21, 240–241.

Pitkanen, A., Savander, V., and LeDoux, J.E. (1997). Organization of intra-amygdaloid circuitries in the rat: an emerging framework for understanding functions of the amygdala. *Trends in Neurosciences*, 20, 517–523.

Quirk, G.J., Repa, C., and LeDoux, J.E. (1995). Fear conditioning enhances short-latency auditory responses of lateral amygdala neurons: parallel recordings in the freely behaving rat. *Neuron*, 15(5), 1029–1039.

Repa, J.C., Muller, J., Apergis, J., Desrochers, T.M., Zhou, Y., and LeDoux, J.E. (2001). Two different lateral amygdala cell populations contribute to the initiation and storage of memory. *Nature Neuroscience*, 4, 724–731.

Rogan, M.T., and LeDoux, J.E. (1995). LTP is accompanied by commensurate enhancement of auditory-evoked responses in a fear conditioning circuit. *Neuron*, 15(1), 127–36.

Rogan, M.T., Staubli, U.V., and LeDoux, J.E. (1997). Fear conditioning induces associative long-term potentiation in the amygdala. *Nature*, 390, 604–607.

Romanski, L.M., and LeDoux, J.E. (1993). Information cascade from primary auditory cortex to the amygdala: corticocortical and corticoamygdaloid projections of temporal cortex in the rat. *Cerebral Cortex*, 3, 515–532.

Schafe, G.E., Atkins, C.M., Swank, M.W., Bauer, E.P., Sweatt, J.D., and LeDoux, J.E. (2000). Activation of ERK/MAP kinase in the amygdala is required for memory consolidation of Pavlovian fear conditioning. *Journal of Neuroscience*, 20(21), 8177–8187.

Schafe, G.E., and LeDoux, J.E. (2000). Memory consolidation of auditory Pavlovian fear conditioning requires protein synthesis and protein kinase A in the amygdala. *Journal of Neuroscience*, 20(18), RC96.

Schafe, G.E., Nader, K., Blair, H.T., and LeDoux, J.E. (2001). Memory consolidation of Pavlovian fear conditioning: a cellular and molecular perspective. *Trends in Neurosciences*, 24, 540–546.

Silva, A.J., Rosahl, T.W., Chapman, P.F., Marowitz, Z., Friedman, E., Frankland, P.W., Cestari, V., Cioffi, D., Sudhof, T.C., and Bourtchuladze, R. (1996). Impaired learning in mice with abnormal short-lived plasticity. *Current Biology*, 6(11), 1509–1518.

Weeber, E.J., Atkins, C.M., Selcher, J.C., Varga, A.W., Mirnikjoo, B., Paylor, R., Leitges, M., and Sweatt, J.D. (2000). A role for the beta isoform of protein kinase C in fear conditioning. *Journal of Neuroscience*, 20(16), 5906–5914.

11 Taste Aversion

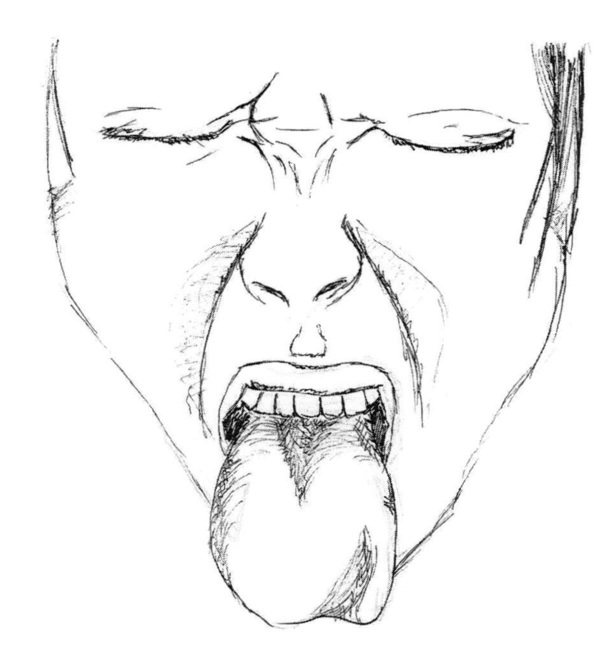

Introduction and Background

Imagine having a meal out in one of your favourite restaurants and ordering a particular dish. Following a most pleasant evening you return home only to spend half the night vomiting in the bathroom. How could this have happened? you ask yourself. We always go to this restaurant; I always have the same meal – what could have made me so ill? We typically try to rationalise our illness; maybe it was the fish, I do not normally have that particular fish; yes, it must have been the sauce – I knew it did not taste quite right. Two months later you are again out for a meal but this time there is no chance that you order the same food again; in fact, you may never even go to the same restaurant ever again. This is despite not knowing why you really got ill or the fact that you actually had quite a pleasant experience while at the restaurant previously. Such a provoked strong reaction has profoundly changed your behaviour. This learning comes about as a result of a single incident but will be remembered for a long, long time.

Learned taste aversion was first described in 1955 by John Garcia. In this paradigm, animals learn to avoid a particular food or taste after it has been paired with an aversive stimulus. In the original experiment, Garcia and his colleagues paired a sweet saccharin solution (which rats quite enjoy) with a dose of gamma radiation, which made the animals ill. Following this pairing, the animals were retested and offered the choice of saccharin solution or water. The animals that had previously been irradiated no longer picked the saccharin solution. Control animals, which had not been irradiated, continued to choose the saccharin solution over water. Figure 11.1 demonstrates the dramatic effect of this pairing. Preradiation exposure, animals picked the saccharin solution approximately 90% of the time. If the animals are given a saccharin solution and at the same time are given a single high dose (57 rads, black dots) or low dose (30 rads, white dots) of radiation, there is a profound decrease in saccharin consumption. Furthermore, this effect lasts for a long time; for example, in Figure 11.1, animals still avoid the saccharin after 30 days.

Behaviour of Taste Aversion

Taste aversion is an interesting type of conditioning and seems to be quite unique compared to the other

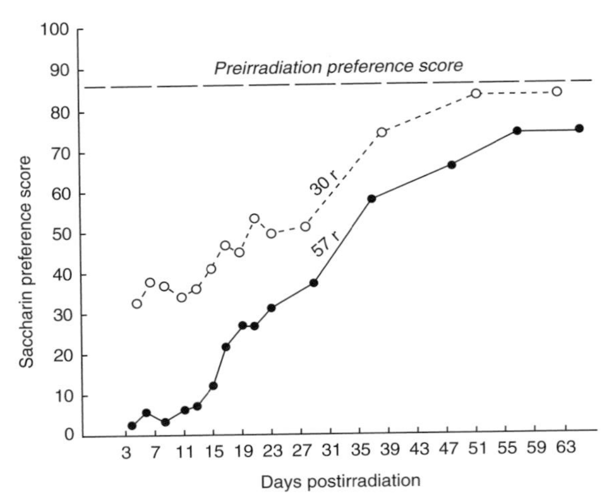

Figure 11.1 Animals avoid saccharin for a long time following exposure to a strong (black dots) or weak (white dots) radiation treatment. Image taken from J. Garcia et al. (1955), Conditioned aversion to saccharin resulting from exposure to gamma radiation. *Science*, 122, 157–158. Reprinted with permission from AAAS.

types of conditioning that have been described. First, it requires very few trials for it to occur; in many taste-aversion paradigms only a single trial is needed. This is very much in contrast with other conditioning tasks, where learning occurs over multiple trials. Second, traditional conditioning tasks generally require the conditioned stimulus–unconditioned stimulus (CS–US) pairing to occur more or less simultaneously, or require the CS (e.g. the bell in Pavlov's classical conditioning) to precede the US (e.g. meat) by a very short interval (e.g. 0.5 second to 2 seconds). This does not seem to be the case for taste-aversion conditioning; the interval between the food presentation and the aversive stimulus can be quite long. For example, in a later paper by Garcia et al. (1966), conditioning still occurred even after a delay of 75 minutes between the pairings. Third, as illustrated in Figure 11.1, taste-aversion conditioning is long-lasting. Fourth, it seems that aversion is selective for taste; for example, if radiation is paired with other nontaste stimuli, e.g. a sound, the effects are not very strong.

Taste aversion has been observed across many species, ranging from the garden slug to humans. Further, while the original experiment used radiation as the aversive stimulus, many other stimuli have been used; for example, lithium chloride (LiCl) is regularly used in current experiments.

From an evolutionary perspective, it makes good sense to avoid tasting substances that might be dangerous for us. For example, rats are foragers and try and obtain nutrients from anywhere. But, as they are unable to vomit and eject poisonous substances from their body, they have to be careful with their selection, often showing an avoidance of unfamiliar food, and rapidly learn to avoid unwanted substances.

Although humans can vomit, we too are often cautious in our approach to new foods and tend to avoid those that disagree with us. Thus it probably does not come as a surprise that novel foods are often much more likely to show a greater conditioned response than familiar foods. Despite this, a number of interesting paradoxes exist in the literature. For example, although many drugs, especially those that produce an emetic response, can be used as an US, drugs that do not produce a vomit response (and in fact may be quite rewarding) are also effective at producing a learned response. In addition, many drugs that do produce an emetic response are not that effective to use as a US (e.g. sodium cyanide). Further discussions of these ideas can be found by Davis and Riley (2010) and Verendeev and Riley (2012).

A Brain System for Taste Aversion

Four basic tastes have been recognised in the literature. These are sweetness (defined by the presence of sugars); sourness (how acidic is a substance, usually compared to hydrochloride acid, or HCl); saltiness (defined by the presence of sodium); and bitterness (how toxic is a substance, usually compared to quinine). Although umami (often described as savoury, the pleasant taste you get when eating meat or other protein-rich food) has been recognised in the East for over a century, it was only scientifically recognised in the mid-1980s.

Taste buds, located on the top of the tongue, respond to different tastes. Information from these buds is then transmitted to the brain via three cranial nerves (facial, vagus and glossopharyngeal nerves). Taste information converges in the brain stem within a structure called the nucleus of the solitary tract (*nucleus tractus solitarius*, NTS, see Figure 11.2). From here, information is sent to the pontine parabrachial nucleus (PBN), and this structure seems to be the first region critical for taste aversion. Indeed, different tastes lead to an increase in neural activation (as measured by the protein c-Fos activation, Yamamoto, 2006) in different parts of the PBN, suggesting that taste is segregated in this region; this chemical separation is often referred to as *chemotopy*. The information is then passed to the insular cortex (IC) via the ventral posteromedial nucleus of thalamus (not included in Figure 11.2). Similar to the PBN, the insular cortex also seems to be chemotopic in nature (Yamamoto et al., 1989; Yamamoto, 2006), with neurons in the more rostral (anterior) part responding more to sweet tastes, whereas neurons that respond to more bitter tastes are more caudally (posteriorly) located. The insular cortex, in turn, projects to the amygdala (Amy in Figure 11.2). As well as receiving projections from the IC, the amygdala also sends information back to this structure (double arrows, Figure 11.2); the amygdala also has a reciprocal relationship with the PBN. Therefore, the amygdala seems to play a privileged role; not only does it project to the nucleus accumbens (NAcb), which is part of the brain's reward system, but it also sends projections to both the ventromedial hypothalamus (VMH, not shown) and lateral hypothalamus (LH); these latter structures are thought to play a direct role in satiety control and feeding behaviour, respectively. Thus, the PBN, IC and amygdala (shaded areas in Figure 11.2) seem to be the key structures in taste aversion and contribute to different aspects of the behaviour.

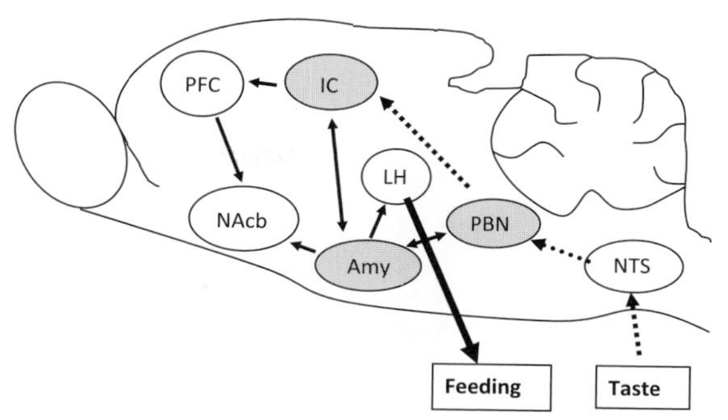

Figure 11.2 A sagittal (side) view of the rat brain, illustrating various structures thought to be involved in taste aversion. Those shaded in grey are considered particularly important. Dashed lines represent input pathways and double arrows represent reciprocal connections. Key: Amy, amygdala; IC, Insular cortex; LH, lateral hypothalamus; NAcb, nucleus accumbens; NTS, nucleus of the solitary tract; PBN, parabrachial nucleus; PFC, prefrontal cortex.

Neural Changes Following Conditioned Taste Aversion

Rather than treating taste aversion as a unitary behaviour, some authors (see Yamamoto and Ueji, 2011 for more details) have categorised it into a variety of behaviours, allowing conditioned taste aversion (CTA) to be studied in more depth. We will also use this approach here. Such behaviours include: (1) the *augmentation* of the conditioned stimulus; (2) a shift from a substance being originally perceived as something pleasant to something that should be avoided (*hedonic shift*); (3) formation of an *association* between the unconditioned stimulus and the conditioned stimulus, with a long-lasting associative memory; and (4) a general *alertness* and wariness of new substances.

Augmentation of the Conditioned Stimulus

As mentioned above, neurons in the PBN (as well as in the IC and amygdala) are taste-specific and respond to different tastes. Shimura et al. (1997) found that following conditioning, the neural response to specific tastes becomes more salient

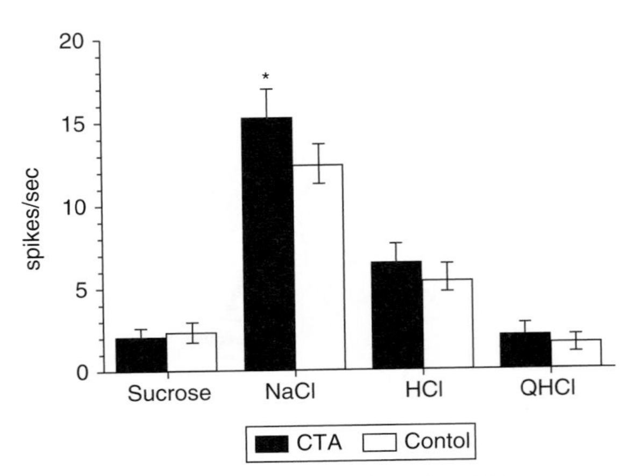

Figure 11.3 Selective activation of parabrachial neurons to NaCl, which is enhanced further following conditioning. Image taken from T. Shimura et al. (1997), Salient responsiveness of parabrachial neurons to the conditioned stimulus after the acquisition of taste-aversion learning in rats, *Neuroscience*, 81(1):239–47, with permission from Elsevier.

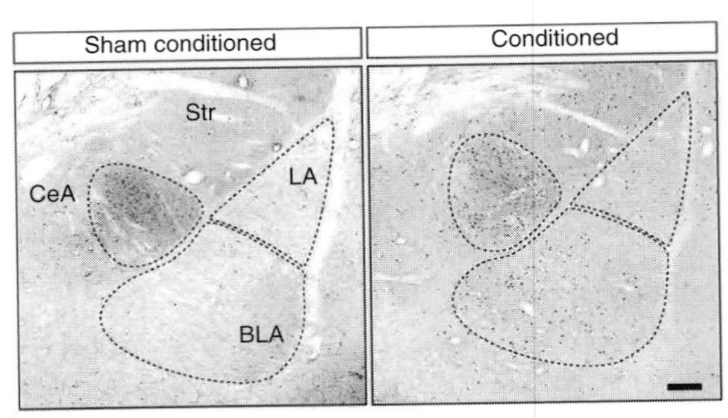

Figure 11.4 Photomicrographs showing c-Fos activation (small black dots) in various subregions of the amygdala following re-exposure to a sucrose solution in the sham (left) and conditioned animals (right). Key: CeA (central amygdala), LA (lateral amygdala), BLA (basolateral amygdala), Str (striatum). Image taken with permission from T. Yamamoto (2006), Neural substrates for the processing of cognitive and affective aspects of taste in the brain. *Arch Histol Cytol.* 69(4):243–255.

(augmented). For example, in one experiment, thirsty animals were presented with a saline (NaCl) solution, which was immediately followed by an injection of lithium chloride (LiCl), an unpleasant toxic substance. Upon further presentation of the NaCl solution alone, there was an immediate decrease in solution intake – the animals were conditioned. However, recordings of neurons from the PBN (Figure 11.3) showed an enhanced neural response in the conditioned group (CTA) compared to animals that had not been injected with LiCl. In addition, this enhanced response was only observed in the sodium-responsive neurons and not in neurons that responded to other tastes (sucrose, HCl, etc.). So, while there was a decrease in the behaviour, there was a concurrent enhancement of the neural response. In other words, as the saline solution has been associated with something unpleasant, there is now an avoidance of the saline solution (decrease in behaviour – drinking) but an actual increase in neural activity (especially in the sodium-related neurons). Furthermore, this enhanced neural response is not confined to the PBN but is observed in the NTS (Chang and Scott, 1984), the IC (Yasoshima and Yamamoto, 1998) and the amygdala (Yasoshima et al., 1995). On a practical level, one of the advantages of an increased neural response is that it may allow the animal to detect a particular substance quicker within a food source. This may provide a warning to the animal, allowing it to avoid that food in the future, or at least to approach it with caution.

Hedonic Shift (Aversion)

A second major feature of taste-aversion learning is that there is a shift in animals' response, from being attracted to something (e.g. sucrose) to a sudden avoidance of the same substance. This is termed *hedonic shift* (Yamamoto, 2006). Can this behavioural shift be represented at the neural level? A number of structures in the brain have been shown to be activated following *retrieval*, the first exposure to the CS (e.g. sucrose) following conditioning. Using the immediate-early gene Fos as a marker of neural activity, researchers found there is increased activity in a number of structures including the supramammillary nucleus, the thalamic paraventricular nucleus, the amygdala and the nucleus accumbens (NAcb) following retrieval. For example, Figure 11.4

shows an increased number of neurons expressing Fos-Li (small black dots) in the amygdala, particularly the basolateral amygdala (BLA), in conditioned animals compared to control (sham-conditioned) animals following re-exposure to a sucrose solution. This suggests that neural structures that had not been active (to sucrose, for example) before conditioning have now become so. There has been a shift in the neural response.

Both the nucleus accumbens and the amygdala are considered to be part of the brain's reward system. The amygdala, primarily the BLA, sends direct projections to the NAcb, which in turn projects to the ventral pallidum (VP), and onto the lateral hypothalamus (LH, see Figure 11.2). For its part, the LH is considered to be the brain's feeding centre. Stimulation of the LH, for example, can cause the desire to eat, while damage or lesions to this region can cause the reduction of food intake. Therefore, the NAcb-VP-LH circuit is considered to be a critical intermediate between the brain's reward system (the desire to eat) and the action of eating, making this an ideal circuit to underlie any hedonic shift.

Is there a mechanism whereby the desire to eat a palatable substance can be reduced? The primary neurotransmitter involved in the projection from NAcb to the VP is GABA (the brain's inhibitory neurotransmitter), and it is this neurotransmitter that is thought to be involved in inhibiting the animal's response to the palatable CS following conditioning. This has been shown in a study by Inui et al. (2007). Here the authors injected the $GABA_A$ receptor antagonist (i.e. blocking the function of GABA) directly into the VP. This had the effect of reversing the aversion to saccharin in animals that had been conditioned. Figure 11.5a shows the low amount of saccharin intake of control animals following conditioning (test, white bar) which is then reversed if given an injection of the $GABA_A$ antagonist bicuculline (test, black bar). This suggests that GABA may be involved in the inhibitory behaviour post-conditioning. It is also possible to show the opposite effect. So, rather than injecting a GABA antagonist, if you injected a GABA agonist (thereby enhancing GABA) into the VP of control animals (unconditioned animals), you would find an increased aversive effect. This effect was shown by Shimura et al. (2006) and is displayed in Figure 11.5b. Following injection of muscimol, a $GABA_A$ receptor agonist, into the VP, animals displayed a reluctance to consume saccharin (no ingestive response, Figure 11.5b). Interestingly these muscimol-injected animals also displayed strong aversive behaviour such as head shaking, chin rubbing and tongue protrusions (striped aversive bar, Figure 11.5b).

Further evidence of the role played by GABA comes from microdialysis techniques (Chapter 3), whereby Inui et al. (2009) have shown that the concentration of GABA increases in the VP region upon presentation of a saccharin solution in the conditioned group (Figure 11.5c) when compared to unconditioned control animals.

So far we have discussed the NAcb-VP-LH circuit, but what role does the amygdala play in hedonic shifting? As mentioned above, there is increased neural activity in the BLA subregion of the amygdala during retrieval. In addition, this area projects directly into the NAcb-VP-LH pathway. However, the role of the amygdala is controversial; many studies indicate that lesions of the central nucleus of the amygdala (CeA) have little impact on taste-aversion learning, whereas lesions to the basolateral amygdala (BLA) do have an effect. However, even with the BLA studies there are many interpretational issues (see Reilly and Bornovalova, 2005 for discussion) and the BLA's exact role is difficult to elucidate. There is some evidence to suggest that neurons in the amygdala show a different response post-conditioning compared to pre-conditioning. For example, Yasoshima et al. (1995) found that a small number of neurons in the amygdala show an enhanced response to sucrose and saccharin post-conditioning (Figure 11.6, top panel, black bars) compared to before conditioning (white bars), while other amygdalar neurons showed an actual decrease in responsiveness (Figure 11.6, bottom panel). Interestingly, the majority of the neurons recorded

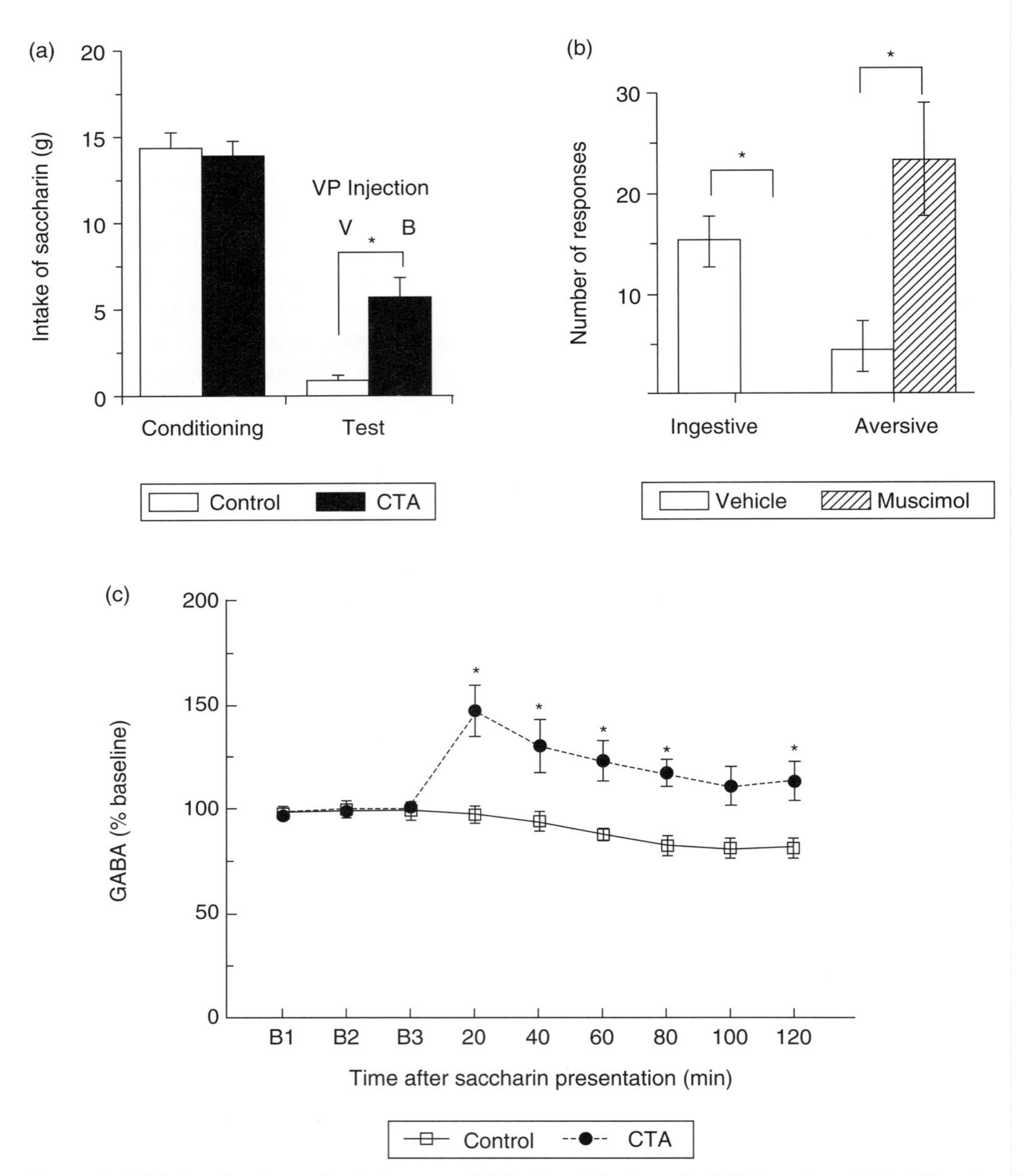

Figure 11.5 (a) Aversion to saccharin is reversed following injection of a $GABA_A$ antagonist in VP region (Test, black bar). (b) Increased aversive effect (striped bar) and decreased ingestive behaviour following injection of a $GABA_A$ agonist. (c) Increased GABA concentration in the VP region upon presentation of a saccharin solution in the conditioned group. CImages taken from T. Yamamoto and K. Ueji (2011), Brain mechanisms of flavor learning. *Front Syst Neurosci.* 5:76 with permission under Creative Commons Attribution License (http://creativecommons.org/licenses/by/4.0/) and modified from T. Shimura et al., 2006 and T. Inui et al., 2007, 2009.

in the BLA have been found to be of facilitatory kind (showing an increase in responsiveness, top panel), but the majority of the neurons recorded in the CeA were of the inhibitory variety (bottom panel). This has led to a debate about whether this neural response is part of the brain's ability to augment the CS (see section above) or whether the amygdala influences hedonic shifting, or both.

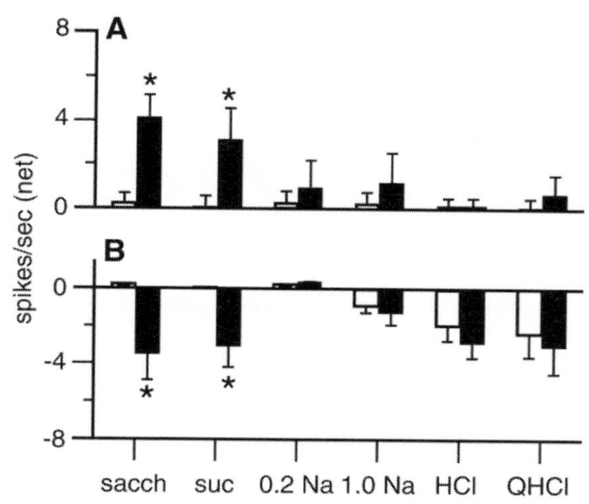

Figure 11.6 Two types of amygdalar neurons exhibiting facilitatory (top panel) or inhibitory (bottom panel) responses to saccharin following conditioning (black bars) compared to before conditioning (white bars). Image taken from Y. Yasoshima et al. (1995), *NeuroReport*, 6, 2424 with permission.

Association of Unconditioned Stimulus and Conditioned Stimulus: Aversion Memory

One of the important features of associative memory is that different types of information can become strongly linked together, for example a person's name and face. If only one element of this information link is subsequently presented (e.g. the face) then the 'combined' information is also activated (the person's name also comes to mind). This convergence and binding of information is also reflected at the neural level, in that 'neurons that fire together, wire together' (see also Chapter 7 on LTP). Taste aversion is no different. If an animal sees or tastes a saccharin solution that had been previously paired or associated with a LiCl injection, the sickness is also recalled. Where in the brain might the site of CS–US convergence take place?

One area that has received much attention is again the amygdala, particularly the basolateral region (BLA). In a recent study, Barot et al. (2008) analysed *Arc* mRNA in the amygdala during different stages of taste-aversion conditioning. *Arc* mRNA is an immediate-early gene (IEG) that is expressed very soon after neural activation, making it an ideal biomarker to see what brain structures are involved in a particular behaviour. Furthermore, as *Arc* mRNA is typically seen in the cell's nucleus 5 minutes after induction before moving into the cytoplasm, this marker is also very useful to examine the time course of activation in various regions. When animals were exposed to the CS only (saccharin), Barot et al. found that *Arc* mRNA was mainly expressed in the cytoplasm of amygdalar neurons. When animals were given the US (LiCl injection) *Arc* mRNA was mainly expressed in the neurons. However, if animals were given the CS and this was followed immediately by the US, *Arc* mRNA was expressed in both the nucleus and cytoplasm of the same amygdalar neuron. Figure 11.7a and b show that the number of cells in the BLA expressing staining in both the nucleus and cytoplasm is significantly higher in the CS–US group compared to the other groups. This suggests that combined information converges onto single neurons of the amygdala. Interestingly, this convergence was not observed in the insular cortex or in other structures. Also, if you reverse the order of conditioning so that you present the US first followed by the CS (Figure 11.7c, US–CS Backward Conditioned), there are a limited number of cells expressing *Arc* mRNA in both the cytoplasm and nucleus compared to 'normal' or forward conditioning.

Although the amygdala itself is clearly an important area in conditioned taste aversion, other research has focused on the interaction between the amygdala and the insular cortex (IC). The IC and the amygdala are highly interconnected. Further, Escobar et al. (1998) have found that stimulation of the BLA leads to LTP in the IC and that this seems to enhance retention of taste aversion over a long time period. As one of the major properties of taste aversion is its long-lasting effect, any neural change must also be long-term. Therefore, in order for memories to be stored, there must be a long-term modification of synaptic strength between brain areas. In a recent review by Guzman-Ramos and Bermudez-Rattoni (2012), there is evidence that long-term changes

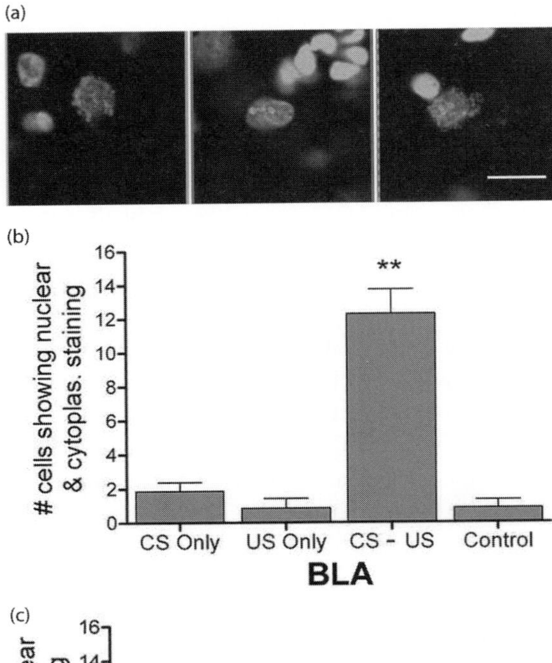

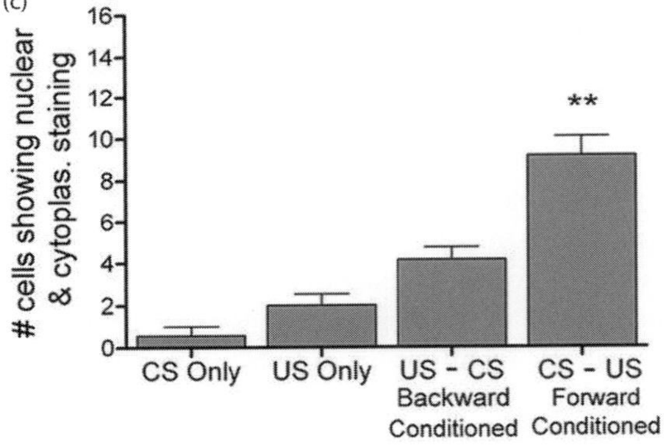

Figure 11.7 (a) Saccharin (CS) produces *Arc* in the cytoplasm (red, left) of amygdalar neurons; LiCl (US) induces *Arc* in the nucleus (middle); CS + US induce *Arc* in both the cytoplasm and nucleus (right). (b) Number of cytoplasm and nuclear stained cells is significantly higher in CS–US paired animals than in all other groups. (c) Forward conditioned animals show significantly more double-labelled neurons than all other groups. Images taken from S.K. Barot et al. (2008), Visualizing stimulus convergence in amygdala neurons during associative learning. *PNAS* 105(52): 20959–20963. Copyright (2008) National Academy of Sciences, U.S.A. For colour version of part (a), please refer to the plate section.

related to aversive memory are done through the interaction of the amygdala and insular cortex. During CS–US pairing, there are changes in both structures that can lead to long-term modification of synapses, allowing memories to remain strong. For example, Guzman-Ramos et al. (2010), and Guzman-Ramos and Rattoni, 2012) have found that saccharin increases dopamine and noradrenaline release in the amygdala and IC, respectively. But, when LiCl is injected after the administration of saccharin, there is an immediate and additional increase in glutamate release in both structures, causing greater modification of synapses and allowing these changes to become more permanent. Guzman-Ramos and Bermudez-Rattoni (2012) offer a possible model of how the increase in neurotransmitters can lead to long-term synaptic modifications, which is outlined in the steps below (see Figure 11.8).

Novel Taste

(1) A novel taste (e.g. saccharin) increases dopamine (DA) and acetylcholine (Ach) release in the IC.
(2) *Simultaneously*, the novel taste increases noradrenaline (NA) release in the amygdala.
(3) In the IC, ACh activates muscarinic receptors activating protein kinase C (PKC).
(4) This, in turn, modifies the NMDA glutamate receptor, leading to short-term memory (STM).
(5) In addition, DA activates D1 receptors in the IC.
(6) This leads to an increase in cAMP through adenylyl cyclase.
(7) In turn, cAMP activates protein kinase A (PKA).
(8) PKA can also modify the NMDA receptor, leading to short-term memory (STM) by increasing the conductance through the channel.
(9) Alternatively, PKA can activate CREB (via MEK and ERK) in the neuron's nucleus, which contributes to the synthesis of new proteins and other gene products, leading to longer term synaptic modifications of the IC and long-term memory (LTM) in this structure.
(10) NA in the amygdala activates beta-adrenergic receptors, which also increases cAMP and activates a similar pathway described above, leading to long-term synaptic modifications in the amygdala.

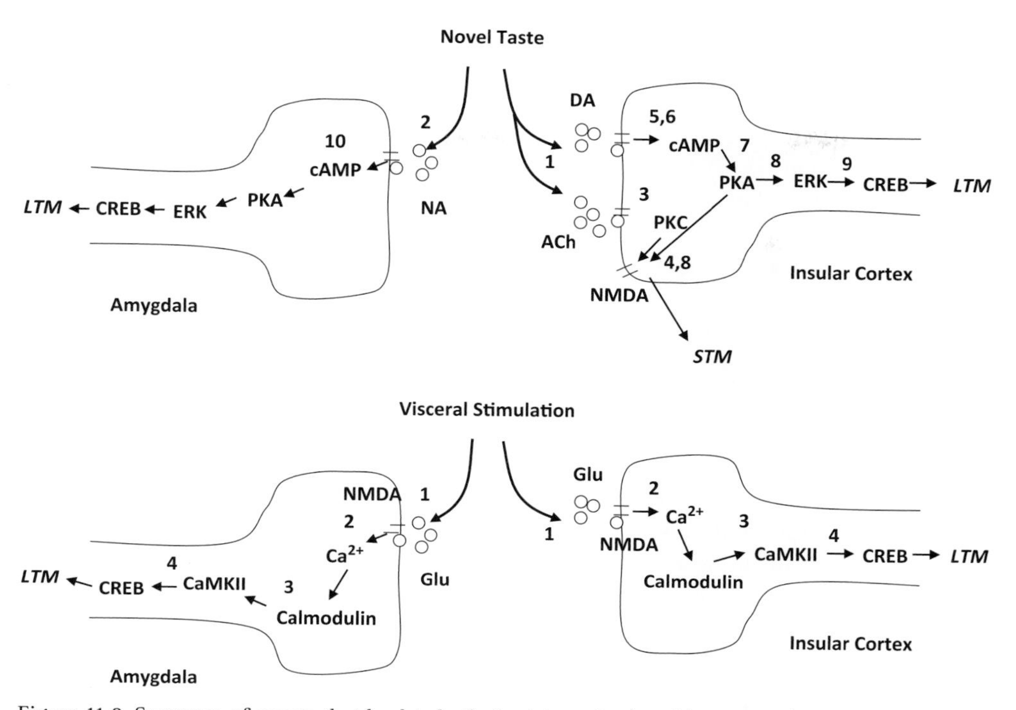

Figure 11.8 Sequence of events that lead to both short-term (top) and long-term (bottom) modifications in amygdalar and insular neurons; see text for details. Model based on K. Guzman-Ramos and F. Bermudez-Rattoni (2012), Interplay of amygdala and insular cortex during and after associative taste aversion memory formation, *Reviews in Neuroscience*, 23(5–6), 463–71.

Visceral Stimulation

(1) Visceral stimulation (e.g. with LiCl) increases the release of glutamate in both the amygdala and IC.
(2) Glutamate activates the NMDA glutamate receptor.
(3) The increase of calcium ions through the receptor activates calmodulin and CaMKII.
(4) This activates cAMP response element-binding protein (CREB), leading to long-term synaptic modifications.

From the steps and schematics, we can observe that neurons in both the amygdala and IC are modified by two molecular pathways. When the novel taste and visceral stimulation co-occur, the modifications may be even stronger and last for longer. Further, even though the amygdala and IC seem to be modified independently, the two regions are being activated/stimulated more or less at the same time. This, in turn, may strengthen the connection between the two areas (see Chapter 7 on LTP) and/or between different neurons within each area. Therefore, as neurons in both the amygdala and IC have been 'primed', when the CS is presented alone this may lead to a greater chance of activation of the 'visceral' neurons as well as those dealing with saccharin, leading to an avoidance response through the lateral hypothalamus (Figure 11.2).

Alertness

A stronger aversion response is acquired when the CS is novel compared to when it is familiar, serving to alert and protect the animal against a substance that might do it harm. Miranda et al. (2000) found that a novel stimulus induces an increased ACh release in the IC compared with the release observed during the presentation of a familiar stimulus. This enhanced ACh response may further activate the IC molecular pathways, leading to enhanced neural modification within the IC and between it and the amygdala.

Summary

The Figure 11.9 (adapted from Yamamoto and Ueji, 2011) presents a simplified model of how the four different processes of alerting, augmentation, aversion and association may interact to lead to the conditioned taste response. In the upper panel the CS (saccharin) activates the amygdala and downstream structures to produce a normal feeding response. If, however, the CS is paired with an unpleasant taste (e.g. LiCl) this enhances the neural response in the amygdala and other structures (middle panel, represented by grey circle neuron). The novelty of the CS may enhance this response even further, via ACh from the IC. The pairing of CS and US also modifies the neurons in the amygdala and forms a strong association that is long-lasting. This changed response is thought to *inhibit* (represented by the minus sign in Figure 11.9, middle) the downstream structures via GABA, shifting the behavioural response from feeding to one of avoidance. As the response of the neurons in the amygdala and other structures remains modified, presentation of the CS alone continues this aversion (lower panel).

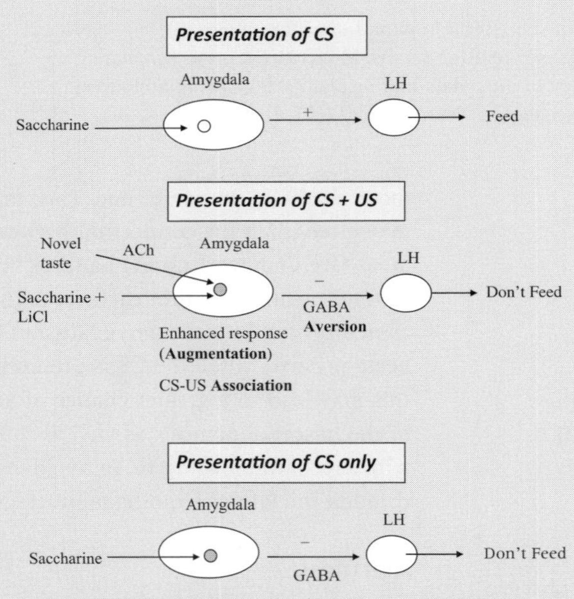

Figure 11.9 A simplified model of how the four different processes of alerting, augmentation, aversion and association may interact to lead to the conditioned taste response. LH = lateral hypothalamus. Adapted from T. Yamamoto and K. Ueji (2011), Brain mechanisms of flavor learning. *Frontiers in System Neuroscience*, 5:76.

Questions and Topics Under Current Investigation

- Consolidation of memories (including taste-aversion memory) is thought to occur in waves, with proteins being expressed many hours after the initial acquisition. What is the molecular mechanism for this consolidation? Is the consolidation process different for taste aversion compared to other types of memories?
- Current research suggests that drugs of abuse are complex compounds that have multiple effects, both positive and negative. Although research on the positive aspects of these drugs has been well documented, the aversive aspects have been generally ignored (see Verendeev and Riley, 2012).
- Why are some drugs, particular rewarding ones, also effective at producing a learned response? Why should such drugs when paired with a CS also produce avoidance?
- Other active areas of research include the exploration of whether loss of appetite and anorexia in humans (e.g. cancer anorexia) can be explained in terms of a taste-aversion model. Although the loss of appetite in these patients may stem from the disease itself or from the medication, the mechanisms responsible for these symptoms are not well understood (Bernstein, 1999).

References

Barot, S.K., Kyonoc, Y., Clark, E.W., and Bernstein, I.L. (2008). Visualizing stimulus convergence in amygdala neurons during associative learning. *Proceedings of the National Academy of Sciences USA*, 105(52), 20959–20963.

Bernstein, I.L. (1999). Taste aversion learning: a contemporary perspective. *Nutrition*. 15(3), 229–234.

Chang, F.C., and Scott, T.R. (1984). Conditioned taste aversions modify neural responses in the rat nucleus tractus solitarius. *Journal of Neuroscience*, 4(7), 1850–1862.

Davis, C.M., and Riley, A.L. (2010). Conditioned taste aversion learning: implications for animal models of drug abuse. *Annals New York Academy of Science*, 1187, 247–275.

Escobar, M.L., Chao, V., and Bermúdez-Rattoni, F. (1998). In vivo long-term potentiation in the insular cortex: NMDA receptor dependence. *Brain Research*, 779(1–2), 314–319.

Garcia, J., Ervin, R.R., and Koelling, R.A. (1966). Learning with prolonged delay of reinforcement. *Psychonomic Science*, 5, 121.

Garcia, J., Kimeldorf, D.J., and Koelling, R.A. (1955). Conditioned aversion to saccharin resulting from exposure to gamma radiation. *Science*, 122, 157–158.

Guzman-Ramos, K., and Bermudez-Rattoni, F. (2012). Interplay of amygdala and insular cortex during and after associative taste aversion memory formation. *Reviews in Neuroscience*, 23(5–6), 463–471.

Guzman-Ramos, K., Osorio-Gómez, D., Moreno-Castilla, P., and Bermúdez-Rattoni, F. (2010). Off-line concomitant release of dopamine and glutamate involvement in taste memory consolidation. *Journal of Neurochemistry*, 114(1), 226–236.

Inui, T., Shimura, T., and Yamamoto, T. (2007). The role of the ventral pallidum GABAergic system in conditioned taste aversion: effects of microinjections of a GABAA receptor antagonist on taste palatability of a conditioned stimulus. *Brain Research*, 1164, 117–124.

Inui, T., Yamamoto, T., and Shimura, T. (2009). GABAergic transmission in the rat ventral pallidum mediates a saccharin palatability shift in conditioned taste aversion. *European Journal of Neuroscience*, 30(1), 110–115.

Miranda, M.I., Ramírez-Lugo, L., and Bermúdez-Rattoni, F. (2000). Cortical cholinergic activity is related to the novelty of the stimulus. *Brain Research*, 882(1–2), 230–2325.

Reilly, S., and Bornovalova, M.A. (2005). Conditioned taste aversion and amygdala lesions in the rat: a critical review. *Neuroscience Biobehavioural Review*, 29(7), 1067–1088.

Shimura, T., Imaoka, H., and Yamamoto, T. (2006). Neurochemical modulation of ingestive behavior in the ventral pallidum. *European Journal of Neuroscience*, 23(6), 1596–604.

Shimura, T., Tanaka, H., and Yamamoto, T. (1997). Salient responsiveness of parabrachial neurons to the conditioned stimulus after the acquisition of taste aversion learning in rats. *Neuroscience*, 81(1), 239–247.

Verendeev, A., and Riley, A.L. (2012). Conditioned taste aversion and drugs of abuse: history and interpretation. *Neuroscience Biobehavioural Review*, 36(10), 2193–2205.

Yamamoto T. (2006). Neural substrates for the processing of cognitive and affective aspects of taste in the brain. *Archives of Histology and Cytology*, 69(4), 243–255.

Yamamoto, T., Matsuo, R., Kiyomitsu, Y., and Kitamura, R. (1989). Taste responses of cortical neurons in freely ingesting rats. *Journal of Neurophysiology*, 61(6), 1244–1258.

Yamamoto, T., and Ueji, K. (2011). Brain mechanisms of flavor learning. *Frontiers in System Neuroscience*, 5:76.

Yasoshima, Y., Shimura, T., Yamamoto, T. (1995). Single unit responses of the amygdala after conditioned taste aversion in conscious rats. *Neuroreport*, 6(17), 2424–2428.

Yasoshima, Y., Yamamoto, T. (1998). Short-term and long-term excitability changes of the insular cortical neurons after the acquisition of taste aversion learning in behaving rats. *Neuroscience*, 84(1), 1–5.

12 Sound Localisation

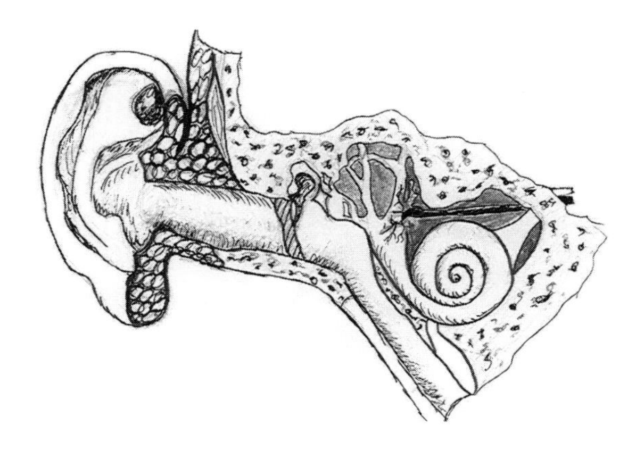

Introduction and Background

Being able to locate the source of a sound is critical. If we want to cross a road, we need to look carefully up and down; we also need be able to tell the direction and location of an oncoming car by listening. Both senses need to be fully tuned. If one is compromised through distraction or other factors, the other needs to keeps us alert to potential danger. Verbal communication also requires us to pay attention to the source of sound, usually the speaker. If we hear a loud noise while talking to someone, we may get distracted and our head immediately turns to the source of the sound, often to the annoyance of the speaker.

In humans, the auditory cortex is an important brain region that is involved in the differentiation of sound and helps us to detect the source of sound. Damage of the auditory cortex, located in the temporal lobes (see Chapter 2), does not cause deafness but does compromise our ability to distinguish between different sounds based on their tones, frequencies and their temporal order – vital functions in communication and language. In addition, damage to the auditory cortex leads to the impairment in the ability to pinpoint precisely the location of a sound. However, humans are visual creatures. Over 55% of our cortex is devoted to visual function, while a mere 3% is thought to be involved in auditory function. In addition, unlike the visual system, about which much is known, the exact role played by the auditory cortex in hearing is not as well-defined.

It is known that the auditory cortex is tonotopic in its organisation – that is, neurons are organised in the cortex according to their response to different sound frequencies. Neurons at one end of the cortex fire in response to high frequencies, while neurons at the other end respond to low-frequency sounds. Unlike the visual and somatic sensory systems, which also have topographical maps, sound is already separated and mapped out prior to reaching the auditory cortex, by the cochlea of the ear. Furthermore, how the auditory cortex in humans can process sound location is still relatively unknown. It is for these reasons that neuroscientists look to an animal that has excellent hearing, an animal in which being able to detect the source of a sound is a matter of life and death. This animal is the beautiful barn owl (see Figure 12.1a).

The owl is one of the most widely distributed bird species, and barn owls are found in most regions of the world, excluding the polar and desert regions. There are nine species of barn owl, which make up their own family (Tytonidae). The common barn owl (*Tyto alba)* is the most numerous of the species. It has a height of approximately 39 cm (15 inches) with a wing span of between 80 and 95 cm (31 and 37 inches).

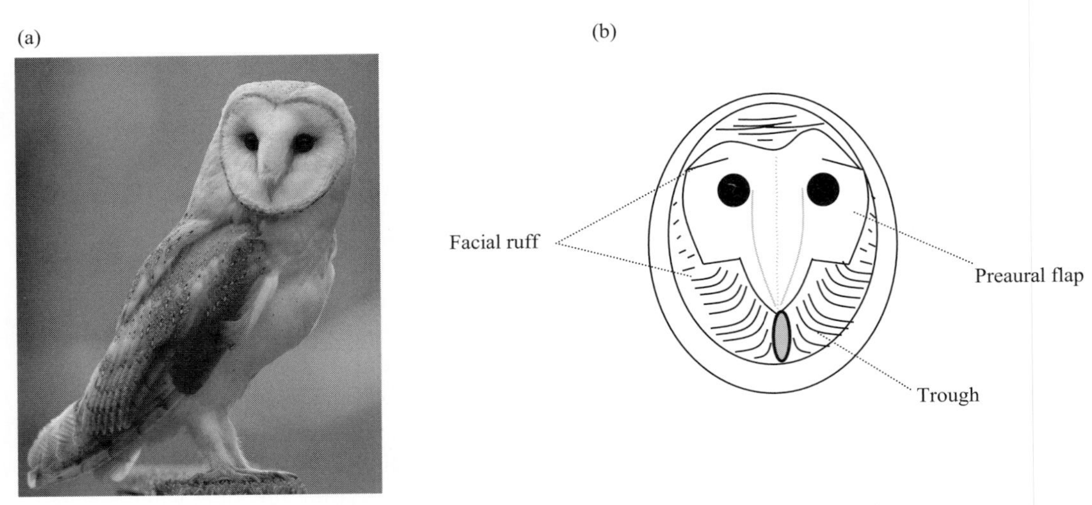

Figure 12.1 (a) Photo of a barn owl by Peter Trimming (England), reproduced under Creative Commons Attribution License (CCA). (b) Important facial features of the barn owl.

Rather than the traditional hooting sound that is often associated with owls, the barn owl produces a long-drawn out shriek. Most barn owls are nocturnal and hunt by night, preying on small mammals, mainly field mice, but they also prey on amphibians and other small birds. The barn owl's skull is relatively small but it has a large round heart-shaped face, typically white. It has dense feathers around its face that are termed the facial ruff (Figure 12.1b). There are two troughs that run through the ruff; they are the owls' equivalent to human ears. These channels collect high-frequency sounds from their environment and funnel them through to the ear canals (Knudsen, 1981).

Behaviour of Sound Localisation in Barn Owls

The barn owl has been studied in depth by Masakazu Konishi since the 1970s. He has used the barn owl to elucidate the behavioural and neural mechanisms underlying sound localisation. In fact, Konishi and his colleagues have, over the last number of years, uncovered almost every step involved in this process, which we will try and explain here.

Evidence suggests that barn owls solely use sound to localise their prey. Field mice are very difficult to see, even during the day; this is mainly due to their colouring but also due to the fact that mice tend to travel through tunnels in grass or snow. At night they are nearly invisible. Roger Payne (1971) clearly demonstrated the owl's dependence on aural information in a set of experiments. In one experiment an owl was trained to catch mice that were released on leaf-filled floor. During the testing phase, the lights were turned off and a wad of paper the approximate size of a mouse was dragged along the floor. The owl snatched at the paper wad and carried it off. The owl was not using odour cues, as paper does not emit any; as both the wad of paper and the leaves were at the same temperature, no infrared cues were emitted. In a second experiment, the room was filled with sand. Mice, trained to drag a wad of paper, were then released and ran silently across the sandy floor. When the owl was subsequently released, it snatched at the noisy paper rather than at the mouse itself.

When owls swoop down on their prey, they turn their heads towards the noise, and align their talons with the body axis of the prey. It occurred to Knudsen and Konishi (1978) that this head-turning response would make an excellent behavioural

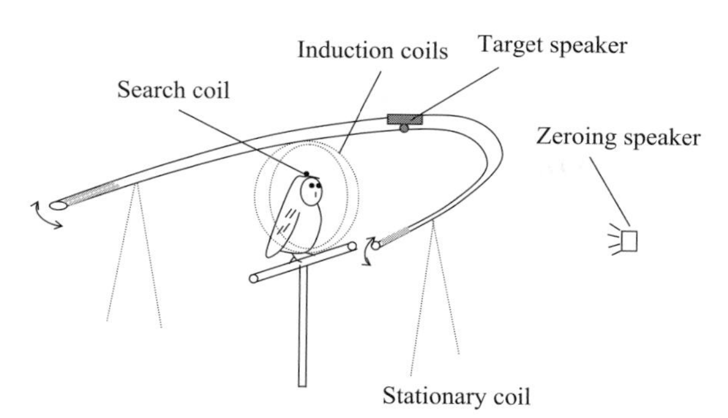

Figure 12.2 Experimental setup used to examine the orienting response in the owl. Adapted from E.I. Knudsen (1981), The hearing of the barn owl, *Scientific American*, 245, 82–91.

measure of sound localisation. As a result they designed a laboratory that would measure the head-turning response in the barn owl. With this new laboratory, they could now accurately measure the ability of owls to locate a sound source, a feat very difficult to do in a free-flight situation.

Head-orientation experiments are relatively simple and allow for the exact measurement of the relationship between head and sound source. In an experimental situation (see Knudsen, 1981, and Figure 12.2) the owl stands on a perch. A ‘search’ coil is mounted on the bird’s head. The owl is placed at the intersection of horizontal and vertical stationary induction coils that create a magnetic field. Any head movement causes a measurable change in the current in the search coil, which can be accurately recorded. The owl’s attention is directed first towards a zeroing speaker that emits a sound. Then a sound is emitted from the target speaker, which causes the owl to direct its head towards the new sound source. This head movement can then be measured. The whole apparatus can also be adjusted with respect to the vertical axis in order that the sound source can be manipulated in three-dimensional space.

One early experiment, conducted by Knudsen and Konishi (1979), has shown that owls respond to high-frequency sounds, typically between 100 Hz and 12,000 Hz, but their preferred range of

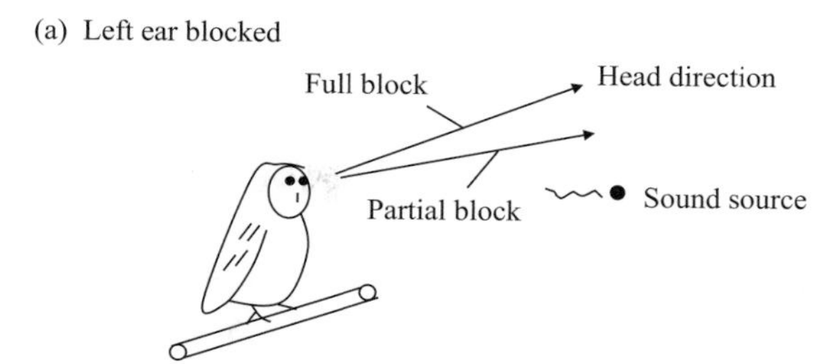

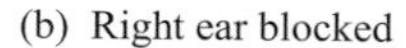

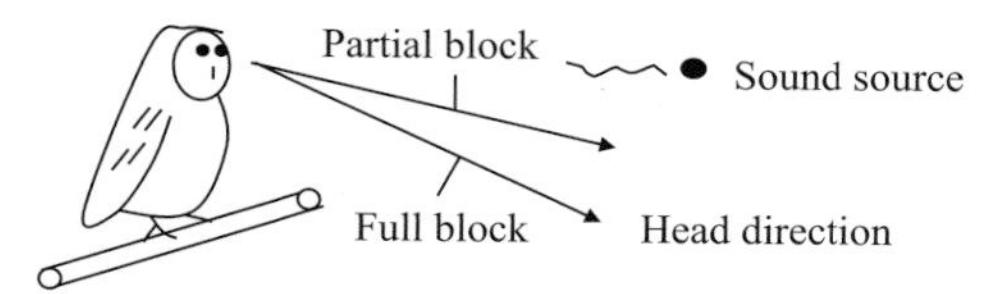

Figure 12.3 (a) A partial or full block of the left ear causes the owl to look upwards in response to a sound. (b) A partial or full block of the right ear causes the owl to look downwards.

sound is between 3,000 and 9,000 Hz. Similar to the auditory cortex of humans, there are neurons in the owl’s auditory cortex that respond to different frequencies of sound in a tonotopic fashion (Cohen and Knudsen, 1999); low-frequency sounds activate neurons at one end and high-frequency sounds activate cells at the other.

In another series of important experiments, Knudsen and Konishi (1979) found that blocking one ear of the owl resulted in the owl having errors in localising a sound source, suggesting that owls require two ears to accurately locate the sound source. Interestingly, partial and complete blockage of an ear led to large errors in localising sound in the elevation plane (vertical axis), although the ability to locate sound in the azimuth plane (horizontal) remained relatively intact. For example, Figure 12.3a shows that if the left ear is blocked and a sound is given, owls tend to move their head upwards, in error. The greater the blockage, the greater the head movement. If the right ear is blocked (Figure 12.3b), owls tend to move their heads downwards. Again, the greater the blockage, the greater the head movement. Blockage of the ears interferes with the sound intensity that reaches

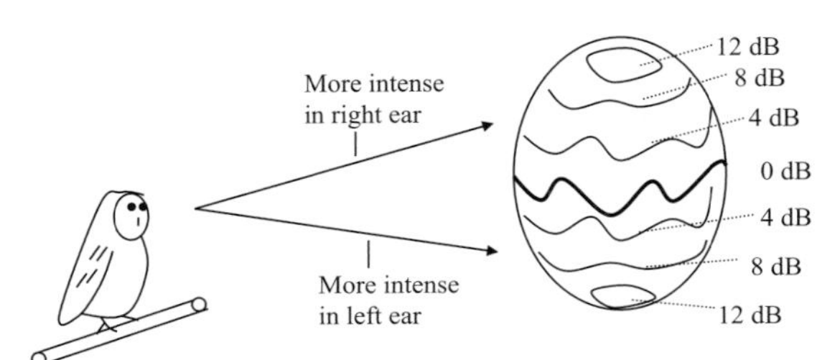

Figure 12.4 Interaural intensity difference between the two ears determines the vertical component of a sound source. If the sound to the right ear was more intense (louder) than to the left ear (e.g. by 12 dB), the owl looked upwards. If the sound to the left ear was more intense (louder) than to the right ear (e.g. by 12 dB), the owl looked downwards. For visualisation purposes, space above and below the eye level of the owl is represented by the oval shape (right). The greater the intensity difference between the two ears, the greater the shift in gaze upwards or downwards. Redrawn from M. Konishi (1993), Listening with two ears, *Scientific American*, 268 (4) 34–41.

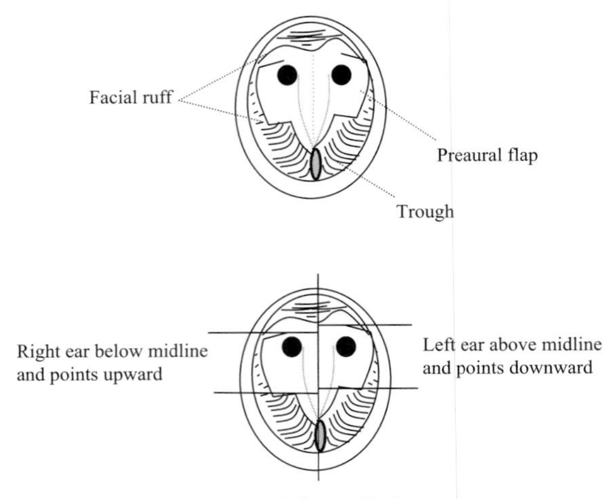

Figure 12.5 Asymmetry of the owl's face contributes to the perceived difference in sound intensities between left and right ears.

the ears, suggesting that the difference in the intensity between the ears is critical for locating sound on the elevation plane. This difference is known as *interaural intensity difference.*

The suggestion that the sound intensity presented in the two ears determines searching along the elevation axis was further confirmed in experiments by Konishi and his colleagues. These authors varied the intensity of the sound presented to the ears (through earphones) while keeping the timing of the presentation constant. Upon presentation, the owl tended to move its head upwards or downwards. The intensity difference and its effect then could be measured. For example, if the sound was presented at 4 decibels higher in the right ear compared to the left ear (while keeping the timing and frequency constant), the owl would turn its head upwards. If sound was presented at 4 decibels higher in the left ear compared to the right ear, the owl would turn its head downward. The greater the intensity difference between the two ears, the greater the shift upwards or downwards (see Figure 12.4).

Amazingly, this directionality along the elevation plane can be directly attributed to the asymmetry of the owls' ears and facial ruff (Figure 12.5). The left ear, for example, is located above the midpoint of the eyes and points in the downward direction. The right ear in contrast is located below the midpoint and points in the upward direction. The facial ruff on both sides of the face is also asymmetrical in arrangement. Below the ruff on both sides are two troughs, which run through the ruff from the forehead to the jaw. Their purpose is to collect high-frequency sounds and channel them to the ears. The left trough is tilted downwards while the right trough is tilted upwards.

Owls can also detect sounds on the azimuth (horizontal) plane. Similar to the experiments presented above, Konishi (1993) varied the timing of the sound presented to both ears (again through earphones) while keeping the intensity of the presentation constant; the owl tended to move its head either rightward or leftward. For example, if the sound is presented to the right ear slightly before that presented to the left ear (e.g. 42 μs, and keeping the intensity and frequency constant), the owl would turn its head rightward. If there is an increased delay of sound presentation (e.g. from 42 μs to 84 μs) the owl would look even further to the right. If, on the other hand, sound is presented to the left ear just before being presented to the right ear, the owl would turn its head leftwards. Similarly, if sound is presented to the right side of the space, it will arrive earlier to the right ear compared with its arrival at the left (Figure 12.6a), as it has a shorter distance to travel. The owl will look to the right. If sound is presented further to the right in space, it will take even longer to arrive to the owl's left ear compared to the right ear (arriving much earlier at the right ear compared to the left ear (e.g. by 84 μs, Figure 12.6b). The owl, in this case, will look even further to the right. Therefore, the greater the timing difference between the two ears, the greater the shift leftward or rightward (Figure 12.6c).

Finally, the frequency at which the sound is presented also seems to affect sound localisation. As suggested earlier, the facial ruff provides directional sensitivity. However, the amount of amplification and directional sensitivity provided by the ruff depends on the frequency of the sound. Frequencies of less than 3 KHz are not well catered for by the ruff. At 3 KHz, the left ear is more sensitive to sounds coming in from 20 to 40 degrees to the left of the owl than to sounds from other areas (Knudsen, 1981); this is represented by the bigger 'left' circle in Figure 12.7a. At higher frequencies, each ear is more sensitive to the sound direction, on both the azimuth plane and the elevation axis. As the frequency increases, the areas of greatest sensitivity move towards vertical alignment (i.e. on the elevation axis), and also the area of sensitivity becomes increasingly smaller and more specific (left and right circles in Figures 12.7b and c are getting smaller and move towards the elevation plane). At 8 KHz, for example, the regions of sensitive hearing are almost directly above and below the animal. The left ear is more sensitive to sounds coming from below the animal and the right ear more sensitive to sounds from above (see Figure 12.7c; and Figure 12.4; Knudsen, 1981).

Neural Circuits Involved in Sound Localisation

The owl's brain includes a number of important nuclei involved in sound localisation dispersed throughout the brain stem, cerebellum and cerebrum. However, sound is processed even before reaching the brain. Sound waves arrive into the ear and reach a receptive structure called the *basilar membrane*. Different parts of this membrane vibrate, depending on the frequency of the sound wave (Figure 12.8a). In addition, the louder the sound, the more this particular region of the basilar membrane will vibrate. Neurons of the auditory nerve are connected along the length of the basilar membrane, therefore a sound at a certain frequency will cause a vibration at a specific area of the basilar membrane, which will then cause neurons in that region to fire action potentials. As the sound grows louder, the neurons will increase their firing rate. The auditory nerve, in turn, sends its projections to two subpopulations of the cochlear nucleus: the magnocellular nucleus (*nucleus magnocellularis*) and the angular nucleus (*nucleus angularis*); see Figure 12.8b.

As well as being organised according to frequency, neurons of the auditory nerve exhibit phase-locking. That is, a neuron tuned to a certain frequency will tend to fire when the sound waveform is at a certain point in its cycle. For example, while a neuron may respond only to a sound at a frequency of 3 kHz, it also only fires an action potential when it reaches the peak of the sound wave (90 degrees, see Figure 12.9a top). Notice too that although this neuron fires when the wave reaches the peak of the cycle, it does not fire at each peak. However, as the noise increases in intensity, the firing rate of the neuron also increases (Figure 12.9a bottom). This neuron still remains phase locked (at 90 degrees). A neuron tuned to a different frequency will tend to fire at a different point along the wave cycle. In the example given in Figure 12.9b, the neuron responds to 5 kHz sound and this neuron fires an action potential at 110 degrees of the sound wave cycle (Figure 12.9b). However, at increasing frequencies many neurons start losing their phase-locking property, but these neurons continue to respond to intensity changes (see Figure 12.9c).

As mentioned, fibres of the auditory cortex send projections to two nuclei: the magnocellular nucleus and the angular nucleus. Collectively, these two nuclei are known as the *cochlear nucleus*. Cells of

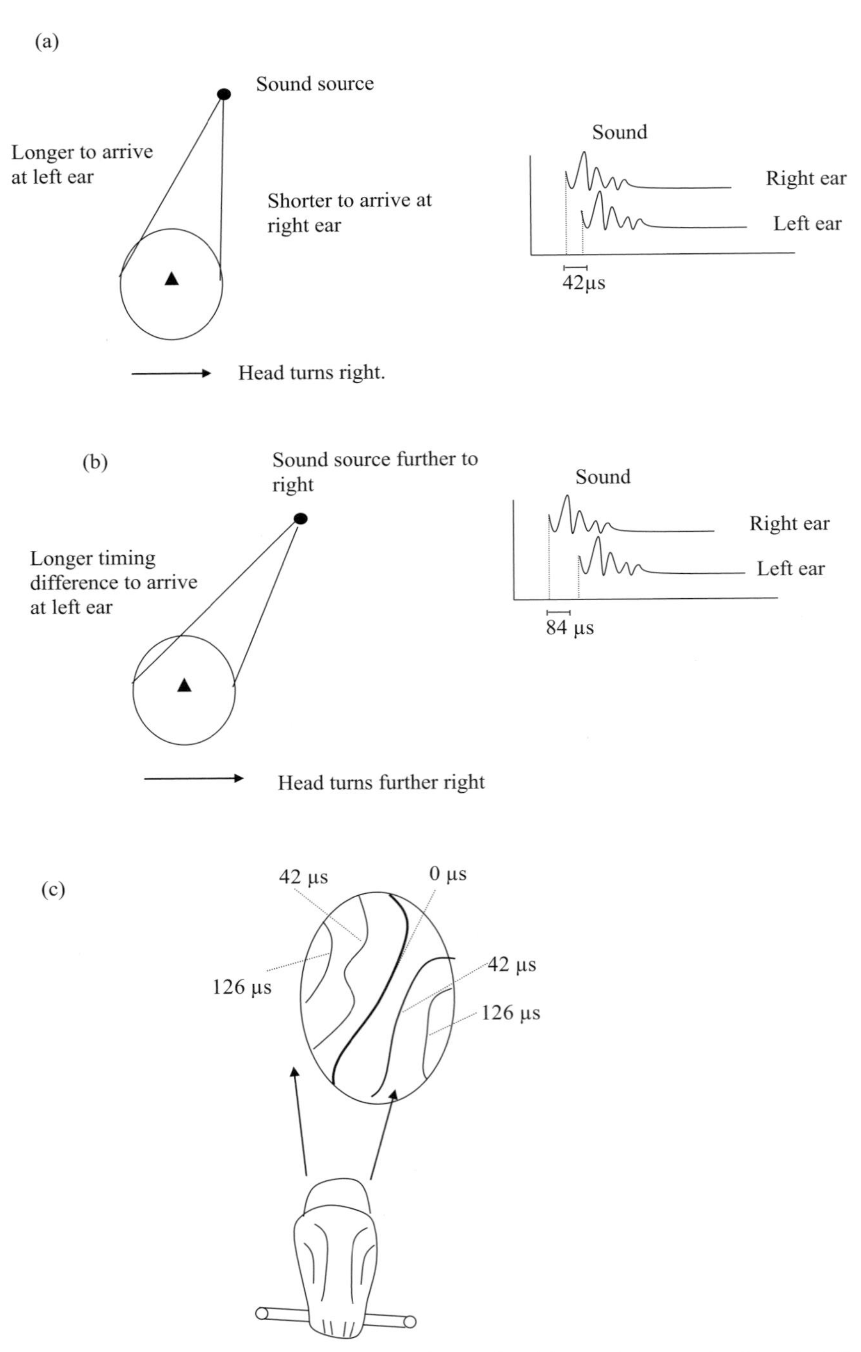

Figure 12.6 (a) If sound is presented in earlier to the right ear, it is perceived by the left ear later (e.g. 42 μs) than by the right ear, and the owl will turn its head towards the right (diagram of owl's head seen from back in all diagrams). (b) If sound is presented even further to the right, the delay to the left ear is even longer, and, the owl will turn its head even further to the right. (c) Interaural time difference between the two ears determines the horizontal component of a sound source. Redrawn from M. Konishi (1993), Listening with two ears, *Scientific American*, 268 (4) 34–41.

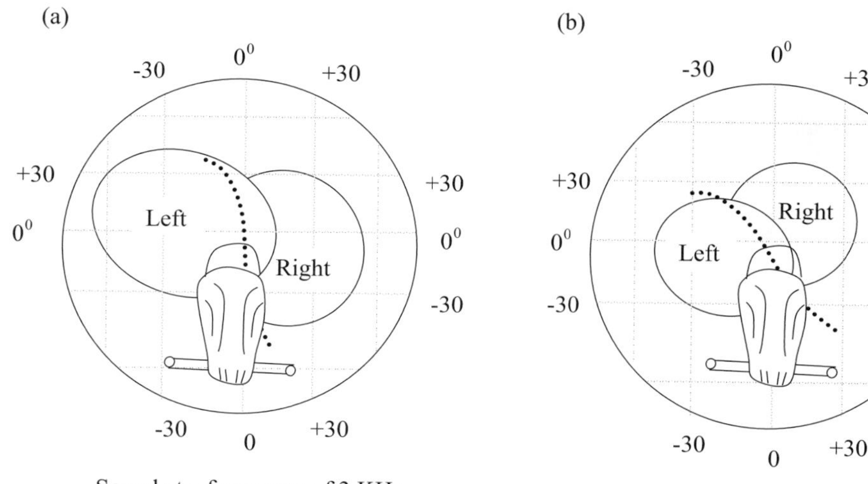

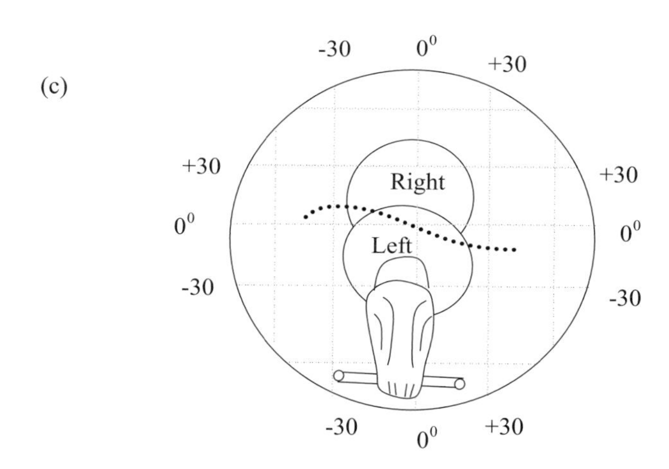

Figure 12.7 The frequency of sound also determines where an owl will look. At 3 KHz the left ear is more sensitive to sounds coming in from 20–40 degrees to the left of space (a), whereas at higher frequencies each ear is very sensitive to the sound direction on both the azimuth plane and the elevation axis (b) and (c). Note again, we are looking at the back of the owl; the owl's right is on our right, its left is on our left. Redrawn from E. I. Knudsen (1981), The hearing of the barn owl. *Scientific American*, 245, 82–91.

the magnocellular nucleus exhibit phase-locking and are also sensitive to intensity changes. However, their range of sensitivity to the intensity changes is not as wide as that of the angular nucleus. Neurons of the angular nucleus, in contrast, are not phased-locked but respond to a wide intensity range. With these two regions we start to see a separation of function. The phase-locking property is useful for determining the timing difference. If the signal reaches one ear slower than it reaches the second ear, the train of phased-locked impulses will be delayed relative to the impulses generated in the opposite ear. This suggests that the owl relies on the *magnocellular nucleus* for measuring *interaural time differences* and the *angular nucleus* for measuring *intensity differences*

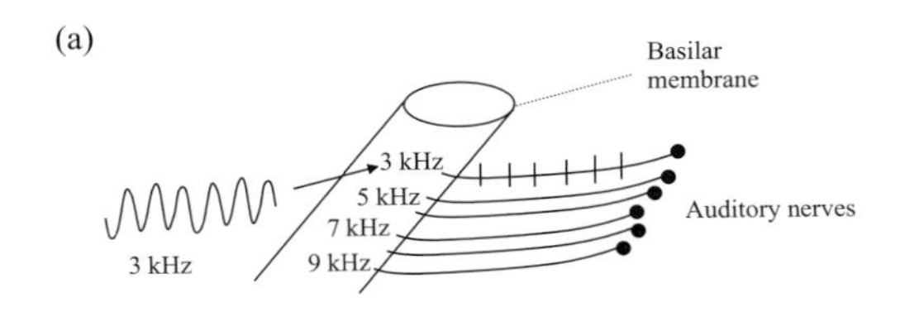

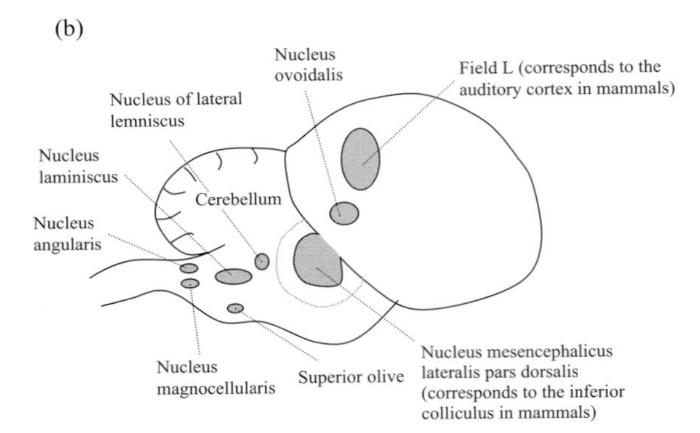

Figure 12.8 (a) Different regions along the basilar membrane of the ear are sensitive to different sound frequencies which, in turn, selectively innervate particular neurons of the auditory nerve. (b) A side view of the owl brain illustrating various structures thought to be involved in sound localisation.

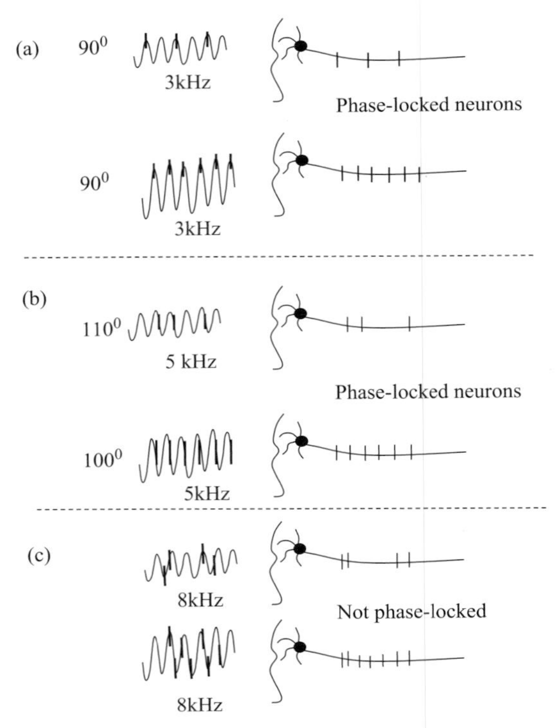

Figure 12.9 At lower frequencies, neurons within the auditory nerves are phase-locked to sound, that is, the neurons fire action potentials at a certain position along the sound wave cycle, e.g. (a) at the top of each cycle, 90°, or (b) on the descending slope 110°. Increasing the sound intensity increases the firing rate while maintaining the phase-lock (lower pictures of a and b). With increasing frequencies, neurons tend to lose their phase-lock but still increase their firing rate with increasing sound intensity (c).

(Konishi, 1993). At low to intermediate frequencies, the animal relies on both nuclei to determine sound localisation, but at higher frequencies the magnocellular neurons gradually become phase-unlocked; therefore, the animal tends to rely more on the angular nucleus and intensity differences to determine the location of a high-frequency sound.

Takahashi et al. (1984) examined the magnocellular nucleus and the angular nucleus in more detail. They found that by adding an anaesthetic to the magnocellular nucleus, thereby preventing the nerve cells from firing, the response to the interaural timing differences was affected. The response of these neurons to intensity differences remained unaffected. Anaesthetising the other cochlear nucleus, the angular nucleus, had the opposite effect. The authors showed that the interaural timing differences of these neurons were unaffected; however, the intensity differences were affected. These findings give further credence to the suggestion that time and intensity cues are processed independently in separate neural pathways before the information is combined further up the system.

How Are the Timing Differences Between the Two Ears Calculated?

While the magnocellular nucleus does tend to respond to timing differences, it is at the next structure along the pathways where such timing difference is more obvious and even stronger. This structure is called the laminar nucleus (*nucleus laminiscus*, Figure 12.8b). The laminar nucleus receives projections from both the left and right magnocellular nucleus; therefore, it is the first structure where signals from both the left and the right ear come together. A key feature of the laminar nucleus is that single neurons within the structure respond to a specific interaural delay. For example, one neuron in the laminar nucleus responds to a delay of 42 microseconds between the sounds arriving at each ear, while another neuron may respond to a 64-microsecond delay. Further, as interaural delay varies with sound location, each neuron of the laminar nucleus, therefore, encodes sound from a particular location of space on the *horizontal* plane (Figure 12.10).

The question of how a particular neuron responds selectively to a delay has vexed neuroscientists for many years but it is now thought that *neural delay lines* formed by axons for both contralateral and ipsilateral inputs make up this time difference. This idea was first developed by Jeffers (1948) but has found experimental support more recently by Smith et al. (1993). Let us consider this in more detail. Figure 12.11a shows that the sound wave is located in the right space and therefore enters the right ear a little earlier than it enters the left. For that reason, the neurons of the right auditory nerve and magnocellular nucleus fire action potentials a little bit faster than the equivalent areas in the left. Axons arrive from both the right and left ear in the laminar nucleus; but if these axons were of similar length, the action potentials arriving from the right would be quicker than those from the left and, therefore, the target neuron in the laminar nucleus will not fire – the target neuron will only fire if the action potentials arrive at the same time (Carr and Konishi, 1988). This is only possible if the action potentials from the right are slowed down somehow. This delay can be achieved if the axons are of different lengths; that is, if the right axon is slightly longer than the left. Now the action potentials can arrive simultaneously and trigger the target neuron (Figure 12.11b). Figure 12.11c shows a schematic diagram of how three single neurons in the laminar nucleus could potentially respond to three different time delays. By simply adding extra lengths onto one of the input neurons, this could create an increase in the delay period.

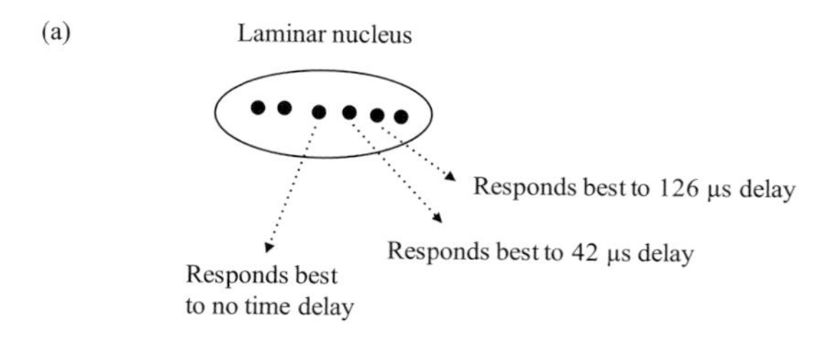

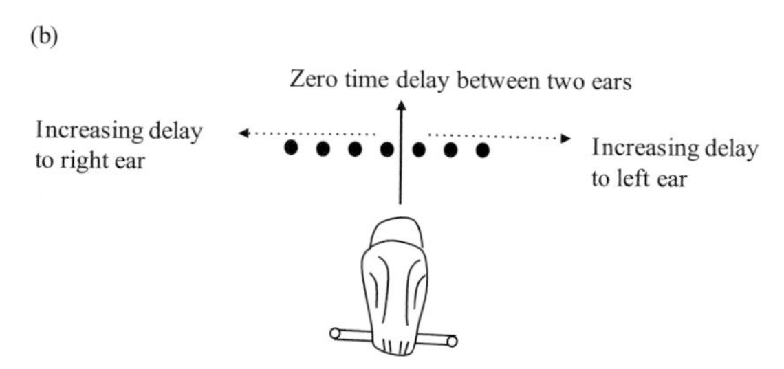

Figure 12.10 (a) Single neurons within the laminar nucleus respond to a specific time delay between the two ears, suggesting that (b) specific neurons within this structure encode sound emanating from a specific region of space along the horizontal plane. Note again, we are looking at the back of the owl; the owl's right is our right, its left is our left.

How Are Intensity Differences Between the Two Ears Calculated?

Intensity information, after being processed through the angular nucleus, undergoes higher processing through a variety of nuclei (Konishi, 1993). One particularly important nucleus is the posterior lateral lemniscal nucleus (nucleus of lateral lemniscus, Figure 12.8b). This is the lowest station to process intensity from both ears. A schematic diagram (Figure 12.12) demonstrates how the lemniscal nuclei of both sides of the brain are connected with the angular nuclei. It is slightly complicated, but the lemniscal nucleus on one side of the head (for example, the left lateral lemniscal nucleus) receives direct input only from the angular nucleus on the opposite side of the brain (right angular

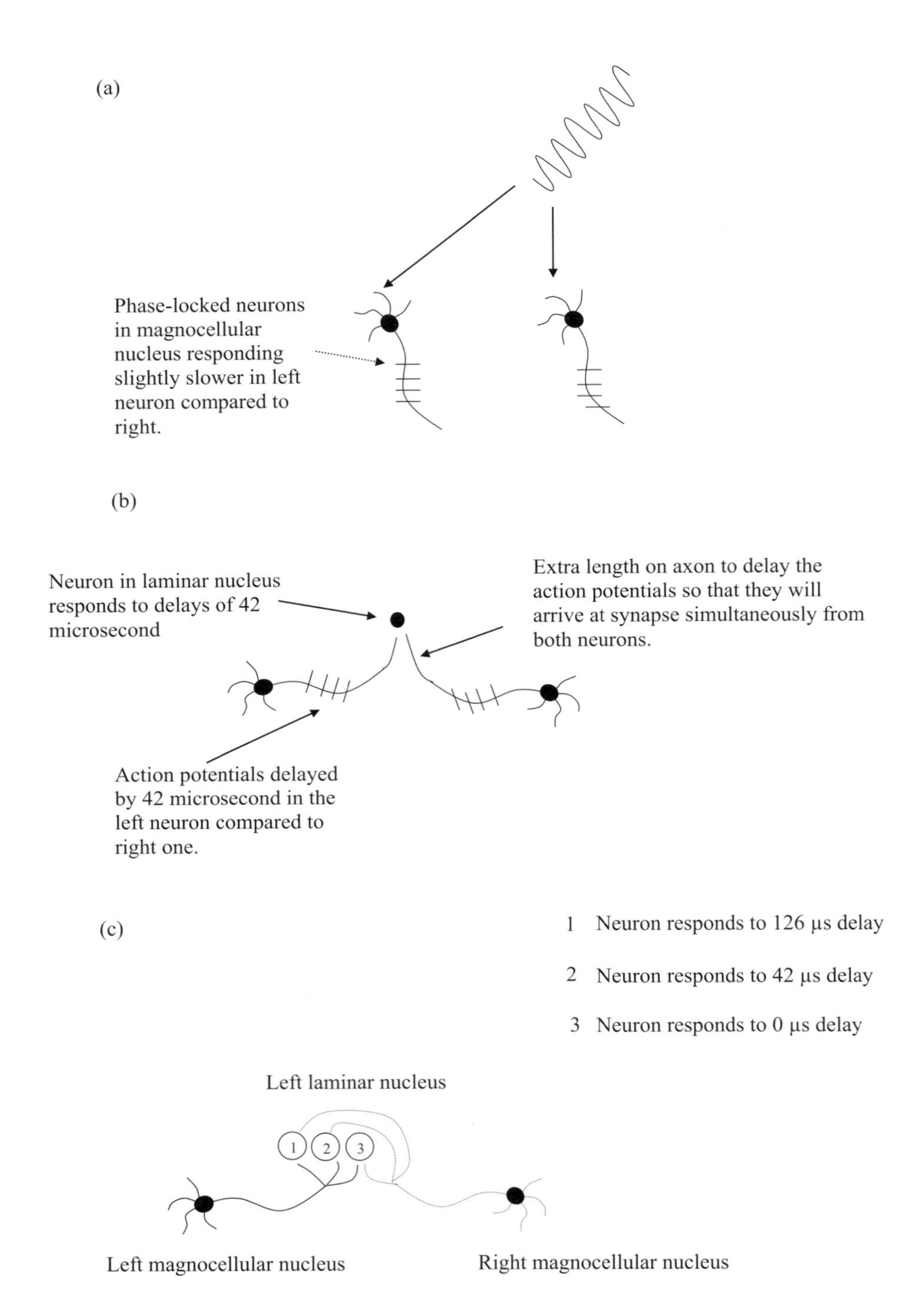

Figure 12.11 (a) If sound enters one ear quicker (e.g. the right) than the other, the neurons in the auditory nerve will also respond quicker. (b) As action potentials in the left are delayed compared to the right, the target neuron in the laminar nucleus will not respond; if, however, the axon of the right neuron is slightly longer than the left, this delays the action potentials so that both sets of action potentials arrive at the exact time, triggering the appropriate neuron in the laminar nucleus. (c) A simplified model of how three single neurons in the laminar nucleus could potentially respond to three different time delays, by simply having axons of different lengths.

nucleus). The left lemniscal nucleus also receives input from the angular nucleus on same side of the head (left angular nucleus), but this indirect input is via the right lemniscal nucleus. The opposite is true for the right lemniscal nucleus. The direct input from the contralateral angular nucleus is excitatory (represented by + in Figure 12.12a), and the input from the contralateral lemniscal nucleus is inhibitory, represented by – (minus sign) in Figure 12.12a. Figure 12.12b represents the excitatory input from the angular nuclei on both sides of the brain as filled circles and solid lines, and the inhibitory input via the contralateral lemniscal nucleus is represented by the open circles and dashed lines. *The difference between the strength of the inhibitory input and that of the excitatory input*

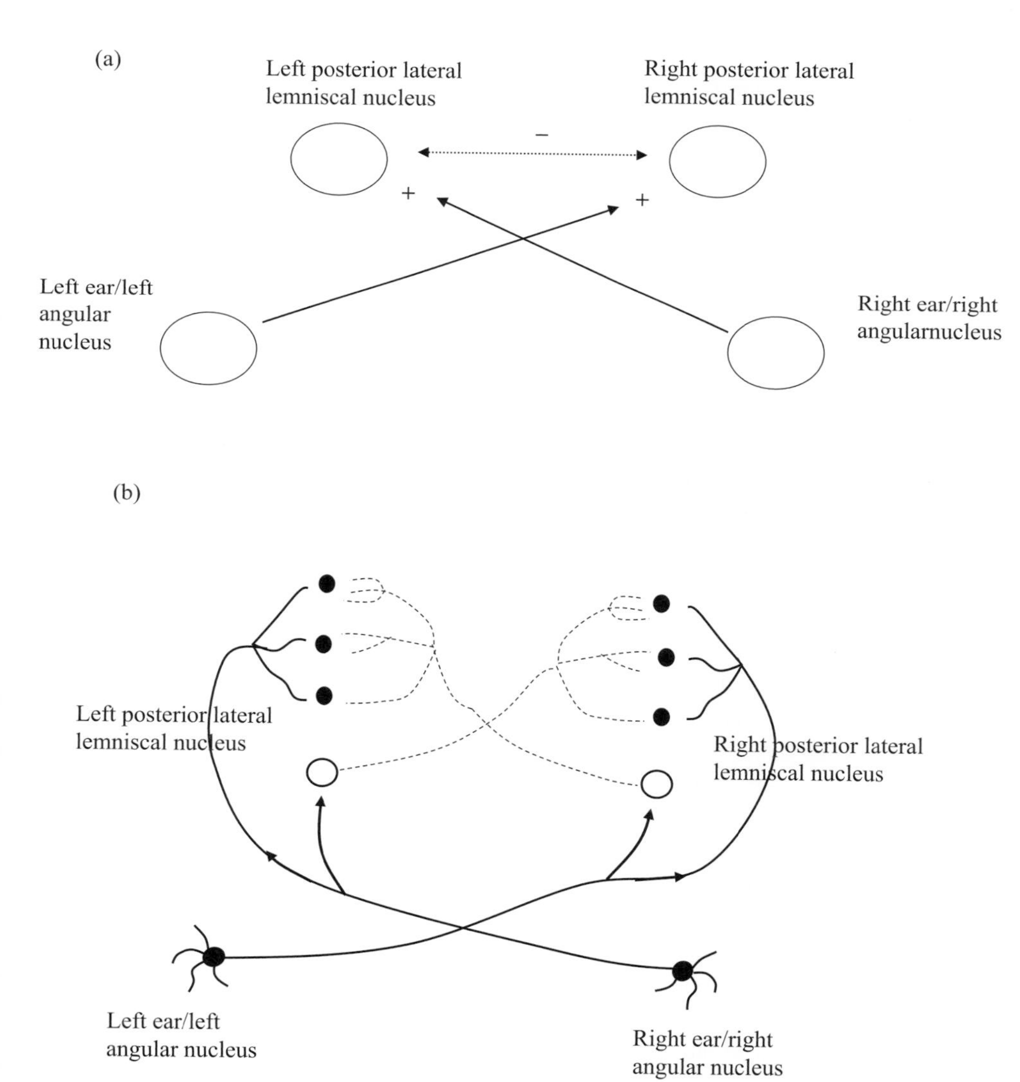

Figure 12.12 (a) A simplified diagram showing the connectivity between the angular nuclei and the posterior lateral lemniscal nuclei (+ and – represent excitatory and inhibitory inputs, respectively). (b) A more detailed circuit diagram showing how the angular nuclei provide excitatory input contalaterally (black neurons and solid lines) to lemniscal nuclei directly and to local inhibitory neurons (white circles and dashed lines). Inhibitory neurons, in turn, innervate the contralateral structure. Therefore, each lemniscal nucleus receives both excitatory and inhibitory inputs.

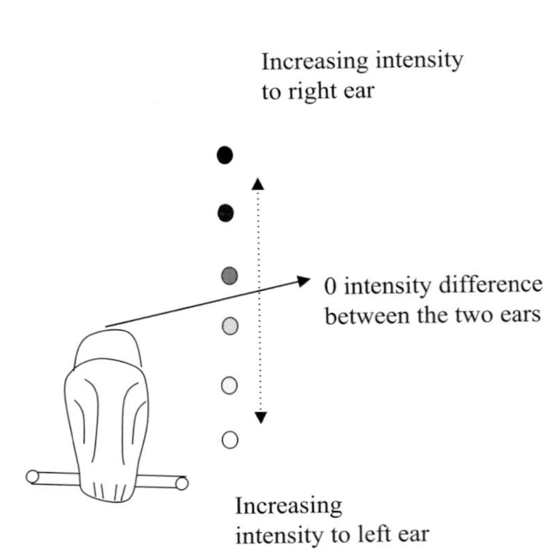

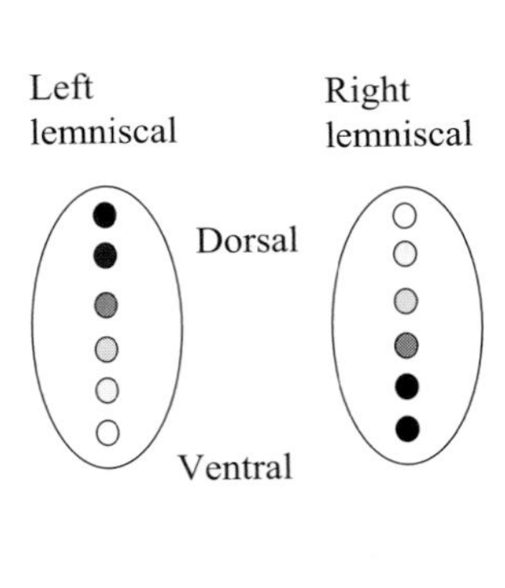

Figure 12.13 Location of sound source on the elevation plane determines the intensity of sound entering the left or right ear; this, in turn, determines the pattern of firing of neurons in the lemniscal nuclei. See text for details.

determines the rate at which the lemniscal neurons fire (Zupanc, 2004). Manley et al. (1988) found that while contralateral stimulation excites and ipsilateral stimulation inhibits the lemniscal nuclei, the strength of the inhibition declines from dorsal to ventral within the nucleus. The number of connections made by the inhibitory nucleus (represented by the white circles and dashed lines) on the contralateral lemniscal nucleus is shown in Figure 12.12b.

What happens when a sound is stronger in one ear compared to the other? A diagram displaying the dynamics of cell firing can be observed in Figure 12.13. In the left lemniscal nucleus, ventral neurons respond most strongly when the sound is louder in the left ear (white circles). On the other hand, dorsal neurons of the left lemniscal nucleus respond most strongly when the sound is louder in the right ear (black circles, Zupanc, 2004). The opposite is true for the right lemniscal nucleus. Therefore, if we imagine that the sound is getting stronger in the left ear, this causes more dorsally located neurons to respond in the right lemniscal nucleus and more ventrally located neurons to respond in the left lemniscal nucleus (increasingly paler circles). However, at the same time the sound is obviously getting weaker in the right ear, therefore encouraging this pattern even further. You will recall that the left and right ear correspond to different areas in space along the *elevation* axis. Therefore, if the sound location is coming from an increasingly upwards direction, there is an increasing intensity to the right ear. However, if the sound location is coming increasingly from a downward direction, there is an increasing intensity to the left ear.

Creating Auditory Maps of the Environment with Higher Brain Structures

So far we have seen specific cells that encode regions of space on the horizontal and the elevation plane independently. Are there cells that encode both bits of information? In the *mesencephalicus lateralis pars dorsalis* (MLD, which corresponds to the inferior colliculus in mammals) and Field L (corresponding to the auditory cortex in mammals), the large majority of neurons respond to the detection and identification of sounds (Knudsen, 1981, and see Figure 12.8). However, Knudsen and Konishi discovered cells in both these regions that specifically respond to sound stimuli coming

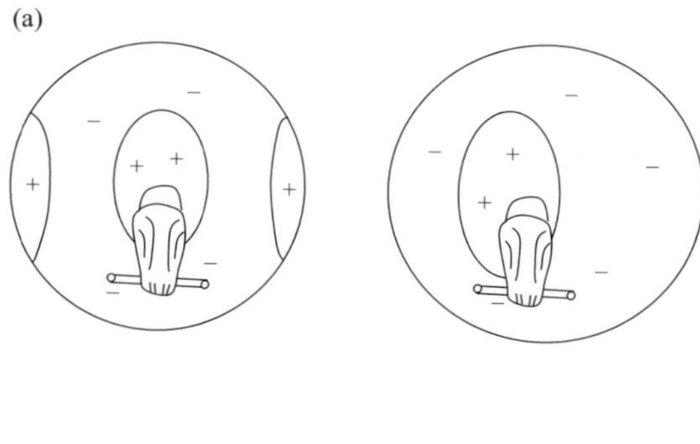

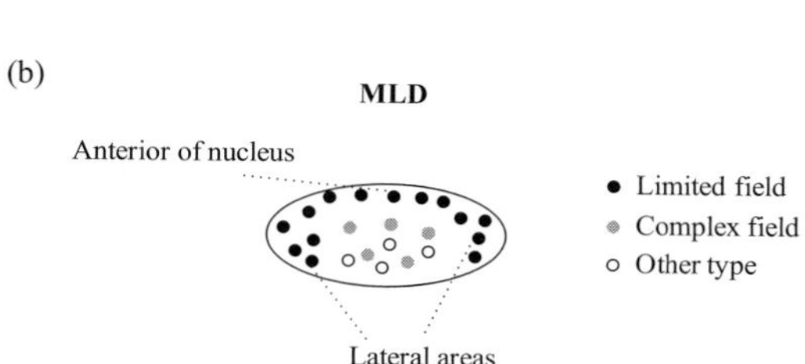

Figure 12.14 (a) Complex field neurons fire strongly when the sound is coming from the extremes of the spatial environment or immediately surrounding the owl (left). Limited field neurons are excited by sounds in a specific region of space (right). (b) Distribution of limited and complex field neurons in the MLD.

from particular areas of space (Konishi, 1993). For example, they found that if a sound was coming from a source directly in front of the owl, a single neuron would fire. However, if the speaker generating the sound was moved away from this position (e.g. to the left of the owl), this neuron would cease firing and a second neuron would then commence firing. This suggests that a single neuron can encode sound emanating from particular region of space – a type of 'auditory space cell'.

Two different types of 'auditory space cells' have been discovered. The first type of cell, called the *complex field neuron*, is found in the MLD, and the second type, called the *limited field neuron,* is found in both MLD and Field L. Complex field neurons fire strongly when the sound is coming from a source surrounding the owl itself, and also when the sound comes from a source located at the extremes of the spatial environment (represented by + in Figure 12.14a, left). In between these regions, the complex field neuron does not respond or indeed is inhibited completely (represented by –). The second cell type, the limited field neuron, is found in both the MLD and Field L. Unlike the complex field neuron, the limited field neurons are excited only by sounds coming from a single region of space (see Figure 12.14a, right).

The limited field neurons are found in both MLD and Field L, but their distribution in each of these brain areas is quite different. In Field L the limited field neurons only make up about 15% of the neuron types in this region and are scattered throughout the structure without an apparent topology. In contrast, in the MLD, limited field neurons are located along the lateral and anterior borders of the MLD (Knudsen and Konishi, 1978), and the medial area of the MLD is composed of complex field neurons (Figure 12,14b) and other types.

Detailed mapping of the limited field neurons in the MLD reveals an interesting pattern of distribution – a distribution of neurons that follows the spatial environment very precisely (Knudsen and Konishi, 1978). Figure 12.15 shows this in more detail. In the left MLD (Figure 12.15, top), neurons in the left lateral region of the structure respond to sound coming from the periphery of the right space (1, 2 and 3) in a systematic fashion. Neurons recorded from the right region of the structure encode sound coming from in front of the space, again in a systematic fashion (4, 5 and 6). In contrast, in the right MLD (Figure 12.15, bottom) neurons in the right lateral region of the structure respond to sound coming from the periphery of the left space (1, 2 and 3). Neurons recorded from the left region of the structure encode sound in more central regions (4, 5 and 6). As we can see, the area located just in front of the owl is represented by two sets of neurons, one set in the left MLD and the other set in the right MLD. The owl is particularly accurate when locating sound in this region of space. Any mouse that makes noise in front of an owl will not last too long.

The next question is: How does the owl construct this auditory map of the environment? A substantial

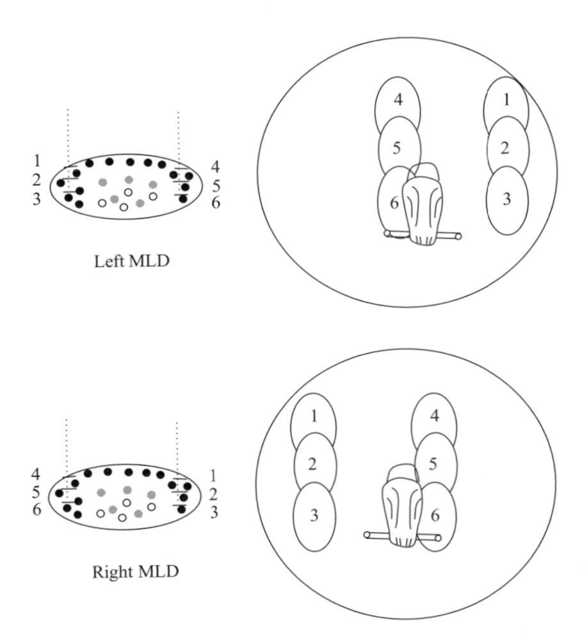

Figure 12.15 Location of cells within the left MLD (top left illustration) and right MLD (bottom left illustration) map very precisely with regions of space. When the electrode is lowered into the structure (represented by vertical line going through the MLD), neuron 1 only fires when sound emanates from region 1 of space, neuron 2 fires when sound comes from area 2 etc. Note again, we are looking at the back of the owl; the owl's right is our right, its left is our left. See text for further details.

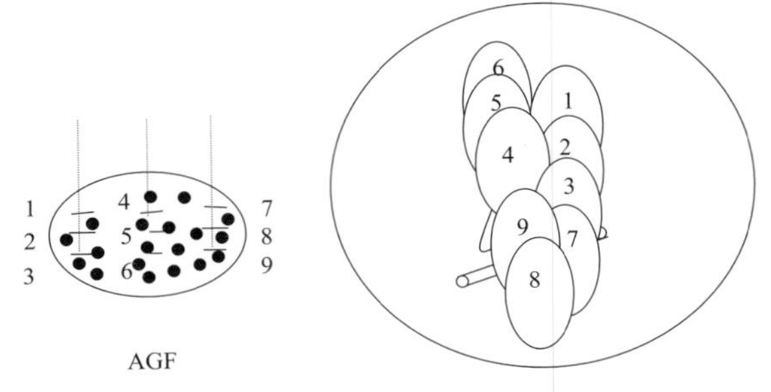

Figure 12.16 Unlike the MLD, the auditory space of the AGF is not organised in a topographical fashion but rather in local clusters.

amount of work in providing the answer to this question comes from Konishi and Moiseeff in the early 1980s. Moiseff and Konishi (1981) recorded a group of these 'space cells' in the mesencephalicus lateralis pars dorsalis (MLD). Earphones were placed onto the anaesthetised owl and sounds were delivered separately to each ear. As researchers recorded the neural activity from the MLD cells (using fine electrode wires), the timing and the intensity of the sounds presented to the ears were varied. It was found that the limited field cells were very sensitive to timing differences of the sounds. Neurons respond to a range of timing differences between 40 microseconds and 90 microseconds. Outside these values, the cells remain quiet. However, each cell fired at a preferred timing, e.g. one cell could fire at 84 microseconds delay, while another at 64 microseconds (similar to what was observed in the laminar nucleus). It was also found that for each limited field neuron there was one difference in intensity of sound that produced maximal firing. Movement away from this preferred intensity led to a decrease in firing of that particular neuron. For example, one neuron may respond only to sounds with differences of 4 dB coming into each ear while another may only respond to a difference of 16 dB. As each neuron approaches its preferred intensity difference, it responds stronger and then weakens its response as it moves away from the preferred intensity difference.

The evidence obtained by recording the space cells in the MLD shows that the auditory map of the environment is created by the same cues that have been observed in downstream structures: interaural time differences and intensity differences. The MLD seems to be the first structure where both time and intensity differences are combined; information from the elevation and horizontal planes converge in this structure.

Finally, an output structure of Field L called the *archistriatal gaze field* (AGF; similar to the mammalian frontal eye fields) is thought to be involved in orienting an animal's gaze in response to both auditory and visual information. Similar to the MLD, neurons in the AGF respond to sounds coming from specific regions of space. However, unlike the MLD, the auditory space of the AGF is not organised in a topographical fashion but rather in local clusters. Neighbouring clusters of neurons may be tuned for different locations (see Figure 12.16). Speculation on why auditory space is mapped this way, in contrast to the midbrain regions' mapping, is ongoing (Cohen and Knudsen, 1999).

Summary

The owl has a remarkable ability to detect a source of sound; it is this feat that allows the animal to precisely capture its prey. In this chapter we describe how sounds on the horizontal plane are detected by a specialised auditory pathway that includes the magnocellular nucleus and laminar nucleus, which is able to distinguish the difference in timing between a sound entering the owl's left ear and one entering its right ear. Sounds on the elevation plane are processed in the brain, via the angular nucleus and posterior lateral lemniscal nucleus, based on the difference in the intensity of sounds reaching each ear. Due to the owls' distinct anatomical features, including the orientation of its facial ruff and ears, sounds from above are detected as stronger by the owl's right ear, whereas sounds from below are detected as stronger by the left ear. Horizontal and elevation information are then combined in structures such as the MLD and Field L, which form a remarkable map of auditory space.

Questions and Topics Under Current Investigation

- How does the auditory system evaluate the significance of sounds and how does it remember auditory targets? These tasks may involve the frontal eye field and higher-order memory, and attentional and emotional areas of the brain.
- Recent evidence suggests that the forebrain (including the primary auditory field and archistriatal gaze field) represents spatial information in clusters rather than as maps (as in the inferior colliculus of the midbrain); see chapter text for details. Why the difference between the various regions?
- Does the mechanism described for sound localisation in this chapter translate to other bird species and to mammals, including humans?
- Does the auditory map change with experience – how plastic is it? How does the auditory map develop from that of a juvenile bird to that of an adult?
- Once the owl has heard its prey and has it localised, it then looks towards it. How does the auditory map interact with the visual system? Research has shown that neurons in the optic tectum respond to both visual and auditory stimuli, and the visual system may help to teach the auditory system to translate such cues (see Knudsen, 2002).
- What are the biochemical and molecular underpinnings of sound localisation?

References

Carr, C.E., and Konishi, M. (1988). Axonal delay lines for time measurement in the owl's brainstem. *Proceedings of the National Academy of Sciences USA*, 85, 8311–8315.

Cohen, Y.E., and Knudsen, E.I. (1999). Maps versus clusters: different representations of auditory space in the midbrain and forebrain. *Trends in the Neurosciences*, 22(3), 128–135.

Jeffers, L.A. (1948). A place theory of sound localization. *Journal of Comparative Physiological Psychology*, 41, 35–39.

Knudsen, E.I. (1981). The hearing of the barn owl. *Scientific American*, 245, 82–91.

Knudsen, E.I. (2002). Instructed learning in the auditory localization pathway of the barn owl. *Nature*, 417, 322–328.

Knudsen, E.I., and Konishi, M. (1978). A neural map of auditory space in the owl. *Science*, 200, 795–797.

Knudsen, E.I., and Konishi, M. (1979). Mechanisms of sound localization in the barn owl (*Tyto alba*). *Journal of Comparative Physiology*, 133, 13–21.

Konishi, M. (1993). Listening with two ears. *Scientific American*, 268 (4) 34–41.

Manley, G.A., Koppl, C., and Konishi, M. (1988). A neural map of interaural intensity differences in the brain stem of the barn owl. *Journal of Neuroscience* 8(8), 2665–2676.

Moiseff, A., and Konishi, M. (1981). Neuronal and behavioural sensitivity to binaural time difference in the owl. *Journal of Neuroscience*, 1(1), 40–48.

Payne, R. S. (1971). Acoustic location of prey by barn owls (*Tyto alba*). *Journal of Experimental Biology*, 54, 535–573.

Smith, P.H., Joris, P.X., and Yin, T.C. (1993). Projections of physiologically characterized spherical bushy cell axons from the cochlear nucleus of the cat: evidence for delay lines to the medial superior olive. *Journal of Comparative Neurology*, 331(2), 245–260.

Takahashi, T., Moiseff, A., and Konishi, M. (1984). Time and intensity cues are processed independently in the auditory system of the owl. *Journal of Neuroscience*, 4(7), 1781–1786.

Zupanc, G.K.H. (2004). *Behavioral neurobiology: An integrative approach*. Oxford University Press, Oxford.

13 Bat Echolocation

Introduction to Bats

Bats are mammals that fly and are found throughout the world. They belong to the order Chiroptera, which is derived from two Greek words; *cheir,* which means 'hand' and *pteron,* which translates as 'wing'. There are over 1,200 species of bat; the majority of them (over 70%) eat insects. The famous vampire bat is the only species that could be termed a parasite, as it feeds on blood. Bats play a crucial role in reducing the insect population and also are pollinators, dispersing fruit seeds. Bats can be classified into two suborders; Megachiroptera and Microchiroptera. The major difference between the two orders is that the Microchiroptera lack fur and a claw on their second toe, but most importantly, at least for our discussion, the Microchiroptera are echolocating whereas the Megachiroptera (with the exception of *Rousettus* species) are not.

Throughout history, humans have had an extreme relationship with bats; they have either treated bats badly or have held them in reverence. For example, in the Roman and Christian tradition, bats have always had a sinister significance, and the bat often has been associated with the devil. Indeed, *Dracula,* written by the Irish writer Bram Stoker, was based on many European mythological folklore stories of vampires and bats; Dracula transforms into a bat many times throughout the novel. Other cultures, however, hold bats in high regard. For example, the ancient Mayans had a bat god, and in China bats are a symbol of long life and happiness.

History of Echolocation

The earliest account of how bats navigate comes from experiments performed by Lazzaro Spallanzani (1729–1799). Spallanzani was an Italian Catholic priest who made many important discoveries on reproduction, digestion and germ theory. He noted that bats were able to navigate in total darkness. Critically, he observed that blinded bats could fly without much difficulty, avoid obstacles and catch prey. Charles Jurine (1751–1819), a Swiss zoologist, read about Spallanzani's work and conducted further experiments on bats. In one experiment, he blocked bats' ears with candle wax and found that these animals would stumble into objects and were unable to catch prey, and would often land in inappropriate locations. Both Spallanzani and Jurine suggested that bats navigate using their sense of hearing.

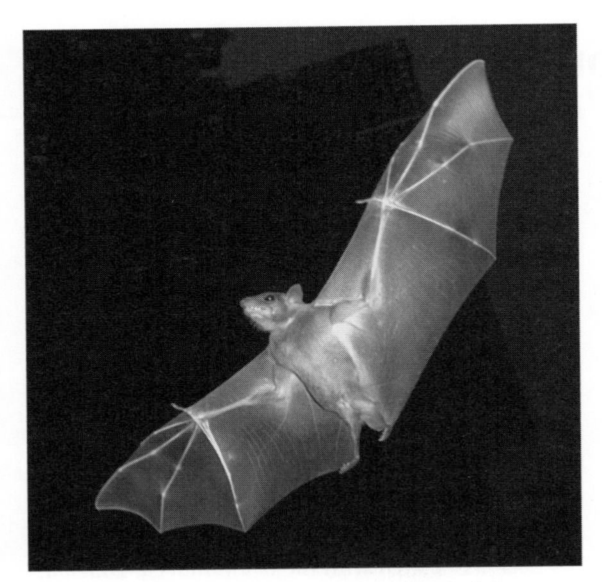

Figure 13.1 Fruit bat (*Rousettus aegyptiacus*), flying in Israel. Original photo by Oren Peles. Reproduced under Creative Commons Attribution License.

However, these experiments were criticised by the French naturalist and palaeontologist Georges Cuvier (1769–1832), who wrote that it was the sense of touch that allowed bats to avoid obstacles and navigate in the dark. As Cuvier was such an influential figure, little research was conducted in the field until the 1900s. Following the sinking of the *Titanic* in 1912, Hiram Maxim, the American engineer and inventor of the machine gun, considered that bats could detect obstacles by sensing the reflections of low-frequency sounds caused by the beating of their wings. He further imagined that it might be possible to replicate this in a machine that could produce sounds and detect far-off obstacles by listening to the echoes produced (sonar).

In the 1930s, as a student at Harvard, Donald Griffin worked with Robert Galambos (a neuroscientist) and George Pierce (a physicist) to develop a device that allowed for the detection of sounds beyond those heard by humans (see Griffin & Galambos, 1941). With this device, the researchers were able to show that bats emitted high-frequency calls that were beyond human hearing. Later, Pierce and Griffin showed that bats increased their rate of high-frequency calls as they approached an obstacle or object. They further demonstrated that bats were unable to navigate effectively if both their mouth and ears were blocked, leading Griffin in 1944 to coin the phrase 'echolocation'. Griffin later showed that sonar was more than simply a mechanism for avoiding obstacles, but could also be used to hunt prey. He demonstrated that the repetition rate of high-frequency calls increased more than ten times in wild bats hunting flying insects compared to the rate he had observed with laboratory bats simply avoiding obstacles. He showed that the repetition rate increased further as the bat captured the insect. He termed this 'a feeding buzz'. Figure 13.2 shows an example of the *Myotis daubentonii* (Daubenton's bat) as it approaches and captures prey. Figure 13.2a demonstrates the flight path before, during and after capturing prey (which occurs at X, point 6). Figure 13.2b demonstrates that upon initial approach, the sounds emitted are quite sparse but as the bat approaches the prey, the emitted sounds become more frequent (Buzz I and Buzz II), culminating in a rapid sequence of pulses, all within a narrow frequency range.

Echolocation Signal Types

There are approximately 800 species of Microchiropteran bats worldwide, which are all thought to use echolocation. All Microchiropteran bats are known to produce calls using vocal cords in their larynges (apart from the fruit bats *Rousettus*, which click their tongues). Although the biosonar pulses can vary from species to species, they can be generally classified into three main types: constant frequency (CF), frequency modulated (FM) and combined CF-FM.

The FM signal is a quick pulse (lasting less than 5 ms) that sweeps downwards through an octave, often sounding like a chirp (Figure 13.3, *Eptesicus*). With the CF, there is a single frequency of tone. But often bats emit a combination of both, with a constant tone that is preceded and/or followed by a downward chirp (Figure 13.3, *Rhinolophus*). This combined CF-FM signal may

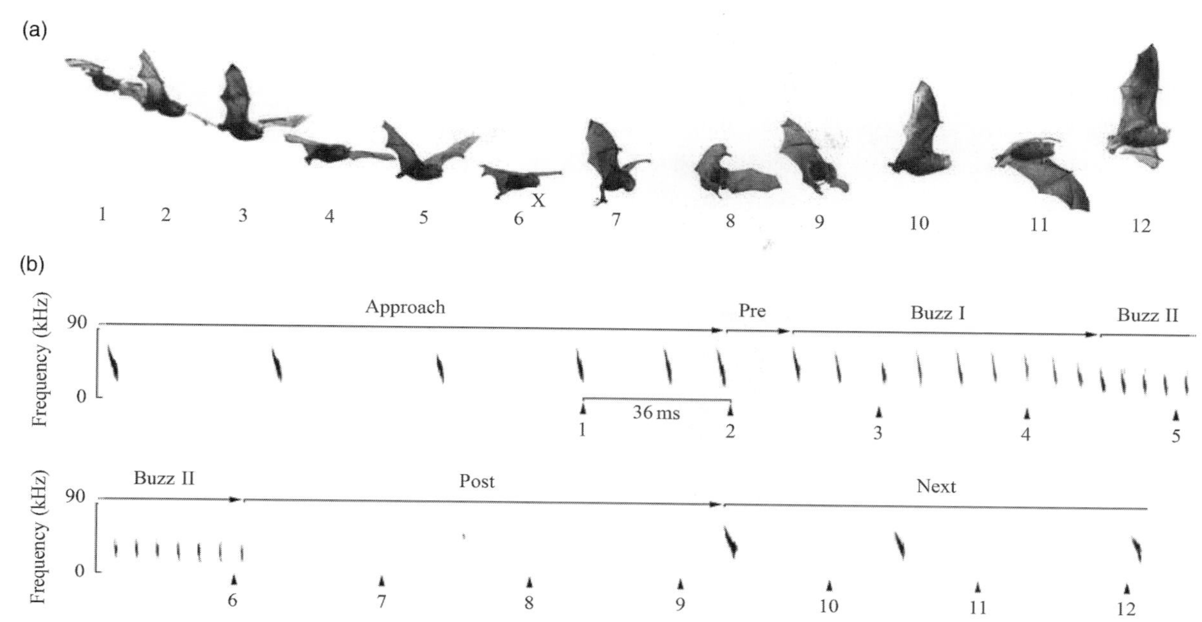

Figure 13.2 (a) A series of picture stills showing *Myotis daubentonii* capturing prey, which occurs at point 6, X. (b) A sonogram featuring the emitted sounds during various phases of capture including: Approach, Pre-buzz, Buzz I, Buzz II, and Post (post-capture). The numbers on the sonogram correspond to the numbers indicated for each still. The figure is republished with permission from A.R. Britton and G. Jones (1999), Echolocation behaviour and prey-capture success in foraging bats: laboratory and field experiments on *Myotis daubentonii*, *J Exp Biol*. 202:1793–801.

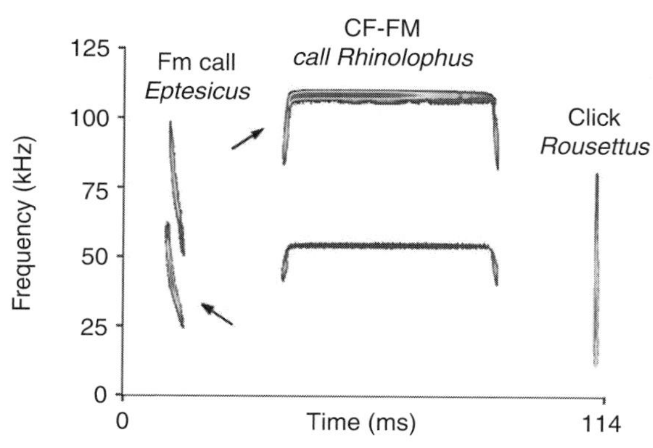

Figure 13.3 Examples of 3 different bats and the calls they emit. Figure taken from N. Ulanovsky and C.F. Moss, What the bat's voice tells the bat's brain. *Proc Natl Acad Sci USA*. 2008, 105(25):8491–8. Copyright (2008) National Academy of Sciences, U.S.A.

be subdivided into a long CF-FM and a short CF-FM. The long CF-FM signal may contain a CF of 10 to 100 ms duration followed by an FM sweep (Figure 13.3, *Rhinolophus*). The short CF-FM may contain a CF of 8 to 10 ms followed by the FM chirp. Most bat species emit only one type of pulse, but this pulse can operate at different harmonics (frequency levels). Notice the call from the *Rhinolophus* operates at two harmonics, one at approximately 100 Hz and another at just above 50 Hz. Table 13.1 gives examples of different bat species and the type of call they emit.

As can be observed in Figure 13.2, irrespective of whether the bat is an FM or a combined CF-FM species, the pulse interval shortens and the frequency range shifts towards the lower frequencies as the bat gets closer and closer to the prey, ending up in what is termed a *feeding buzz*. In some species, the tone type can change, depending upon the situation in which the bat finds itself. For example, the fish-catching bat *Noctilio leporinus* uses the CF and CF-FM pulse type when cruising but the FM when hunting.

Long CF pulses are particularly good for measuring Doppler shifts, which allow bats to detect an object while it is on the move. Although this signal is not that good for discerning details or locating targets precisely, it is useful when

Table 13.1 **Examples of Different Bat Species and Call Types**

Bat species	Call type	Frequency range
Rhinolophus rouxii - Rufuous Horseshoe Bat	FM-CF-FM	
Rhinolophus ferrumequinum – Greater Horseshoe	CF-FM	83 kHz
Hipposideros speoris - Schneider's Leaf-nosed Bat	CF-FM	135 kHz
Pipistrellus mimus - Least Pipistrelle Bat	FM	60–30 kHz
Megaderma - False Vampire Bat	FM with 4–6 harmonics	20–120 kHz
Rousettus leschenaultii - Leschenault's Rousette	Click-like FM pulse but doesn't have a specific structure.	
Myotis lucifugus – Little Brown Bat	FM 0.5–3 ms	30 kHz
Pteronotus parnellii – Parnell's Mustached Bat	CF-FM long CF 5–30 ms and 2–4 ms FM sweep	30.5 kHz, 61.2 kHz

hunting in dense vegetation to get a general idea of objects' locations. In contrast, many FM bats forage in the open (e.g. the Brown Bat), using signals of shorter duration that are particularly useful for estimating three-dimensional details. FM bats can differentiate between a pulse and returning echo of less than 60 microseconds apart, thus their distance estimation is excellent.

Echolocation Behaviour

The key behaviour involved in echolocation is that a bat emits a sound and then picks up on the echo of this sound. Essentially, bats can compare their emitted pulse to the returning echo to determine not only distance and direction but also the identity of the prey. With the FM signal, the emitted pulse is of a very short duration. This allows for a time delay between the call and the returning echo. This delay is very important, because bats don't want to be deafened by their own call, and the delay allows time for their ear muscles to relax and pick up the returning echo. This 'low-duty cycle' is especially important for distance estimation. For example, if you know the speed of sound through air (343.2 m/sec, which is a constant) and you know the time for the echo to return, it is then easy to calculate the distance between the bat and an object. The bat seems to be able to perform these calculations. With continuous calls (CF-FM), the bat is able to separate the call from the echo through frequency differences. With this 'high-duty cycle', bats do not have to worry about the pulse and echo overlapping with each other, as the pulse is emitted at one frequency but the echo returns at another frequency.

Bats' ears and auditory cortices are particularly well tuned to different frequencies. When flying, bats also take advantage of the Doppler shift. This allows them to detect motion, velocity and the location of prey. For example, when we hear the siren of an emergency vehicle passing, the sound seems to change in pitch as its distance from us changes. This phenomenon is known as the Doppler effect and was named after the Austrian physicist Christian Doppler in 1842. In essence, as a police car approaches the listener, the sound waves coming from the car's siren are compressed, but as the car moves away the sound waves stretch out. In Figure 13.4a, we can see the change in frequency as the police car approaches and moves away from the person. In this example the frequency increases from 3.5 Hz to 4 Hz as the vehicle approaches and then it decreases from 3 Hz to 2.5 Hz as the vehicle moves away. Similarly, as a bat approaches a stationary object it emits a pulse, but the echo that the bat receives in return is Doppler-shifted.

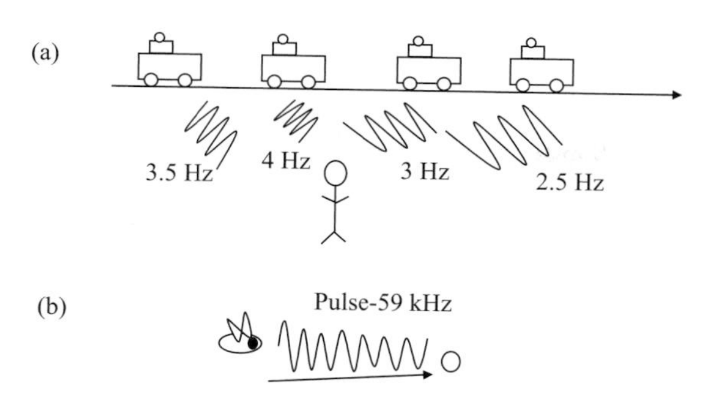

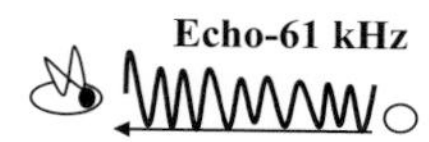

Figure 13.4 (a) The Doppler effect, how frequency changes with movement. (b) How the bat can use the Doppler effect to tell how fast the prey is going. If the echo is at a higher frequency than the pulse the bat emitted, then the bat is gaining on the prey.

Thus, if the frequency of the return echo is slightly higher than the pulse it had originally emitted (Figure 13.4b), the bat is gaining on the object/insect. If, however, the frequency of the returning echo is lower than the pulse the bat emitted, it knows that the insect is moving away from it.

The Doppler shift, however, can cause problems for the bat because often the returning echo may lie just outside its optimal acoustical frequency range. For example, one of the frequencies emitted by the Mustached Bat is around 59 kHz, if the returning echo is at 61 kHz. This frequency may lie slightly outside the bat's optimal detection range. In order to deal with this, the bat can reduce the frequency of its next emitted pulse by approximately 2 kHz; thus the echoes are now within the bat's optimal hearing range. This is termed *Doppler shift compensation*, which was first observed in 1968 by Hans-Ulrich Schnitzler. The ability to change frequencies is very useful when attempting to capture an insect. Typically, the insect is either stationary with its wings fluttering (often against a stationary background, wall, vegetation etc.) or the insect is flying. As the insect's wings are beating, the fluttering adds an extra ripple on the returning echo. The bat can detect the small Doppler shift given off by the fluttering, as well as minute changes in amplitude that are superimposed on the returning echo. From these tiny changes, the bat can determine the size of the insect.

Neural Basis of Echolocation

It is within the bat's auditory system that the specialisation of echolocation occurs. Similar to the way all mammals' ears function, sound waves are transmitted down the auditory canal of the bat's ear and reach the eardrum (Figure 13.5). The eardrum then vibrates at the same frequency as the sound wave. Sound then passes through three small bones (malleus, incus and stapes) of the middle ear, which help the transmission and amplification of sound vibrations. The main structure of the inner ear is the spiral-shaped, fluid-filled cochlea. Contained within the cochlea is the basilar membrane, which vibrates when sound waves pass through. The basilar membrane is thicker and wider at the cochlear apex compared to the base, where it is a lot thinner. If we think of guitar strings, there are thick ones (wide) and narrow ones. The thick strings vibrate at low pitches and the thinner strings vibrate at high pitches. Similarly, because the thickness levels of the basilar membrane change along the length of the cochlea, the basilar membrane at the apex vibrates to low frequencies and the membrane at the base vibrates to higher frequencies (see Figure 13.5). Hair cells along the basilar membrane therefore respond to different frequencies and transmit this information to the brain via auditory nerves.

If we think about a sound wave, there are a number of features that need to be encoded by the brain and its neurons: the amplitude, the timing and the frequency of the signal. The amplitude of the sound wave is encoded by the firing rate of the auditory neurons; the higher the amplitude, the greater the neural discharge rate. The timing of the signal maps onto the interval between discharges of the neuron, and the frequency depends on the location of firing along the basilar membrane, as discussed earlier.

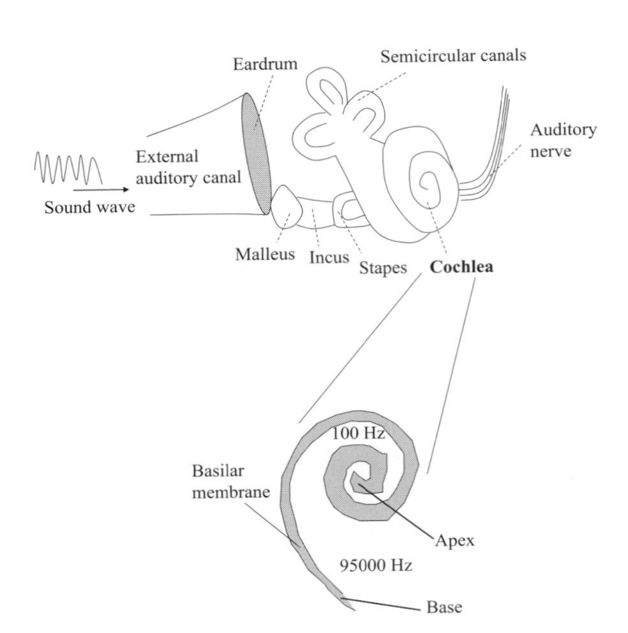

Figure 13.5 An illustration of the workings of the inner mammalian ear, including an enlarged view of the basilar membrane, which is located inside the cochlea.

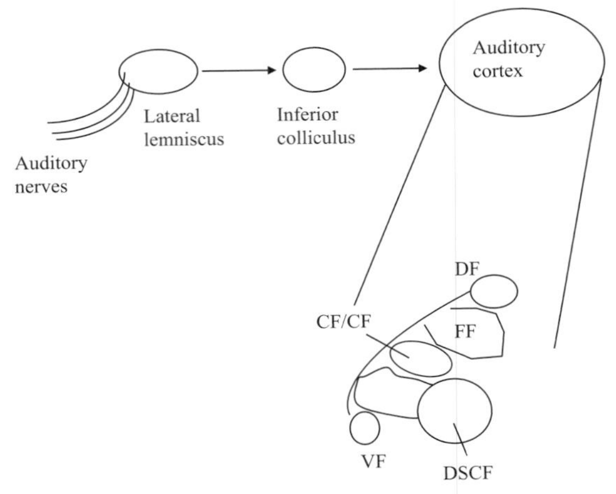

Figure 13.6 A schematic representation of the anatomical pathway from the cochlea to the auditory cortex, with an enlarged view of various subregions of the auditory cortex. Key: CF-CF, constant frequency-constant frequency area; DF, dorsal fringe; DSCF, Doppler-shifted CF region; FF, frequency modulation–frequency modulation area; VF, ventral fringe.

The range of frequency that is encoded along the basilar membrane is not uniformly represented. There is often an overrepresentation of certain frequencies, depending on the bat in question. This feature is similar to what is observed in the somatosensory cortex of humans (see Chapter 2), where, for example, the tongue and fingers are over-represented in the brain due to their constant use. Similarly, the spiral ganglion cells (of the cochlea) of the Mustached Bat are extremely sensitive to frequencies of between 61 and 61.5 kHz (the echo frequency that has been Doppler-shifted). Likewise, the echo of the *Rhinolophus* is overrepresented at 83 kHz. This overrepresentation of frequencies in the basilar membrane and activity of the neural output is mirrored in higher-order brain structures also, such as the auditory cortex.

How Is Distance Information Calculated?

From the cochlea, auditory information is encoded in a neural system that starts at the cochlear nucleus and proceeds through the lateral leminscus and inferior colliculus before ending up in the auditory cortex. We have encountered some of these processes and structures in the previous chapter. The auditory cortex consists of a number of subregions, and these areas contain specialised neurons that respond to different elements of the sound (Figure 13.6). Distance information, represented by the delay between the emitted pulse and the returning echo, is encoded in three areas of the auditory cortex: the FF (FF for frequency), dorsal fringe (DF) and ventral fringe (VF). The FF area, as well as the DF and VF areas, contain one type of specialised cells called FM-FM neurons. FM-FM neurons respond not to the emitted pulse or to the returning echo but, amazingly, to the *delay* between the pulse and echo of the FM signal (see Figure 13.3 and 13.8). Figure 13.7 shows a typical neural response, recorded from the VF region of the auditory cortex of the Mustached bat. As can be observed, the neuron responds somewhat to the pulse (P) and the echo (E); however this neuron particularly likes the delay between the pulse and echo (responding most strongly to delays of 1 millisecond).

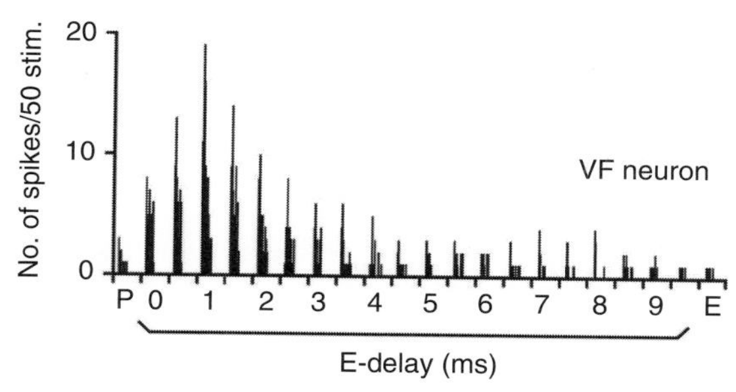

Figure 13.7 An illustration of how a particular neuron in the VF region responds preferentially to a delay of 1 sec between the pulse (P) and echo (E). Republished with permission of Society of Neuroscience from from J. Tang and N. Suga (2009), Corticocortical interactions between and within three cortical auditory areas specialized for time-domain signal processing, *J. Neurosci.* 29(22):7230–7; permission conveyed through copyright Clearance Centre, Inc.

Because each FM-FM neuron in the FF region is tuned to a particular echo delay, most FM-FM neurons therefore respond to a target at a particular distance. The distance detected can be very precise, with some species of bat detecting distances of 15 mm, which roughly translates as detecting a delay of approximately an 85-millionth of a second! These so called delay-tuned neurons are located in many structures of the auditory system, such as the medial geniculate and the inferior colliculus, as well as in the auditory cortex. In addition, these delay-tuned neurons have been found in many bat species, including *Pteronotus parnellii, Myotis lucifugus, Rhinolophus rouxii* and *Eptesicus fuscus.* While in most bats neurons respond to the delay between the pulse and the echo of the FM component occurring at the same frequency, in the *Pteronotus parnellii* (Mustached Bat) delay-tuned neurons respond to the FM components of the pulse and the echo of different harmonics. For example, neurons may respond best to the delay between the FM component of the pulse emitted at 24–29 kHz harmonics (FM_1) and the FM component of the echo of a higher harmonic (FM_2, 48–59 kHz; FM_3, 72–89 kHz or FM_4, 96–119 kHz). Thus, the neurons do not respond well to the emitted pulse, the echo, the CF or the FM sound presented individually; rather, the neurons respond best to the delay between the pulse and the echo of different harmonics. For example, the recorded neuron in Figure 13.8 responds best to delays of 4 ms between the pulse FM harmonics of 1 (thick black line, Figure 13.8 top) and echo FM harmonic 3 (dashed thick black line, Figure 13.8 top).

Why do bats emit pulses at different frequencies (i.e. at different harmonics)? The first harmonic is often the weakest, so weak that other bats often don't detect it. Although the bat can just about hear its own emitted pulse, it may not be able to hear the returning echo, as the echo is often even weaker than the pulse and may be below that bat's own hearing threshold. Thus the FM_1 pulse acts as a baseline to which the returning echoes of higher harmonics can be compared. Further, as other bats cannot hear the emitted first pulse, it saves a lot of confusion in a crowed colony. Interestingly, neurons only respond to combination of the *first* harmonic and others, and not to the combination of different higher harmonics, that is, they respond to FM_1/FM_3 and FM_1/FM_2 but not FM_2/FM_3.

As well as neurons responding to delays between the pulse and echo (either within one harmonic or between different harmonics), areas FF, DF and VF are beautifully organised so that neurons located in the more rostral (anterior) region of, for example, the FF respond to very short delays (e.g. 2 ms), while neurons located more posteriorly respond to longer delays. Further, there seems to be additional segregation along harmonic lines (Suga, 1990; see Figure 13.9). To calculate the distance from the delay information, we can apply the formula: Best range = Best delay × 17.2 cm. So delays of between 0.4 ms to 18 ms correspond to a range of between 7 cm and 310 cm. This is extremely accurate! The FF, DF and VF are well connected and the response of neurons in DF and VF are very similar to those in FF. The major difference among the regions is length of the echo delay that is represented along the anterior–posterior axis. The delay is up to 18 ms (see Figure 13.9) in the FF region, up to 9 ms in DF region and 5.5 ms in the VF area, thus, the VF region is particularly important during

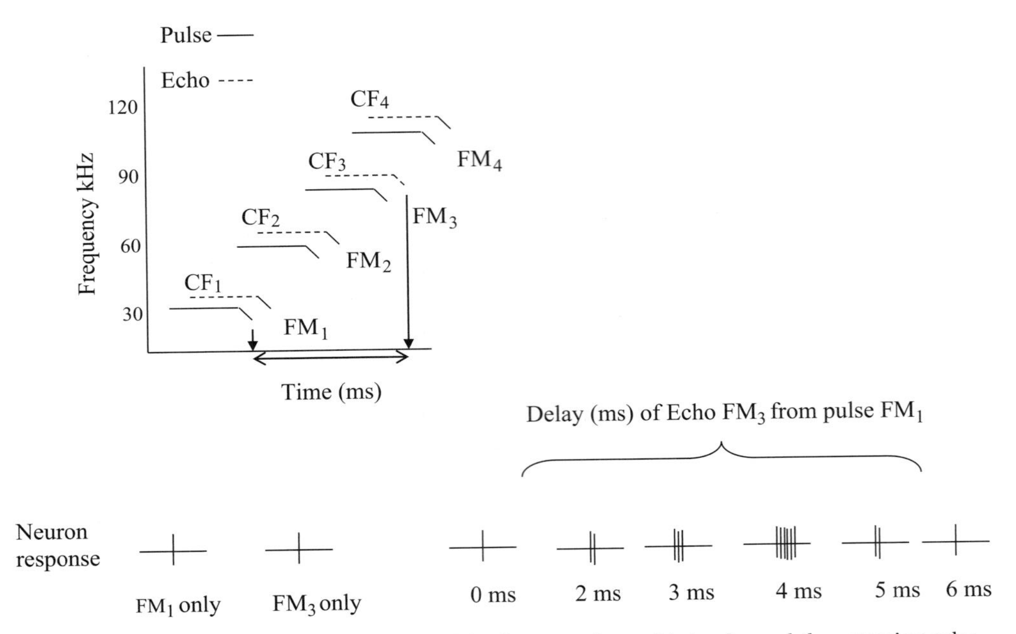

Figure 13.8 The neuron responds best to the delay between the emitted pulse and the returning echo operating at a different harmonics, in this case the delay between the FM part of the pulse at 30 Hz and the FM_3 part of the returning echo at approximately 90 Hz. Figure adapted from J.J. Wenstrup and C.V. Portfors, C.V. (2011), Neural processing of target distance by echolocating bats: functional roles of the auditory midbrain, *Neuroscience Biobehavioral Review*, 35(10), 2073–83.

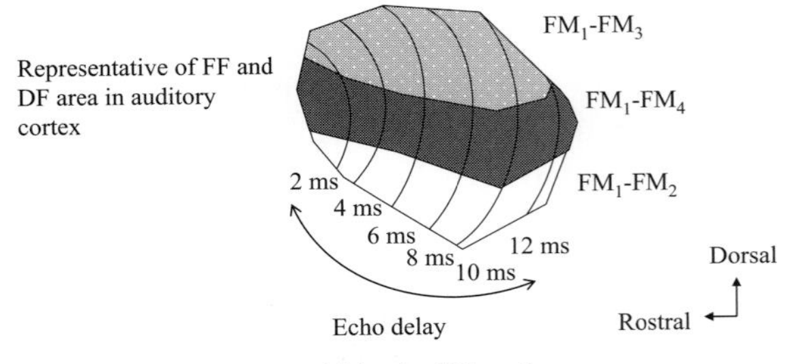

Figure 13.9 Neurons within the FF region are segregated. Those neurons located more rostrally (anterior) respond to very short delays compared to those located more caudally (posterior). Segregation along frequency lines is also observed in this region.

the terminal phase (Edamatsu and Suga, 1993), when the bat closes in on its prey and accuracy is paramount. Although segregation is clear in the auditory cortex, similar segregation is seen in other auditory regions, such as the inferior colliculus (IC), but in the IC no organisation of best delays is observed (Wenstrup and Portfors, 2011).

Delay Lines

As described in the previous section, there are neurons in the FF region that respond to delays – not the pulse or the returning echo, but the delay between the pulse and echo. How can a single neuron do this? One possibility is the presence of ‘delay lines’ (see also Chapter 12). If we consider that when the initial pulse is emitted, a neural response is generated (represented by the star on the Pulse FM_1 neuron in Figure 13.10), then, after a certain period of time (e.g. 12 ms) the echo activates a second neuron (see Figure 13.10, top right). However, as the ‘pulse’ action potential has already started its journey along its axon, there has to be a method whereby this action potential is delayed in order that the pulse action potential and the echo action potential converge on a single neuron at the same time. One mechanism that can be used to slow down the journey of an action potential is the presence of interneurons – that is, there are more neurons for the pulse action potential to travel through before reaching its destination (illustrated by the presence of the white-circle neuron in Figure 13.10). Alternatively the pulse neurons could have extra-long axons (see Chapter 12), or even longer delays could be caused by some form of inhibition. These mechanisms, known as 'delay lines', are thought to occur in the inferior colliculus, before the convergence of action potentials can occur at a single neuron in FF (Figure 13.10, bottom right).

How Is Velocity Information Processed and Calculated?

A second prominent region of the auditory cortex is known as CF-CF (see Figure 13.6), and it is thought that this region represents velocity

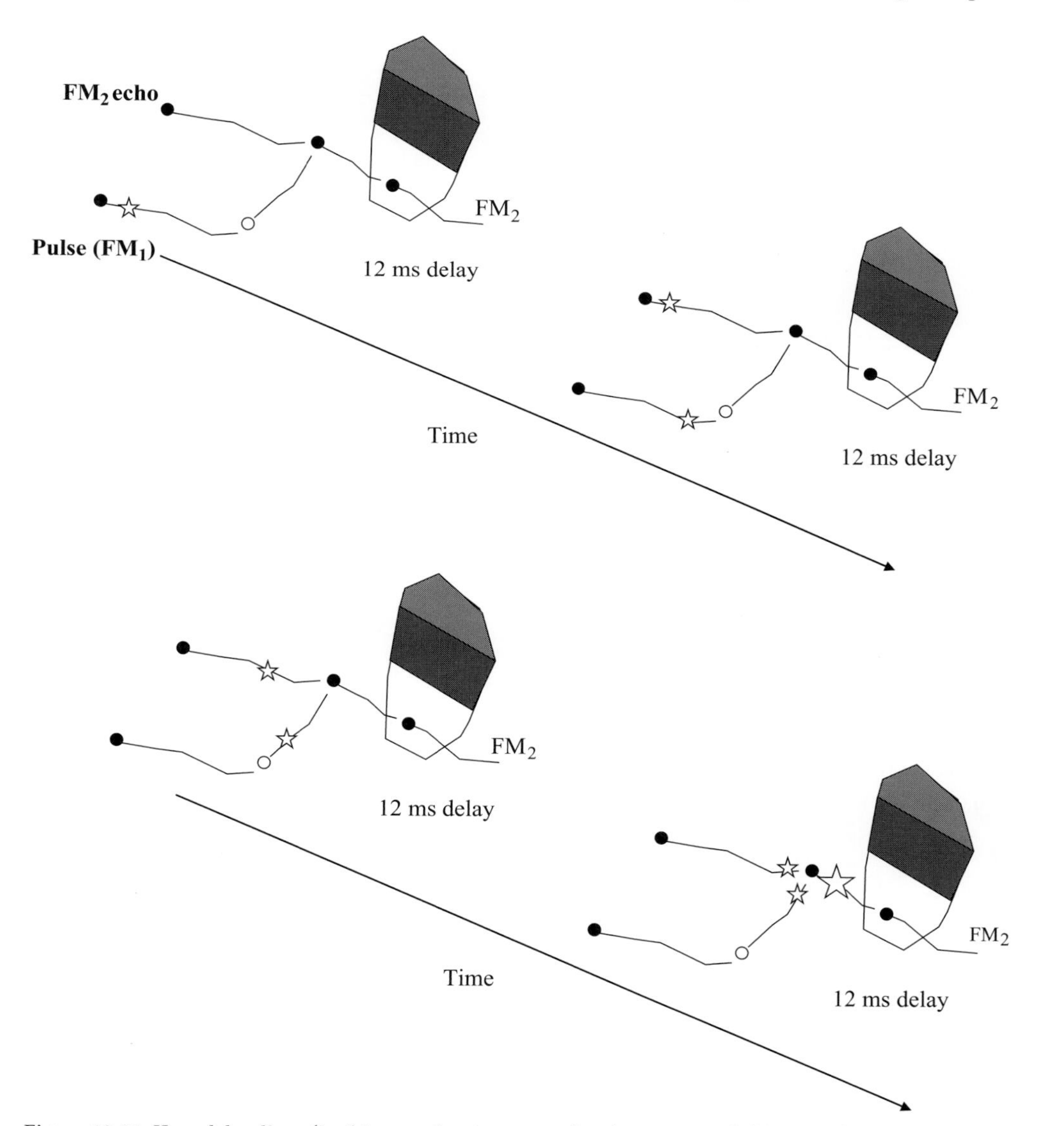

Figure 13.10 How delay lines (in this case the presence of an interneuron (white circle) may help the convergence of information onto a single neuron in the FF region. Small stars represent action potential travelling down each axon until they eventually converge (large star).

information. The CF-CF region of the auditory cortex is composed of neurons that respond to relationships between CF components of different harmonics. Like the FF neurons, CF neurons do not respond to the CF part of the pulse or echo presented alone, but to the *relationship* between the pulse CF (thick black line, Figure 13.11a) and the echo CF of a higher harmonic (dashed thick black lines, Figure 13.11a). Interestingly, as in the FF region, there is segregation within the CF-CF

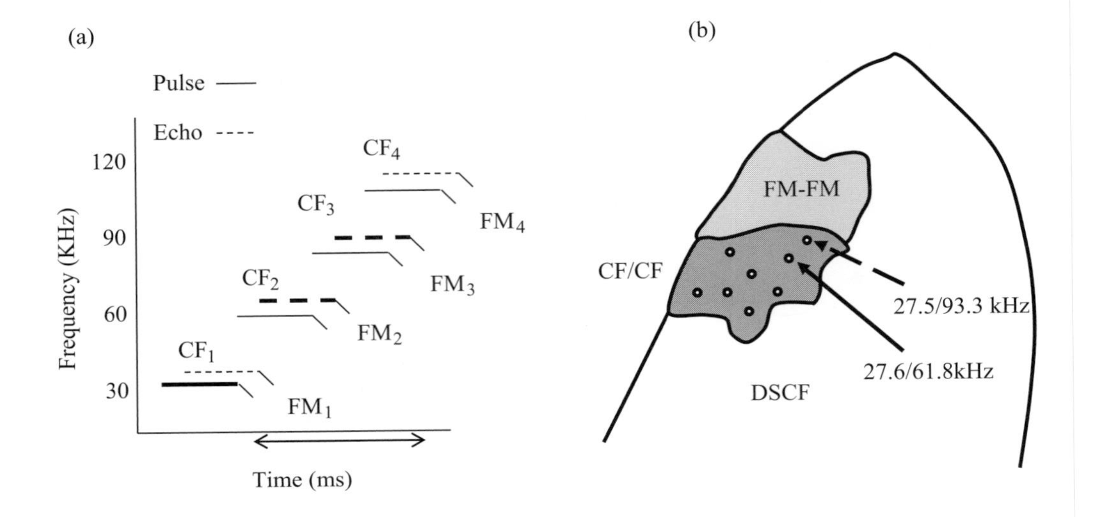

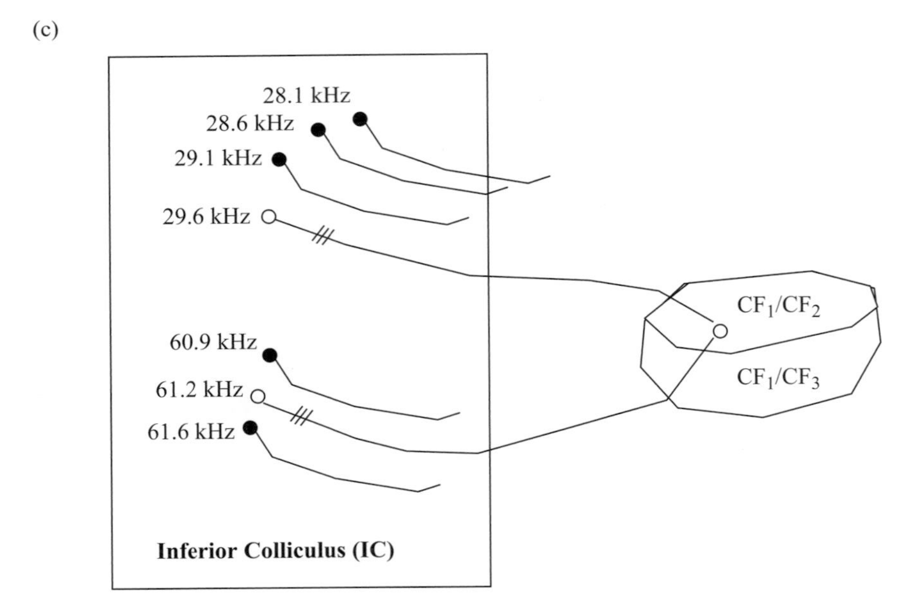

Figure 13.11 (a) Neurons in region CF-CF respond best to the delay between the CF component of the emitted pulse of harmonic 1 (thick black line) and the CF component of returning echo operating at one of the higher harmonics (thick black dashed lines). (b) An illustration of recorded neurons in the CF-CF region that respond to the combination of CF_1:CF_2 (black arrow) and CF_1:CF_3 (dashed arrow). (c) An illustration of how a neuron in CF-CF responds when two frequency-dependent neurons in the IC are activated.

region. There are two clusters of neurons. One set of neurons responds to CF_1/CF_2 relationships and the second cluster responds to CF_1/CF_3 relations. As the CF_1 harmonic has a frequency 28 and 30 kHz, the CF_2 is around 61 kHz and CF_3 is 92 kHz (see Figure 13.11a), there is a nice relationship between the various harmonics. CF_2 is more or less twice that of CF_1, and CF_3 is more or less three times that of CF_1. Notice I say 'more or less' – this relationship is not exact. Due to the Doppler effect, the ratio is slightly higher. Figure 13.11b illustrates this better. In the CF-CF (dark grey) region, each white circle corresponds to a recorded neuron. Some neurons respond best to the CF_1/CF_2 combination, for example 27.6/61.8 kHz (black arrow). This ratio is slightly higher than 2 times at 2.23. Likewise, other neurons respond best to the CF_1/CF_3 combination, for example 27.5/93.3 (dashed arrow). This ratio is again slightly higher than 3 times at 3.39. So these neurons are very exact, allowing for precise velocity to be calculated. Indeed, Suga (1990) suggested that a 1-kHz shift would arise if the target was moving at 2.8 m/s, so a 2.23-kHz shift would arise if the target was moving at 6.2 m/s. Furthermore, Suga (1989, 1990) has found that there are clusters of neurons in the CF-CF region that represent velocities up to 9 m/s. Interestingly, there is a disproportionately higher amount of CF-CF area that is devoted to velocities of between 0 to 4 m/s. These particular speeds are critical to a bat when capturing its prey or coming into land.

How can the CF-CF region respond to such relationships between the various harmonics? It is thought that neurons downstream to the auditory cortex, in the inferior colliculus, respond to single frequencies (e.g. CF_1, CF_3 or CF_2). Therefore, in order for a particular neuron in the CF-CF region to become activated, it requires the combined activation of two frequency-dependent neurons in the downstream structure (Figure 13.11c).

The third major area in the auditory cortex that plays an important role in echolocation is the Doppler-shifted CF region (DSCF, see Figure 13.6). This is one of the largest areas, occupying up to 30% of the auditory cortex. Although neurons in

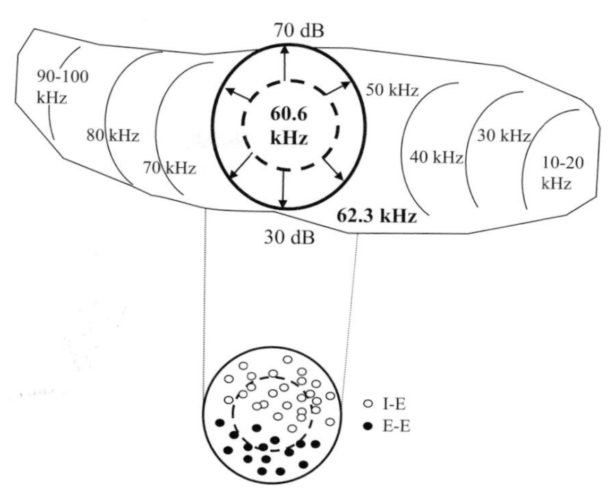

Figure 13.12 Neurons within the DSCF region are segregated according to frequency and sound amplitude. There is an overrepresentation of 61 kHz (to account for the Doppler shift from the echo). In addition, DSCF contains specialised I-E and E-E neurons, which respond to sounds presented to different ears.

the DSCF region respond to different frequencies along its length (see Figure 13.12), they primarily respond to echo CF_2 (Fitzpatrick et al., 1993; Kanwal et al., 1999) at least in the Mustached Bat. Sixty-one kHz is overrepresented in this area (as well as 1 kHz above, which takes the Doppler shift from the echo into account). This suggests that the DSCF region is primarily involved in the precision of the Doppler shift: How fast is the insect moving; is the bat gaining on the insect? (See Figure 13.4.) Neurons responding to the CF_2 echo follow a topographical map whereby as one moves outward (from dashed to thick black circle in Figure 13.12), the frequency changes. As one moves out at different angles from the centre (represented by the black arrows), neurons respond to different sound amplitudes, from 13 to 98 dB (Suga, 1989), the range 30–50 dB is overrepresented, and more-intense sounds are located dorsally and less-intense sounds located are ventrally (Manabe et al., 1978). Echo amplitude depends on the size and distance of the target: the closer the target, the larger the evoked amplitude.

In addition to this topographical organisation, within the area of the DSCF (Figure 13.12), a number of different types of neurons have been identified and classified (Manabe et al., 1978). Some are termed I-E (or ipsilateral inhibitory and contralateral excitatory neurons) and others are called E-E neurons (ipsilateral excitatory and contralateral excitatory). These neurons respond to tones and to the ear to which sound is presented (contralateral or ipsilateral). It is thought that E-E neurons are good for general target detection (e.g. to tell that there is an insect nearby), but I-E neurons are better at detecting direction and more precise localisation (e.g. the insect is 40 cm away at an angle of 30 degrees). Much more research is needed regarding the precise details of such specialised neurons, but knowledge regarding how sound is localised, paralleling the work described in the barn owl (Chapter 12), will help in this respect.

Summary

Bats are amazing creatures. They can navigate through an environment, avoiding obstacles and detecting prey as it moves. All of this is achieved through using auditory information. More than just listening to detect a target, bats emit a pulse and analyse the returning echo. This echo provides information on distance, direction, relative speed and even size of the target. Scientists are now beginning to understand how these feats are possible by delving into the inner workings of the bat's auditory cortex and surrounding regions. Within the various subregions of the auditory cortex, there are neurons that respond to the delay between the pulse and the echo, others that respond to the frequency, more that respond to the tone and others that can calculate the Doppler shift. All of these neurons work together in order for the bat to function effortlessly.

Questions Not Addressed That Are Active Research Topics

- Do the different neurons combine to form an overall map of the environment?
- What role do learning and memory play in echolocation?
- How does the auditory map translate into a motor map?
- Are the mechanisms found in the bat translatable to other echolocating species?
- What role does genetics play in echolocation?

References

Britton, A.R, and Jones, G. (1999). Echolocation behaviour and prey-capture success in foraging bats: laboratory and field experiments on Myotis daubentonii. *Journal of Experimental Biology*, 202 (Pt 13):1793–801.

Edamatsu, H., and Suga, N. (1993). Differences in response properties of neurons between two delay-tuned areas in the auditory cortex of the mustached bat. *Journal of Neurophysiology*, 69(5), 1700–1712.

Fitzpatrick, D.C., Kanwal, J.S., Butman, J.A., and Suga, N. (1993). Combination-sensitive neurons in the primary auditory cortex of the mustached bat. *Journal of Neuroscience*, 13(3):931–40.

Griffin, D., and Galambos, R. (1941). The sensory basis of obstacle avoidance by flying bats. *Journal of Experimental Zoology*, 86(3):481–505.

Kanwal, J.S., Fitzpatrick, D.C., and Suga, N. (1999). Facilitatory and inhibitory frequency tuning of combination-sensitive neurons in the primary auditory cortex of mustached bats. *Journal of Neurophysiology*, 82(5), 2327–2345.

Manabe, T., Suga, N., and Ostwald, J. (1978). Aural representation in the Doppler-shifted-CF processing area of the auditory cortex of the mustached bat. *Science*. 200(4339), 339–342.

Schnitzer, H.-U. (1968). Die Ultraschall-Ortungslaute der Hufeisen-Fledermäuse (*Chiroptera-Rhinolophidae*) in verschiedenen Orientierungssituationen. [The ultrasonic locating sounds of Horseshoe bats (Chiroptera-Rhinolophidae) in different orientation situations.] *Journal of Comparative Physiology* 57(4):376–408. [in German]

Spallanzani, L. (1794). Lettere sopra il sospetto di un nuovo senso nei pipistrelli [Letters on the suspicion of a new sense in bats]. Torino (Turin), Italy: Stamperia Reale. [in Italian]

Suga, N. (1989). Principles of auditory information-processing derived from neuroethology. *Journal of Experimental Biology,* 146, 277–286.

Suga, N. (1990). Biosonar and neural computation in bats. *Scientific American, June,* 60–68.

Tang, J., and Suga, N. (2009). Corticocortical interactions between and within three cortical auditory areas specialized for time-domain signal processing. *Journal of Neuroscience*, 29(22), 7230–7237.

Ulanovsky, N., and Moss, C.F. (2008). What the bat's voice tells the bat's brain. *Proceedings of the National Academy of Sciences USA*, 105(25), 8491–8.

Wenstrup, J.J., and Portfors, C.V. (2011). Neural processing of target distance by echolocating bats: functional roles of the auditory midbrain. *Neuroscience Biobehavioral Review*, 35(10), 2073–2083.

Recommended Reading

Zupanc, G.K.H. (2004). *Behavioral neurobiology: An integrative approach*. Oxford University Press, Oxford.

14 Spatial Navigation

Introduction

Being able to get from a to b is a critical skill for all animals. To survive, animals need to be able to explore their environment and find food, shelter and mates. Getting to a particular location and remembering where to go is an impressive feat matched only by the ability to return home. Some animals travel thousands of kilometres, even crossing oceans and deserts, before reaching their destination. Then, a few months later they make the return journey home, often ending up in the exact field, loft, cave or river from which they originally set off. It is hard not to be impressed with such migratory feats, which require not only stamina and knowledge but precision too. Many bird species, including storks, Manx shearwaters, swifts and swallows (Figure 14.1, left) make such journeys. The Arctic tern holds the record for the longest distance travelled. These birds leave their Arctic breeding grounds to reach the opposite end of the earth in the Antarctic, a round trip of some 70,000 km!

Generally, migratory birds have distinct routes and are very faithful to such routes; indeed even subspecies of birds may have their own distinct route. For example, European blackcaps tend to spend the summer in central Europe, particularly in Germany and Austria; however one population (located more to the west) tends to migrate towards Spain and onwards to western Africa, while the more eastern population migrates towards eastern Mediterranean and onwards to eastern Africa. Impressive migratory distances are not just the preserve of birds. Some turtles, including leatherback and loggerhead turtles, can also migrate large distances searching for food (jellyfish), mates and nesting beaches (see Figure 14.1, right). Some leatherbacks, for example, can swim up to 10,000 km, crossing the Pacific Ocean from Indonesia to California in the United States.

How do animals perform such amazing feats? There are a number of mechanisms that are thought to aid their navigation, including the use of sun, the use of the earth's magnetic field and various landmarks. The use of the sun as a navigational tool was demonstrated elegantly by the Austrian ecologist Karl von Frisch (1886–1982) while studying the behaviour of bees. When bees returned to the hive after foraging for food, von Frisch (1953) noticed that the worker bee performed an intricate dance. This dance consisted of the bee moving in a certain direction, shaking its abdomen, returning to a starting point and repeating the directional movement. This sequence was repeated multiple times. Von Frisch discovered that with

Figure 14.1 Left: Photograph of barn swallow by Malene Thyssen (Denmark) reproduced under Creative Commons Attribution License. Right: Green turtle *Chelonia mydas* coming up for air; photo by Brocken Inaglory, reproduced under Creative Commons Attribution License.

this dance the bee communicated both the distance and the direction of the location of food to the other bees. The direction moved by the bee is directly related to the position of the sun. If the bee moves upwards, then the food is located in the direction of the sun. If the bee moves downwards, the food is in the opposite direction to the sun. The rate of abdomen shaking relates to the distance the bee must travel in the particular direction.

Migratory birds also use the sun to determine the direction in which they should travel. This was first discovered by Gustav Kramer (1910–1959), a German zoologist who showed that migratory birds trapped in cages got very restless, particularly close to the time for migration. They tended to flap about the cage, pointing in the direction that they needed to travel. Kramer (1952) noticed that on a cloudy day the restless behaviour decreased and birds did not tend to show any directionality. He hypothesised that such behaviour was governed by the sun. To test this, he arranged mirrors to be fixed to the windows of the cage so that sunlight shone into the cage at a particular angle rather than directly. He observed that the birds' directionality also shifted in relation to the angle shift.

Subsequent experiments have shown that birds, bees and other animals can detect polarised light. As such, they can determine the position of the sun, even on a cloudy day, as long as there is a small patch of blue sky. Although the sun is a very useful directional cue, many animals can also detect the earth's magnetic field, which provides both directional and positional information. The ability of the European robin to detect the earth's magnetic fields was demonstrated by Wolfgang Wiltschko, who observed that the direction of restless behaviour could be changed by manipulating the local magnetic field using a magnetic coil (Wiltschko and Wiltschko, 1996).

More recent studies seem to suggest that birds can actually visualise magnetic fields (Biro, 2010). Although it is generally agreed that the sun, stars (in the case of nocturnal migration) and magnetism all contribute to long-distance migration and play a particularly important role in the initial period of the journey, other factors including landmarks are also critically important, perhaps even more important. For example, Tim Guilford and colleagues recently demonstrated that homing pigeons use individualised routes to get home and seem to learn various landmarks along the way (Meade et al., 2005). This demonstrates a route memory that is built up with experience. Interestingly, if the pigeons are released from a new site away from the learnt path, they will try to make their way back to this learnt route. The landmarks used tend to include natural features in the landscape such as rivers and hedgerows, but even motorways and railroad tracks are also used.

Large-scale migration is difficult to study, but with the increasing use of satellite and miniature GPS tracking systems, large amounts of detailed information about various routes and behaviours can be now obtained and examined. However, studying spatial navigation and memory in the laboratory is much easier. Although the laboratory setting does not allow for the examination of long-distance navigation, it does allow more controlled experiments to be conducted and allows researchers to tease apart the various mechanisms underlying navigation. Maze learning has been an important aspect of psychology throughout its history (see Chapter 4 for more details), and with the development of sophisticated neural electrophysiological recording and behavioural tracking systems, the neural correlates underlying navigation can also be elucidated.

A Brain System for Navigation: The Role of the Hippocampus

One structure of the brain that seems to be critically involved in spatial navigation and spatial memory is the hippocampus. In humans and non-human primates, the hippocampus (red) and the entorhinal cortex (blue), often referred to as the *parahippocampal complex*, are located deep within the temporal lobes. These are also prominent structures in rodents (Figure 14.2 top). The hippocampus can be divided into a number of subregions, including the dentate gyrus (DG), area cornu ammonis 3 (CA3), area CA2 and area CA1. In the macaque and human brains an area CA4 is also recognised. Figure 14.2, bottom, shows a coronal section through the brain of the mouse, macaque and human hippocampi, and indicates the different subregions and their relative locations. The entorhinal cortex is located beside the hippocampus

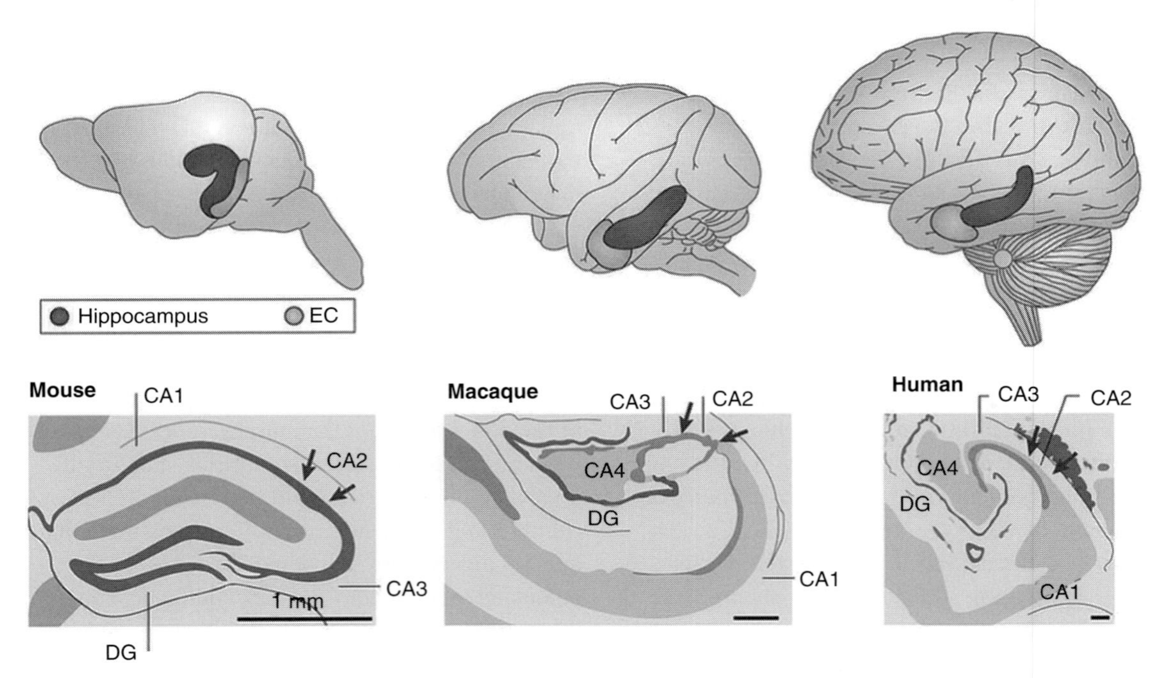

Figure 14.2 (top) Location of hippocampus and entorhinal cortex (EC) in mouse, macaque and human brains. (bottom) Section through the hippocampal region of each species, showing the various subregions. DG, dentate gyrus; CA1–4, cornu ammonis 1–4. From Strange et al. (2014), Functional organization of the hippocampal longitudinal axis, *Nature Reviews Neuroscience*, 15, 655–669. Reprinted by permission from Macmillan Publishers Ltd. For colour version, please refer to the plate section.

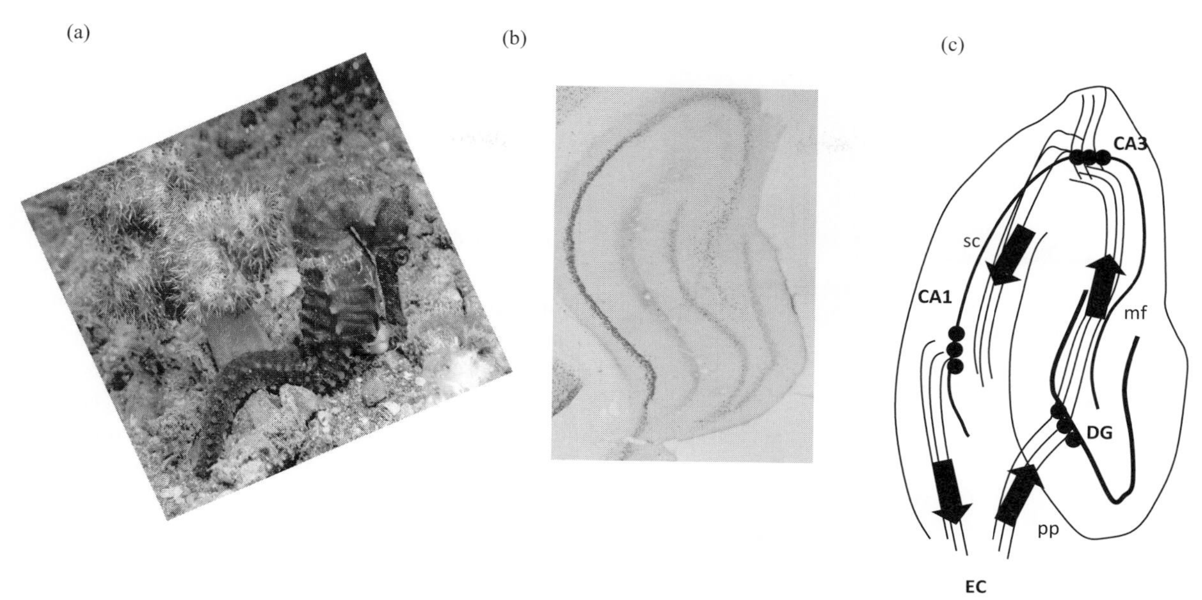

Figure 14.3 (a) Photograph of a seahorse thanks to Sarah McCready and Mark Theunissen. (b) Section through hippocampus showing its close resemblance to a seahorse. Image thanks to Drs Francesca Farina and Daniel Barry. (c) Pattern of projections through the hippocampus (see text for description).

and is generally divided into two: the lateral entorhinal cortex (LEC) and medial entorhinal cortex (MEC). We have encountered these structures previously with our discussions of LTP and LTD.

The word *hippocampus* derives from the Greek for 'seahorse', named for the close resemblance of the brain structure to the beautiful marine creature (see Figure 14.3a,b). Information is generally thought to flow through the hippocampus in a unidirectional manner (Figure 14.3c), but see Commins et al. (2002) for the possibility of back projections. Information from the sensory cortices including visual, auditory and olfactory information converge onto the hippocampus through two parallel pathways that innervate the lateral and medial entorhinal cortex. The EC fibres, termed the *perforant pathway* (pp), innervate small granular cells in the dentate gyrus. From here axons (mossy fibres, mf) extend into area CA3 which, in turn, send axons to CA1 pyramidal cells via the Schaffer collaterals (sc). Projections from area CA1 return to deep layers of the EC through the subiculum (not shown). This is a much-simplified picture with many other projections not included, such as reciprocal projections from the EC to CA3, CA1 and subiculum, the projections to the thalamus and other subcortical structures through the fornix and the anterior commissure connecting the left and right hippocampus.

There is much evidence to suggest that the hippocampus is involved in spatial navigation and memory. Lesions of the hippocampus in rats and mice impair the ability of animals to find a hidden platform when doing the Morris water maze task (Morris et al., 1982; Diviney et al., 2013; see also Chapter 4). In this task, animals are required to escape a pool of water by finding a hidden platform located somewhere in the arena. As they cannot see the platform directly, they must use the landmarks in the environment to locate it. Lesions of the hippocampus do not have an effect on animals' visual or motor abilities, but do seem to impair their memory of the platform's location.

Further evidence for the role of the hippocampus in spatial navigation and memory comes from studies that show food-storing birds (e.g. the

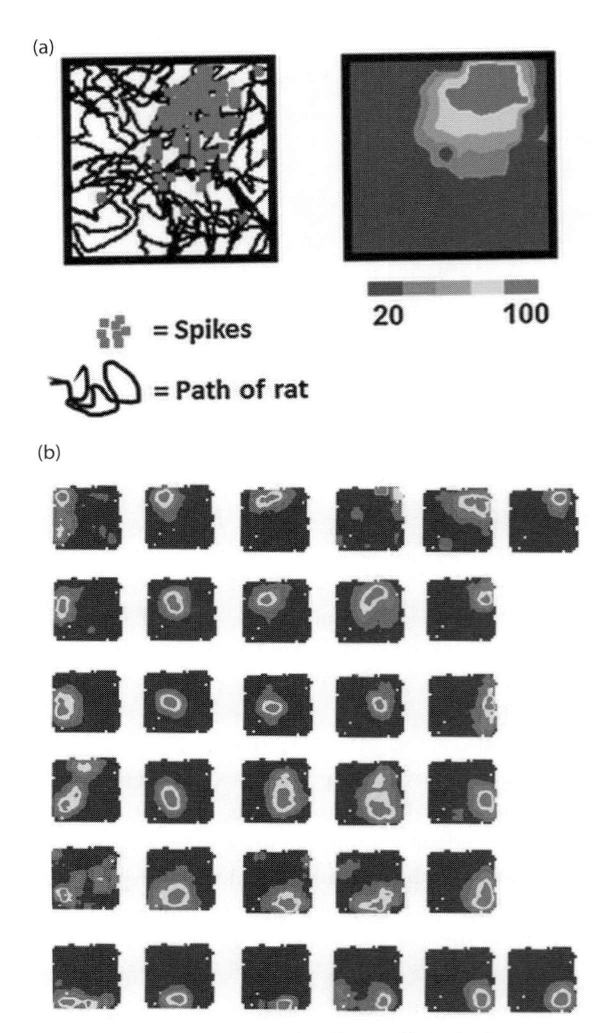

Figure 14.4 (a) Example of a place cell. A single neuron in the rat hippocampus fires (red dots, left image) when an animal is at a specific location in the square arena. Image taken from R.M. Grieves and K.J. Jeffery (2017), The representation of space in the brain, *Behav Processes*, 135:113–131, with permission from Elsevier. (b) Recordings of multiple place cells show that the entire environment is encoded. Image from Professor John O'Keefe Nobel lecture. Copyright © The Nobel Foundation (2014); Source: http://Nobelprize.org. For colour version, please refer to the plate section.

marsh tit, *Parus palustris*) have larger hippocampi compared to those birds that do not store food (e.g. the great tit, *Parus major*; Krebs et al., 1989). Food-storing birds must remember where the food is stored in order to retrieve it at a later date. Further, food-storing behaviour tends to increase in autumn and winter, and an increase in the size of the hippocampus and the rate of hippocampal neurogenesis is also observed in autumn and winter in many food-storing birds (Sherry and Hoshooley, 2010).

Similar findings have been observed in humans. For example, London taxi drivers are required to spend two years learning the thousands of streets in London before being given a license. It has been shown that qualified taxi drivers have a larger hippocampus (posterior part) compared to non-taxi drivers, and that the longer the time spent as a taxi driver, the larger the hippocampus (Maguire et al., 2000). Having to navigate the streets and recall multiple routes requires the continuous activation of the hippocampus and its subsequent growth. Further evidence for the role of the human hippocampus in spatial memory comes from patient studies, and particularly from those suffering from Alzheimer's disease (see Schröder and Pantel, 2016 for details).

Neural Basis of Navigation: Place Cells, Grid Cells and Head-Direction Cells

In previous chapters we were able to take a particular behaviour and try to map the neural circuitry directly onto this behaviour. Spatial navigation is very complex, requiring multiple sensory systems (visual, auditory and olfactory) interacting with a motor output. Included within this interaction are attentional, motivational and memory systems. Unfortunately, direct mapping of spatial navigation onto a specific neural system that includes all of these processes is not yet possible. However, over the last 40 years, much has been discovered about the neural underpinnings of navigation, culminating in the awarding of the Nobel Prize in Physiology or Medicine in 2014 to John O'Keefe, and Edvard and May-Britt Moser for their discoveries of very specialised cells located within the parahippocampal complex.

In order to move from one location to another, there are a number of essential things that we need to know. For example, we need to know where we are at the moment (our current location). We also need to know the intended destination (our goal), the direction in which we have to travel and the distance we have to travel. Further, as we move towards our goal, the direction and distance need to be continuously updated. Some of these essential features of navigation are encoded by very specialised cells within the hippocampal region.

Knowing the Current Location

In 1971 John O'Keefe began to record electrical activity from single cells in area CA1 of the rat hippocampus (O'Keefe and Dostrovsky 1971). The animals were allowed to freely explore a small square arena as recording took place. Over the course of a few months, a distinct firing pattern emerged. It was found that cells within area CA1 were not interested in what the animal was doing, but rather *where* the animal was located within the arena. Such cells were subsequently termed *place cells*. Figure 14.4a shows that when an animal explores the square arena, the hippocampal cell only fires when it is in the upper right position of the environment; it does not respond to any other location. So here we have a single cell firing because the animal is in a particular location. What about the rest of the environment? If we record a second cell, this will fire in a different location. Indeed, Figure 14.4b demonstrates that different hippocampal cells encode different locations. In this experiment, when 32 cells are recorded simultaneously each cell encodes a specific location. Therefore, with just 32 cells the entire environment can be encoded.

Since the discovery of place cells, many of their properties have emerged. It has been shown that place cells are dependent on the external landmarks in the environment. When the landmarks are rotated, place cells also rotate in proportion. Place cells can continue to fire in the dark, suggesting a memory component. Place cells can discriminate between two environments; for example, if an animal explores a square arena and a place cell emerges, the same cell will not fire in a circular environment. Other interesting features emerge when the environment is manipulated. If the environment is stretched, for example from a square to a rectangle, the place cell also stretches. Such cells clearly depend on the subject's location relative to cues and other features in the environment.

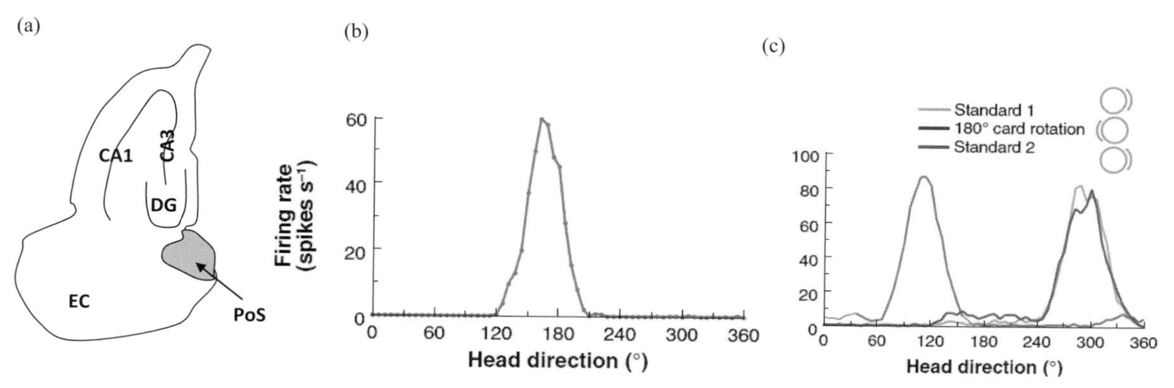

Figure 14.5 (a) Location of the postsubiculum (PoS), where head-direction cells were originally discovered. EC, entorhinal cortex; DG, dentate gyrus; CA1/3, cornu ammonis 1/3. (b) Example of a head-direction cell that has a preferred direction of 160°. (c) Rotation of visual cues in the environment causes the concurrent rotation of the head-direction cell. Image taken from J.S. Taube (2007), The head direction signal: origins and sensory-motor integration, *Annu Rev Neurosci.* 30:181–207. Reproduced with permission of *Annual Review of Neuroscience,* Volume 30, © by annual reviews, http://www.annualreviews.org.

Knowing the Direction

A second major finding was the discovery of cells that fired when an animal's head was facing in a particular direction, irrespective of where the animal was located in the arena. These cells, originally described by Jim Ranck in 1984 and characterised more fully in the early 1990s, are called *head-direction cells.* These cells were originally found in an area lying between the hippocampus and entorhinal cortex called the *postsubiculum* (PoS, see Figure 14.5a). Such cells have now been described in many other regions of the brain, the most prominent of which is the anterodorsal nucleus of thalamus; 60% of cells in this region are thought to be head-direction cells. Other regions that contain head-direction cells include the lateral mammillary nucleus (25% of cells), the lateral dorsal thalamus (30% of cells), the retrosplenial cortex (10%), the striatum (6%), and more recently the entorhinal cortex and rostral thalamus (see Taube, 2007 for details and Giocomo et al., 2014, and Jankowski et al., 2015). Similar to place cells, removal of a familiar visual cue or turning the lights off does not affect the firing rate greatly. In addition, rotation of a visual cue leads to the similar rotation of the firing of head-direction cell. Figure 14.5b shows a particular cell recorded from the postsubiculum that only fires when the animal's head is in the 160^0 direction. Figure 14.5c shows that when the visual cue card is rotated 180^0 anticlockwise (Figure 14.5c inset) the preferred head-direction firing shifts from 300^0 to approximately 120^0. When the cue card is placed back in its original location, the preferred firing direction also shifts back.

Knowing the Intended Destination

Although reward-related behaviour has long been associated with the nucleus accumbens, more recent analysis suggests that this region plays more of a role in action selection that helps in goal-directed behaviour (Floresco, 2015). Interestingly, it has been discovered that the hippocampus may also play a role in this regard. In an experiment conducted by Hok et al. (2007), the authors required animals to go to a certain location within a large circular arena, where they had to wait before obtaining food. As animals performed this task, place cells were recorded from hippocampal area CA1. As well as finding cells that encoded for a particular location in the arena, the same cell would also fire, to a lesser degree, in the goal location. Figure 14.6a demonstrates place cell firing from 4 recorded hippocampal cells (NW, E, NE and SE positions of the arena for cells 1, 2, 3 and 4, respectively). With each of these cells, there was also some firing in the vicinity of the goal (located in W and represented by the black circle). This suggests that the same cell can fire in different locations within the arena and that it may encode goal-like information.

In another task, conducted by Dudchenko and colleagues (Ainge et al., 2007), animals were required to make two choices in order to obtain food at the end of double Y-maze task. The authors found that place cells fired much more in the start box compared to other areas of the maze (Figure 14.6b). In addition, the authors found that the place cell did not fire to the same extent for each journey, but rather the cell seemed to fire more before heading towards a particular goal. For example, Figure 14.6b shows more firing of the place cell in the start box before the animals went to box 1, compared to when they travelled to the other boxes. This suggests that the place cell is encoding not only its current location but its intended destination as well.

Knowing the Distance to Travel

In 2005 Edvard and May-Britt Moser discovered cells in the medial entorhinal cortex (EC) that fired in multiple locations in the environment. So rather than a single cell encoding for a particular location, here we have a single cell that encodes multiple locations (Hafting et al., 2005). These cells do not fire in random locations, but rather fire in a beautiful systematic gridlike pattern (Figure 14.7a), thus these cells have been termed *grid cells.* Similar to both place cells and head-direction cells, these

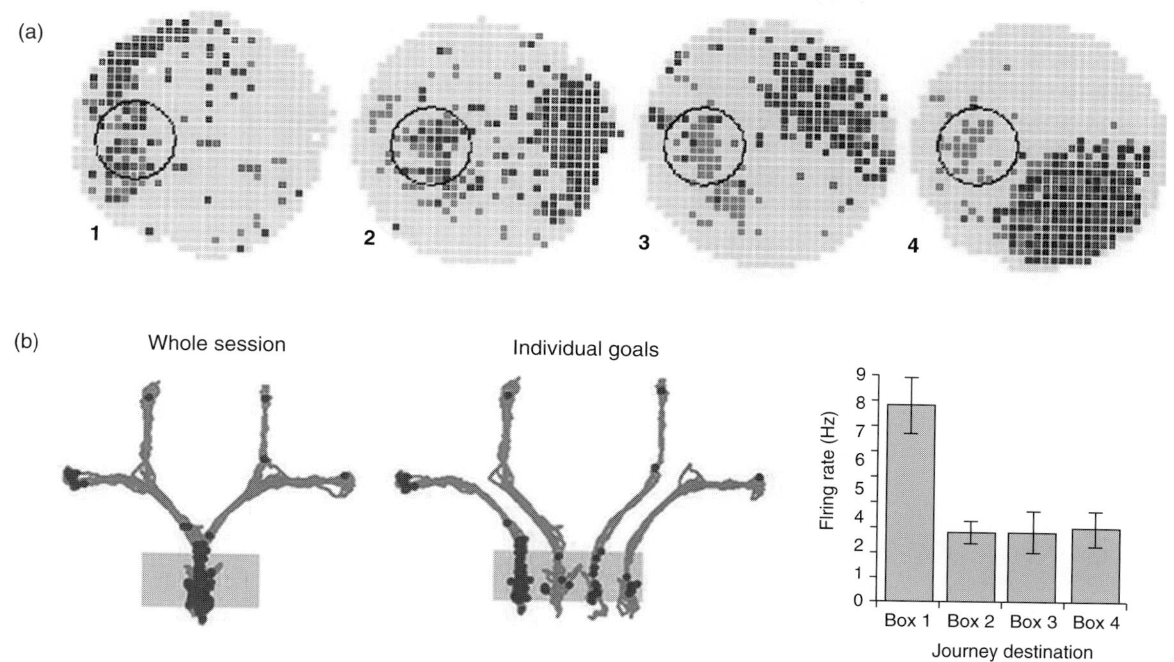

Figure 14.6 (a) Firing of hippocampal place cells with additional firing at the goal location (black circle). Republished with permission of Society of Neuroscience from V. Hok et al. (2007), Goal-related activity in hippocampal place cells, *J Neurosci.* 27(3):472–482; permission conveyed through copyright Clearance Centre, Inc. (b) Firing of cells at the start of the task seem to predict the animal's current location and intended destination. Republished with permission of Society of Neuroscience from J.A. Ainge et al (2007), Hippocampal CA1 place cells encode intended destination on a maze with multiple choice points, *J Neurosci.* 27(36):9769–9779; permission conveyed through copyright Clearance Centre, Inc.

cells continue to fire in the dark (Figure 14.7b) and also rotate their firing pattern in relation to the rotation of environmental cues. Another feature of grid cells is that different scales can be encoded and represented. As the electrode is lowered deeper into the EC (from dorsal to ventral), the spacing between cells increases. A particularly interesting feature that distinguishes grid cells from place cells is that they preserve their orientations and scale from one environment to another. For example, Figure 14.7c shows recordings from three separate grid cells as the animal is placed from Room 1 into a smaller arena (Room 2) before being returned to its original environment. Whereas place cells change depending on the configuration of the arena and environment, grid cells do not. This suggests that grid cells operate more like a universal map and do not care about the details of a particular environment.

Having cells that fire at regular intervals as they travel through space allows animals to maintain an idea of distance. For example, the animal may simply have to calculate how many times a particular cell has fired on the journey from point a to point b.

So far we have mentioned the main types of cells that have been found within various regions of the parahippocampal complex. Other cells such as border or boundary cells, which fire in relation to the arena's wall, also exist, and these have been found in the subiculum (S, Figure 14.8) and the EC (Lever et al., 2009). In addition, there are cells termed *conjunctive cells* that encode both head-direction and grid properties; these have been found primarily in the deep layers of the EC (Sargolini et al., 2006). More recently, *speed cells* have also been discovered in the medial EC (Kropff et al., 2015); such cells may encode for

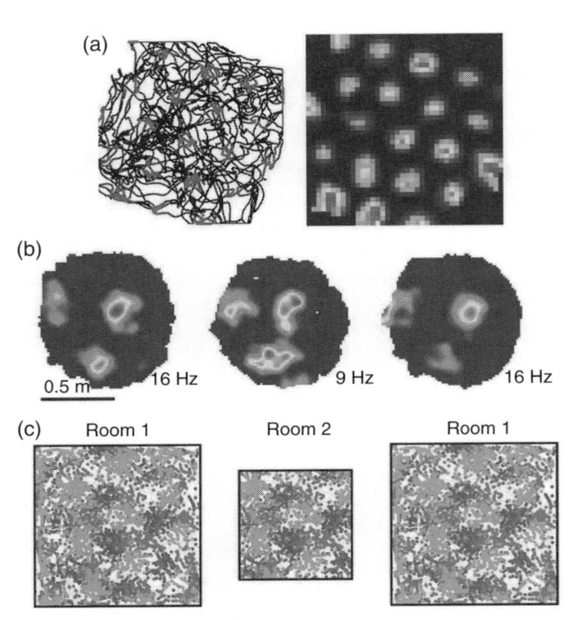

Figure 14.7 (a) Example of a grid cell. A neuron in the medial entorhinal cortex fires in multiple areas in a gridlike pattern (as represented by the red dots, left, and the red/yellow colours, right) as the animal traverses a square arena. Image taken from R.M. Grieves and J.K. Jeffery (2017), The representation of space in the brain, *Behav Processes* 135:113–131, with permission from Elsevier. (b) Grid cells maintain their firing in the dark. Light condition (left), Dark condition (middle), Light condition (right). (c) Grid cells maintain their firing pattern across different environments. Images (b) and (c) from Professor Edvard Moser's Nobel lecture. Copyright © The Nobel Foundation (2014); source: http://Nobelprize.org. For colour version, please refer to the plate section.

the velocity of an animal and play an important role in determining where and how far an animal has travelled, independent of the environment. Such a system may not rely on visual, auditory or olfactory input but on the animal's own movements and sense of location through the vestibular and proprioceptive system, a mechanism termed *path integration*. Indeed, it is thought that many insects, such as ants, rely on this mechanism to navigate. See Figure 14.8 for a summary of the various cell types discovered in the hippocampal region.

Cells Representing Space Across Species

As mentioned at start of the chapter, the hippocampus is an important structure for spatial navigation

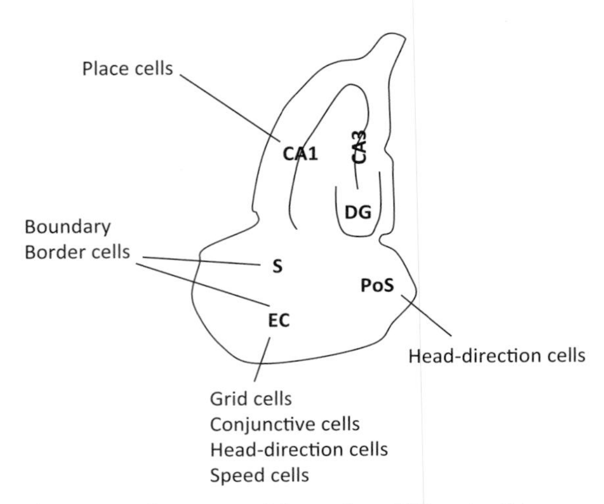

Figure 14.8 Summary of the various different cell types and where they have been found within the hippocampal region. CA, cornu ammonis; DG, dentate gyrus; PoS, postsubiculum; EC, entorhinal cortex; S, subiculum.

not only for rodents but also for humans. Are any of the previously discussed cells also seen in the human hippocampus? Due to ethical considerations, it is difficult to insert electrodes directly into the human hippocampus and begin recording. However, some patients suffering from temporal lobe epilepsy may need to undergo surgery. In order to determine the focus of the seizures, an electrode may have to be inserted into the hippocampal region. This allows for an electrical readout from the structure and provides important information to the physicians. Under very strict conditions, the patients may also be willing to undergo a variety of tasks when the electrodes have been inserted.

Under such circumstances, Itzhak Fried and colleagues (Ekstrom et al., 2003) directly recorded over 300 neurons from the hippocampus, parahippocampal area, amygdala and frontal lobes of epileptic patients. The patients were required to explore and navigate through a virtual-reality town. The authors found that 11% of cells responded to a specific place in the virtual world, and these were located primarily in the hippocampus, i.e. they were *place cells*. A number of other cells, termed *view cells*, responded to a single object during the navigation task and included items such as a particular shop or a passenger; these cells were located primarily in the parahippocampal region. More cells (21%) responded to the patient's goal – for example, a particular shop that the participant had to find. These cells were found in all regions sampled.

In another study, it was shown that cells in the human EC responded to whether a patient was moving clockwise or anticlockwise around a virtual environment (Jacobs et al., 2010). Both these studies suggest that there are cells within the human hippocampal complex and surrounding structures that are involved in navigation. These cells may have similar but not identical properties to those found in rodents. Furthermore, the representation of space may very much depend on the species involved. For example, view response cells, as described earlier in the human parahippocampal area, have also been described in the non-human primate (Rolls and O'Mara, 1995), perhaps emphasising the greater role of the visual system in primates compared to rodents (see Chapter 2). Likewise, place cells have also been discovered in the bat (Yartsev and Ulanovsky, 2013), but these fields are three-dimensional in nature, with cells firing as the animal flies through a particular location in space. Similar to rodent place cells, with just a small number of cells recorded, the bat's entire environment can be covered.

Beyond Spatial Navigation

Finally, the role of the hippocampus has long been associated with memory in general and not just spatial memory or indeed spatial navigation. The famous patient HM (Henry Molaison- see also Chapter 7), who had both hippocampi removed due to severe epileptic seizures in 1953, was unable to form any new declarative memories (memories for facts and events). He was unable to remember anything after his operation right up until he died in 2008, nothing of world events and nothing regarding his own personal history. His short-term memory remained relatively intact and he was able to carry on a conversation, but as soon as the person he was speaking with left the room, HM had no recollection of ever having met that particular person. Some memory systems remained intact; for example, he was able to form new habits and skills. The importance of HM lies in the idea of identifying the hippocampus as a critical structure in memory and showing that there are multiple memory systems. More recently, the role and functioning of the hippocampus has expanded further, with many authors suggesting that it is also involved in future planning (see Figure 14.6, top), i.e. the role of memory is not simply to recall past events but to prepare and imagine the future (Addis et al., 2007). For example, it has been demonstrated that patients with bilateral hippocampal damage find it very difficult to construct scenes using their imagination (Mullally et al., 2014). Such findings indicate that so much more has yet to be discovered about the hippocampus and its function.

Summary

Unlike the subjects of previous chapters, the behaviour of spatial navigation cannot be directly mapped onto a single circuit at this time. Despite this, navigation can be broken into various constitutive parts, including knowing where one is currently located, and knowing the distance and direction one has to travel. These behaviours are represented by the firing pattern of different cells within structures of the parahippocampal complex, including the hippocampus itself and the surrounding entorhinal cortex. Such cells are not unique to rodents, but have their equivalents in a variety of different species, including humans, non-human primates and bats. In humans (and possibly in rodents) the hippocampus may play an expanded role in general memory and even in imagination. As such, understanding the structure and cellular components is critical for understanding many disorders such as Alzheimer's disease, which have memory and spatial navigation deficits at their heart.

Questions and Topics Under Current Investigation

- Do migratory birds and animals also have place cells and other specialised navigational cells? How do these cells respond in enlarged naturalistic environments?
- How do all the various navigational cells interact with each other?
- How are the various navigational cells formed?
- How does the hippocampal navigation system interact with other memory systems and the cells contained within?
- How do attention, motivation, planning and other systems interact with the navigation system?
- Sleep is known to consolidate memory, and place cells are reactivated during sleep (see e.g. de Lavilléon et al., 2015). How are the processes connected and what are the exact mechanisms?

References

Addis, D.R., Wong, A.T., and Schacter, D.L. (2007). Remembering the past and imagining the future: common and distinct neural substrates during event construction and elaboration. *Neuropsychologia*, 45, 1363–1377.

Ainge, J.A., Tamosiunaite, M., Woergoetter, F., and Dudchenko, P.A. (2007). Hippocampal CA1 place cells encode intended destination on a maze with multiple choice points. *Journal of Neuroscience*, 27(36), 9769–9779.

Biro, D. (2010). Bird navigation: a clear view of magnetoreception. *Current Biology*, 20(14), R595–596.

Commins, S., Aggleton, J.P., and O'Mara, S.M. (2002). Physiological evidence for a possible projection from dorsal subiculum to hippocampal area CA1. *Experimental Brain Research*, 146(2), 155–60.

de Lavilléon, G., Lacroix, M.M., Rondi-Reig, L., and Benchenane, K. (2015). Explicit memory creation during sleep demonstrates a causal role of place cells in navigation. *Nature Neuroscience*, 18(4), 493–495.

Diviney, M., Fey, D., and Commins, S. (2013). Hippocampal contribution to vector model hypothesis during cue-dependent navigation. *Learning & Memory*, 20(7), 367–378.

Ekstrom, A.D., Kahana, M.J., Caplan, J.B., Fields, T.A., Isham, E.A., Newman, E.L., and Fried, I. (2003). Cellular networks underlying human spatial navigation. *Nature*, 425(6954), 184–188.

Floresco, S.B. (2015). The nucleus accumbens: an interface between cognition, emotion, and action. *Annual Review of Psychology*, 66, 25–52.

Giocomo, L.M., Stensola, T., Bonnevie, T., Van Cauter, T., Moser, M.B., and Moser, E.I. (2014). Topography of head direction cells in medial entorhinal cortex. *Current Biology*, 24(3), 252–262.

Grieves, R.M., and Jeffery, K.J. (2017). The representation of space in the brain. *Behavioral Processes* 135, 113–131.

Hafting, T., Fyhn, M., Molden, S., Moser, M.B., and Moser, E.I. (2005). Microstructure of a spatial map in the entorhinal cortex. *Nature*, 436(7052), 801–806.

Hok, V., Lenck-Santini, P.P., Roux, S., Save, E., Muller, R.U., and Poucet, B. (2007). Goal-related activity in hippocampal place cells. *Journal of Neuroscience,* 27(3):472–482.

Jacobs, J., Kahana, M.J., Ekstrom, A.D., Mollison, M.V., and Fried, I. (2010). A sense of direction in human entorhinal cortex. *Proceedings of the National Academy of Sciences USA*, 107(14), 6487–6492.

Jankowski, M.M., Passecker, J., Islam, M.N., Vann, S., Erichsen, J.T., Aggleton, J.P., and O'Mara, S.M. (2015). Evidence for spatially-responsive neurons in the rostral thalamus. *Frontiers in Behavioral Neuroscience*, 9, 256.

Kramer, G. (1952). Experiments on bird orientation. *Ibis*, 94, 265–285.

Krebs, J.R., Sherry, D.F., Healy, S.D., Perry, V.H., and Vaccarino, A.L. (1989). Hippocampal specialization of food-storing birds. *Proceedings of the National Academy of Sciences USA,* 86(4), 1388–92.

Kropff, E., Carmichael, J.E., Moser, M.B., and Moser, E.I. (2015). Speed cells in the medial entorhinal cortex. *Nature*, 523(7561), 419–424.

Lever, C., Burton, S., Jeewajee, A., O'Keefe, J., and Burgess, N. (2009). Boundary vector cells in the subiculum of the hippocampal formation. *Journal of Neuroscience*, 29, 9771–9777.

Maguire, E.A., Gadian, D.G., Johnsrude, I.S., Good, C.D., Ashburner, J., Frackowiak, R.S., and Frith, C.D. (2000). Navigation-related structural change in the hippocampi of taxi drivers *Proceedings of the National Academy of Sciences USA*, 97(8), 4398–403.

Meade, J., Biro, D., and Guilford, T. (2005). Homing pigeons develop local route stereotypy. *Proceedings of Biological Science*, 272(1558), 17–23.

Morris, R.G., Garrud, P., Rawlins, J.N., and O'Keefe, J. (1982). Place navigation impaired in rats with hippocampal lesions. *Nature*, 297(5868), 681–3.

Mullally, S.L., Vargha-Khadem, F., and Maguire, E.A. (2014). Scene construction in developmental amnesia: an fMRI study. *Neuropsychologia*, 52, 1–10.

O'Keefe, J., and Dostrovsky, J. (1971). The hippocampus as a spatial map. Preliminary evidence from unit activity in the freely-moving rat. *Brain Research*, 34(1), 171–175.

Ranck, J.B. Jr. (1984). Head direction cells in the deep layer of dorsal presubiculum in freely moving rats. *Society for Neurosciences Abstracts*, 10, 599.

Rolls, E.T., and O'Mara, S.M. (1995). View-responsive neurons in the primate hippocampal complex. *Hippocampus*, 5(5), 409–424.

Sargolini, F., Fyhn, M., Hafting, T., McNaughton, B.L., Witter, M.P., Moser, M.B., and Moser, E.I. (2006). Conjunctive representation of position, direction, and velocity in entorhinal cortex. *Science*, 312(5774), 758–762.

Schröder, J., and Pantel, J. (2016). Neuroimaging of hippocampal atrophy in early recognition of Alzheimer's disease—a critical appraisal after two decades of research. *Psychiatry Research*, 247, 71–78.

Sherry, D.F., and Hoshooley, J.S. (2010). Seasonal hippocampal plasticity in food-storing birds. *Philosophical Transactions of the Royal Society of London, Series B*, 365(1542), 933–43.

Strange, B.A., Witter, M.P., Lein, E.S., and Moser, E.I. (2014), Functional organization of the hippocampal longitudinal axis. *Nature Reviews Neuroscience*, 15, 655–669.

Taube, J.S. (2007). The head direction signal: origins and sensory-motor integration. *Annual Review of Neuroscience*, 30, 181–207.

Von Frisch, K. (1953). *The dancing bees: An account of the life and senses of the honey bee.* Harvest Books, New York.

Wiltschko, W., and Wiltschko, R. (1996). Magnetic orientation in birds. *The Journal of Experimental Biology,* 199, 29–38.

Yartsev, M.M., and Ulanovsky, N. (2013). Representation of three-dimensional space in the hippocampus of flying bats. *Science*. 340(6130), 367–72.

15 Birdsong Learning

Introduction and Background

Being able to communicate is one of the great achievements of biology. Indeed, without communication, the ability to interact with other members of one's species would be severely hampered, which could have serious consequences. For example, communication serves in many critical activities in the animal kingdom, such as mating rituals, finding food and shelter, asserting dominance and signalling warnings. Many animals have very sophisticated and varied forms of communication that go beyond human capabilities. These include the amazing ability of honeybees to communicate about food sources through very intricate dances, the use of clicks and whistles of many dolphin and whale species, the use of bioluminescence of some fish and invertebrate species and the distinct warning calls of many species of monkeys.

Language is a distinctively human form of communication. Language has not only allowed humans to communicate effectively with each other, but has allowed us to express and contemplate abstract ideas and thoughts. In a simple conversation, you can speak an average of 180 words a minute. The words used are selected from a mental dictionary of between 60,000 and 120,000 words. This continuous flow of language is not only effortless, but is almost perfect. We do pause at times when speaking and sometimes have to search for the correct word, but it is close to being perfect. This performance in generating language is matched only by the way we comprehend language, its grammar, meaning, sound forms, linguistic structure and even emotional content.

Language holds a special place in the history of neuroscience. Late in the nineteenth century, the famous French physician Paul Broca (1824–1880) reported on a case of a man named Leborge who slowly lost the ability to speak. In fact, by the time Broca saw him in 1861, the man's speech had deteriorated so much that he was only able to utter the word "tan"; the man was given the nickname Tan on account of this. Following the death of Tan, Broca performed an autopsy and removed his brain. What he discovered was a large amount of damage and tissue loss in the frontal lobe of his left hemisphere. This area is now known as *Broca's area* and is critically involved in speech production (see Chapter 2).

Around the same time, a German physician named Carl Wernicke (1848–1905) discovered another important brain region for speech. This region, now referred to as *Wernicke's area*, is located in the superior part of the temporal lobes and plays an important role in speech comprehension (rather than speech production). These discoveries, amongst others, led to the idea of *localisation of function*. This important concept suggests that different parts of the brain are specialised for different functions, an idea that has been repeatedly demonstrated throughout this book.

Infants are born without language, but after a short period of time (approximately 10 months) they begin to babble. Rudimentary language begins to emerge. Typically, a baby hears sounds emanating from its parents' mouth; these sounds are then repeated and remembered. With time, the sounds, in the form of words, match exactly those of their parents. As well as hearing sounds from others (parents, etc.), children depend on hearing the sounds that they themselves emit. Indeed, in cases where children become deaf early in early life, there is often a rapid decline in speech behaviour. How do humans develop language? The closest model we have that resembles language acquisition is birdsong learning.

An Introduction to Songbirds

Half of all bird species belong to the suborder Oscines (of the order Passeriformes, perching birds); the term 'oscine' is derived from the Latin word *oscen*, meaning 'singing bird'. Well-known examples of oscines include the zebra finch (see Figure 15.1), canaries and white-crowned sparrows; these birds are most commonly used in the study of birdsong learning. The male bird tends to sing more often than the female, usually as part of the mating ritual. Many of the brain structures that are thought to underlie singing, as well as the general song system, are better developed in males than in females. For example, one particular brain structure, the higher vocal centre (HVC), is three times the size in male canaries as in their female counterparts; in zebra finches, it is actually eight times larger (Nottebohm and Arnold, 1976).

Figure 15.1 Photograph of zebra finch. Image thanks to Beth A. Vernaleo and Catherine Dooling.

Studies of birdsong learning began with William H. Thorpe (1954) and Peter Marler (1955) in the 1950s. These authors found that if wild birds were taken as hatchlings and tutored with songs of adult birds that were not related, these young birds ended up producing songs that resembled the tutor's songs, and not the songs of their biological parents. This finding would suggest that birdsong is learned, rather than having a strong genetic component. Further evidence posits that hatchlings brought up in isolation produce very abnormal songs. Likewise, children who have grown up in the wild and without any human contact have also demonstrated language issues (for example, see Newton, 2002).

How Songbirds Learn to Sing

The key concept of birdsong learning is that songbirds must hear themselves sing in order to be able to vocalise. If birds are deafened after exposure to tutor songs (typically from the male parent) but before they begin vocalising themselves, they produce an abnormal song. Song learning starts at a stage (early sensorimotor period) that could be considered equivalent to babbling in infants, where low-amplitude sounds that are noncommunicative are produced. This early production is called a *subsong*. During the rehearsal stage (later sensorimotor period) birds do not need a tutor any longer but rely on an 'internal representation' or a memorised template of the song. During this later period, birds use their own auditory feedback and compare it to this internal template. After this, the song is said to be *crystallised*. Therefore, song learning can be divided into a number of distinct behaviours, each having its own distinct neural circuitry:

- The production of motor commands that give rise to sounds of a song.
- The learning and memorisation of sounds, where sound information from the tutor is stored in long-term memory during the sensitive sensorimotor period.
- The evaluation and comparison of the bird's own song (BOS) to the memorised information. Initially, this is a very rough draft, but over time, with rehearsal, this output is refined and matches the tutor song. The song, at this stage, is said to be crystallised.

Figure 15.2a gives an excellent example of how the subsong develops into a crystallised song. Initially, the song doesn't resemble anything like the tutor song – compare the earliest sonograph panel (40 days) to the top tutor panel – but with time (100 days in Figure 15.2a), the juvenile song matches the song of the tutor (compare 100-day panel to top panel). At this stage, the song has become crystallised. How the song develops and becomes crystallised depends very much on the species involved. For example, in sparrows the development of the song takes close to one year, with a clear separation between the sensory stage, the sensorimotor stage and the stage when the song is crystallised (Figure 15.2b, top). However, in the zebra finch the acquisition period is much shorter, approximately 100 days, with the sensory period overlapping with the sensorimotor phase (Figure 15.2b, middle). Canaries develop their song over a two-year period, with multiple overlapping stages (Figure 15.2b, bottom).

Birdsong System in the Brain

Within the bird brain a number of different regions have been identified as playing a critical role in both song production and learning. These regions are often referred collectively as *the song system*. The structures are displayed in Figure 15.3 and may be equivalent to many of the areas involved in language in humans. The HVC and the robust nucleus of the archistriatum (RA) may be considered as part of the motor system. Projections to hypoglossal nucleus (which controls the bird's vocal organ) arise from area RA and originate in the HVC (see Figure 15.3, grey regions). This set of projections is also known as the *posterior descending pathway* (solid black lines) and may be homologous to the motor pathway in humans starting in the motor cortex and projecting to the brain stem. Another set of projections, known as the *anterior forebrain pathway* (dashed lines involving black structures in Figure 15.3) also arises in the HVC but involves area X (which is considered to be similar to the basal ganglia in humans). This, in turn, sends afferent projections to the medial nucleus of the dorsolateral thalamus (DLM) and onto the lateral magnocellular nucleus of the anterior neostriatum (LMAN). The pathway then converges with the posterior descending pathway in the RA. The anterior forebrain pathway may be homologous to projections through the basal ganglia and thalamic structures in the human brain.

Neurophysiology of Song Production

Using the zebra finch as an example, the birdsong typically begins with a number of very

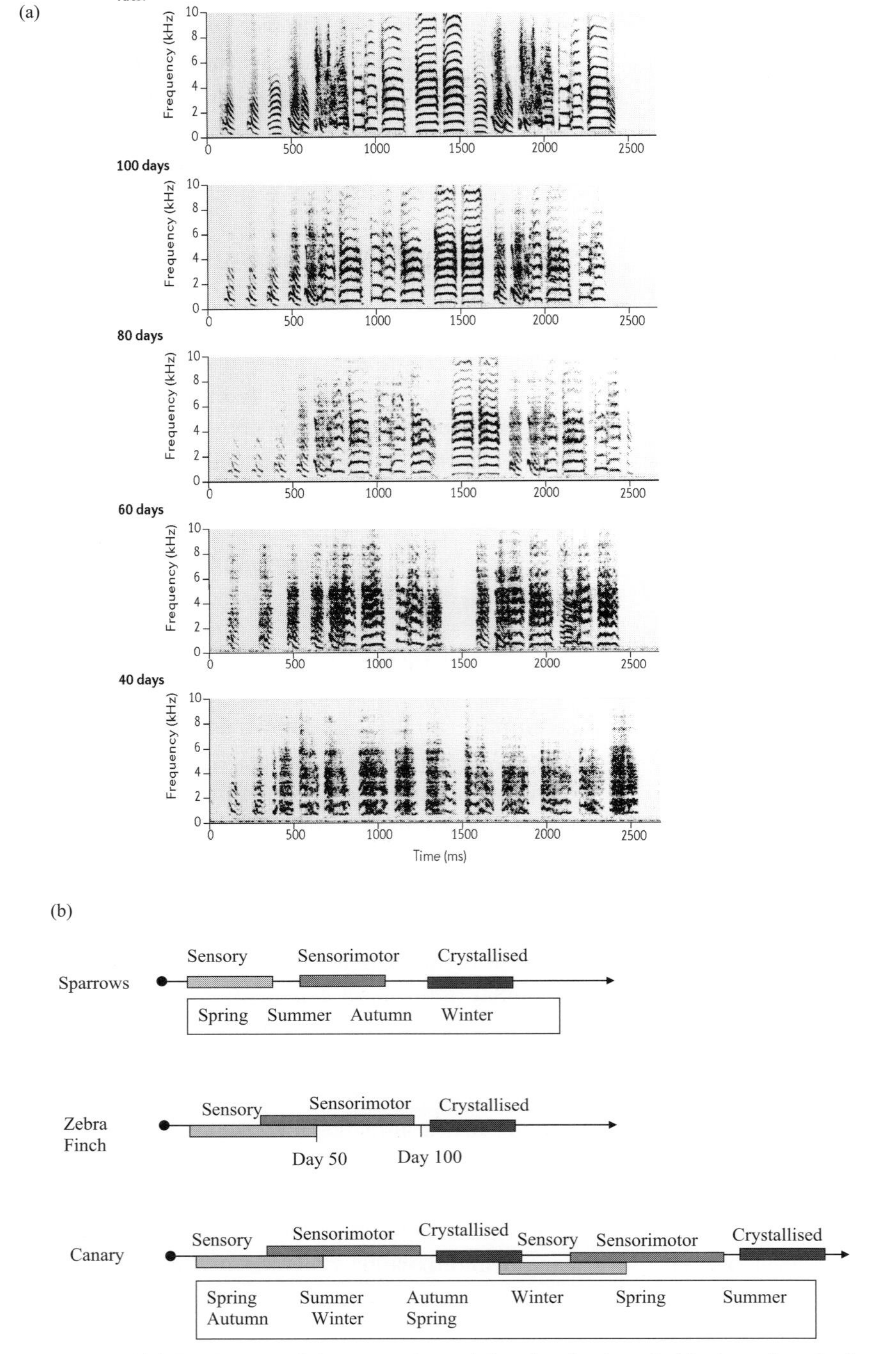

Figure 15.2 (a) Sonograms of the song of an adult zebra finch male (the 'tutor') and of a juvenile at different stages of song learning. First vocalisations (at 40 days after hatching) are called subsong. Gradually over time, the song crystallises and becomes like the tutor's. From J.J. Bolhuis and M. Gahr, Neural mechanisms of birdsong memory, *Nat Rev Neurosci.* (5):347–357, copyright (2006). Reprinted with permission from Macmillan Publishers Ltd. (b) Timeline for song learning across different bird species. From M.S. Brainard and A.J. Doupe, What songbirds teach us about learning, *Nature* 417(6886):351–358. Copyright (2002). Reprinted with permission from Macmillan Publishers Ltd.

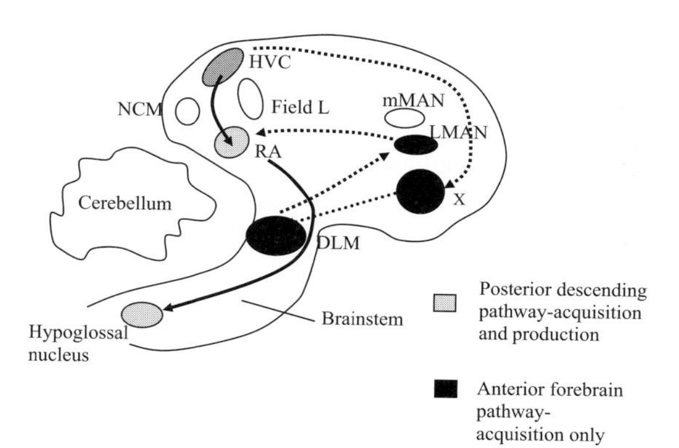

Figure 15.3 Neural circuit of the birdsong system (see text for abbreviations). Key: HVC, higher vocal centre; DLM, medial nucleus of the dorsolateral thalamus; LMAN, lateral magnocellular nucleus of the anterior neostriatum; mMAN, medial magnocellular nucleus of the anterior neostriatum; NCM, nidopallium caudal medial nucleus; RA, robust nucleus of the archistriatum; X, Area X. Diagram adapted from F. Nottebohm (2005), The neural basis of birdsong, *PLoS Biol.* 3(5):e164.

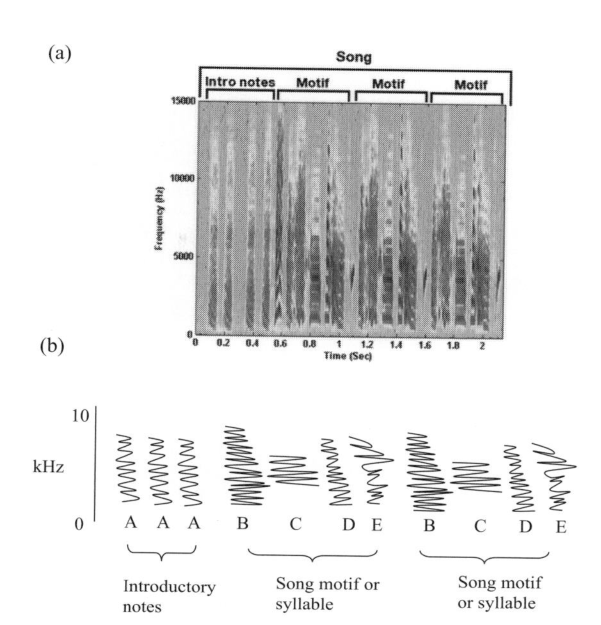

Figure 15.4 (a) A zebra finch song showing introductory notes and song motifs as recorded by Beth A. Vernaleo. (b) A simplified illustration of the song showing the repetition of song patterns.

simple syllables that are identical to each other. These are composed of just one or two notes (introductory notes). This is then followed by what is termed a *song motif* or a sequence of multi-note syllables. These motifs or syllables are repeated a number of times before the song ends. Figure 15.4a shows a song composed of 4 identical introductory notes, which is followed by song motifs. Notice that the notes that make up each motif are quite different from each other in terms of their duration and frequency, but the sequence of notes is repeated exactly three times. Figure 15.4b hopefully illustrates this idea.

The two key structures in song production are the HVC and the RA. There is good evidence demonstrating that HVC projects to the RA, which in turn projects to motor neurons of the vocal cord (see Figure 15.3, grey pathway), which ultimately causes the bird to sing. If you block neural firing in the HVC it abolishes firing in the RA and causes song impairments. During a song, RA neurons fire in a typical bursting fashion of approximately 10 ms in duration. Interestingly, the firing of RA neurons is closely coupled to parts of a song syllable. In the example illustrated in Figure 15.5a, the firing of one RA neuron correlates with motif C and E, while the firing of a second RA neuron correlates with D. HVC cells also fire during a song, but they fire infrequently and do so as bursting spikes; these bursts last approximately 6 ms and are composed of 4 to 5 action potentials per burst. However, HVC neurons fire at a single precise time during the RA firing sequence when they do fire (Hahnloser et al., 2002 and Figure 15.5b). Lesions of the HVC cause worse song impairments compared to lesions of RA. Furthermore, stimulation of the HVC causes disruption of the entire song, whereby the bird must start again. In comparison, stimulation of the RA results in disruption of the structure of syllables only (Vu et al., 1994). This might suggest that the HVC neurons may be seen as a type of 'grandmother' cell, i.e. a cell encoding the entire song rather than only elements of it. It is also thought that HVC may be involved in the timing of the song, perhaps triggering the initiation or termination of syllables (Mooney, 2009).

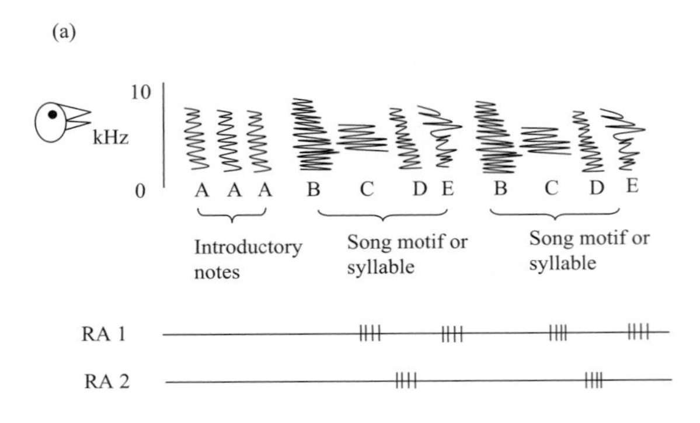

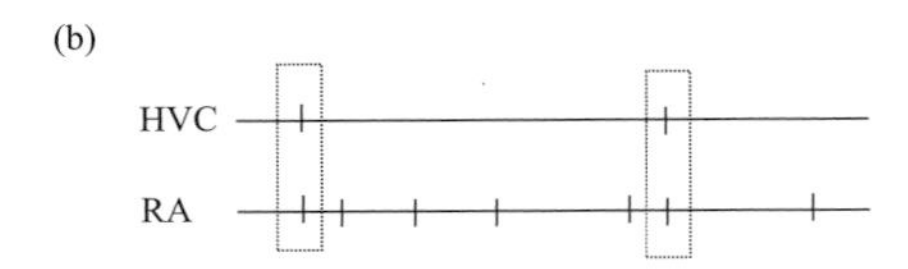

Figure 15.5 (a) Single neurons in RA correlate with specific parts of a song syllable. (b) HVC neurons fire less frequently but are timed with firing in the RA.

Neurophysiology of Song Learning

Songbirds, like children learning a language, learn their songs through trial and error. The juvenile songbird's vocal output is very variable and riddled with mistakes. However, over time, by comparing their own output to that of their tutor, the young songbird's output converges on a stable song. How does this happen? We saw earlier that the posterior descending pathway (black arrows in Figure 15.3) is responsible for motor production, but it is considered that the anterior forebrain pathway (AFP; dashed arrows in Figure 15.3) is critical for the acquisition and learning, and not for the production of a song. For example, lesions of structures along the AFP of juvenile zebra finches have major effects on song development, whereas similar lesions done to adult birds show little long-term effects (Bottjer et al., 1984). One structure thought to be critical in both production and acquisition of song is the RA. The RA, as well as receiving a direct input from the HVC, also receives input from the lateral magnocellular nucleus of the anterior nidopallium (LMAN, part of the AFP stream). It is thought that this input serves to both *control* and *modify* the motor output (see Figure 15.6). Therefore, the LMAN-RA pathway may modulate the juvenile song so that the true adult song becomes encoded and strengthened along the motor pathway. Lesions of the LMAN do not affect ongoing songs, and stimulation of this region does not elicit vocalisation in a quiescent bird. However, Kao and Brainard (2006) demonstrated that LMAN stimulation can modulate the structure of the individual syllables without changing the order. Such modulation can take the form of increases or decreases in sound amplitude or frequency of an individual syllable. Further, Olveczky et al. (2005) demonstrated that inactivation of the LMAN reduced the variability in the songs of zebra finches (Figure 15.6b). The LMAN may therefore serve to add jitter or variability to a song – this is necessary for learning to occur. Olveczky et al. (2005) also showed that this input depends on the NMDA receptor, which is known to play an important role in synaptic plasticity (see also Chapter 7). But how does this added variability help to match the bird's own song to that of the tutor? One suggestion is that the bird retains the sounds from the variable pattern that most resembles the tutor's song.

Song Stored Stored in Memory

Songbirds use both tutor song memories and their own song to modify their vocal output, before the songs become crystallised. However, as the sensory period and the sensorimotor period can be separated by up to 10 months (e.g. with sparrows, Figure 15.2), this suggests that the tutor song must be stored in long-term memory. There is good evidence that one structure, the nidopallium caudal medial (NCM) nucleus, which lies outside the traditional songbird system, is a good candidate for such a storage site. An early study by Mello et al. (1992) showed that exposure of zebra finches or canaries to a conspecific (e.g. tutor) song led to increased immediate-early gene activation (IEG, *zif-268*, a marker of neural activity)

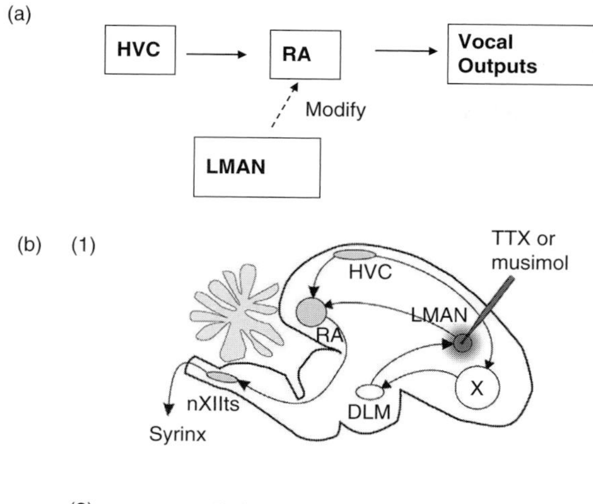

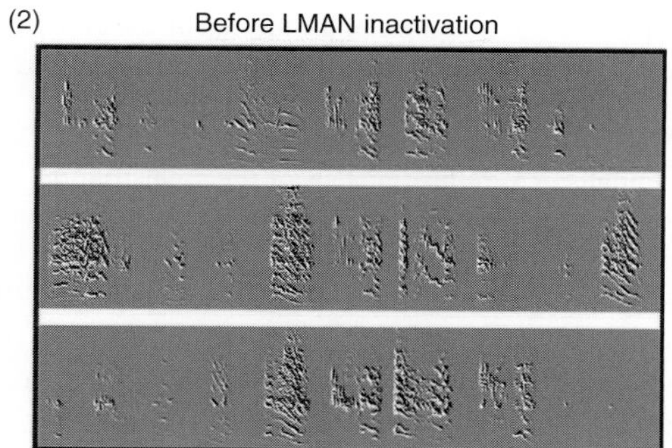

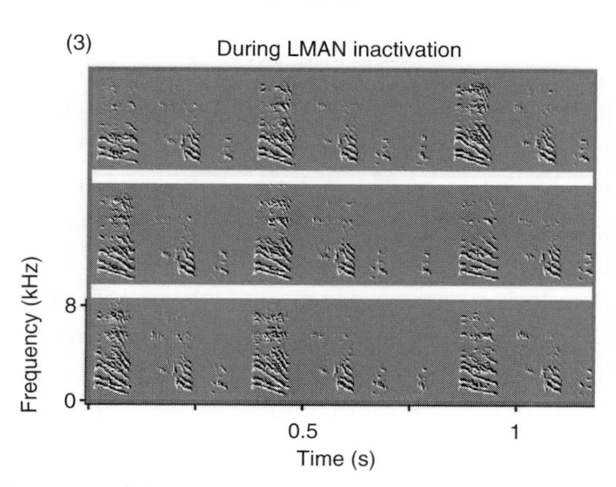

Figure 15.6 (a) LMAN may serve to modify the RA. (b) Inactivation of the LMAN by TTX or other pharmacological treatments causes the song to lose its jitter and variability. Notice how (3) is more structured than (2). Image from B.P. Olveczky, et al. (2005), Vocal experimentation in the juvenile songbird requires a basal ganglia circuit, *PLoS Biol.* 3(5):e153 with permission under Creative Commons Attribution License (http://creativecommons.org/licenses/by/4.0/).

in the NCM. This increase was not observed if the birds were presented with either pure tones or heterospecific songs. Further evidence using IEG activation demonstrates that there is a good correlation between the strength of song learning (as measured by the proportion of song elements copied from the tutor song) in the finch and Zif-268 expression in the NCM (Bolhuis et al., 2000). More recently, Gobes and Bolhuis (2007) showed that lesions of the NCM impaired recognition of the tutor song, as measured by the amount of time spent by the bird near the speaker broadcasting the tutor song compared to a novel one (Figure 15.7a). Similar lesions, however, did not affect the production of the bird's own song (Figure 15.7b). This provides strong evidence that the NCM may store the memory of the tutor song. Interestingly, Terpstra et al. (2004) showed an increase in IEG activation in the NCM of an adult following exposure to songs it heard as a juvenile, again pointing to a role for this structure in memory.

Bird's Own Song Compared to Auditory Feedback

As well as comparing the bird's own song (BOS) to a memory of the tutor song, the bird must be able, over time, to modify its output by using its own auditory feedback. However, differentiating whether a bird is modifying its output in accordance to its own auditory feedback or to a memorised template is very difficult. It is generally accepted that many structures of the birdsong system respond to the auditory presentation of the bird's own song, suggesting a strong auditory input. For example, neurons in the HVC (as well as in Area X and the RA) respond stronger to the BOS than to the tutor song (Solis et al., 2000) or to the song of a conspecific bird. Further, this selectivity of BOS increases from Field L (the equivalent to the primary auditory structure in humans) to HVC, whereby Field L shows little selectivity but the HVC shows very strong selectivity. Field L, in contrast, responds more strongly to conspecific songs rather than to

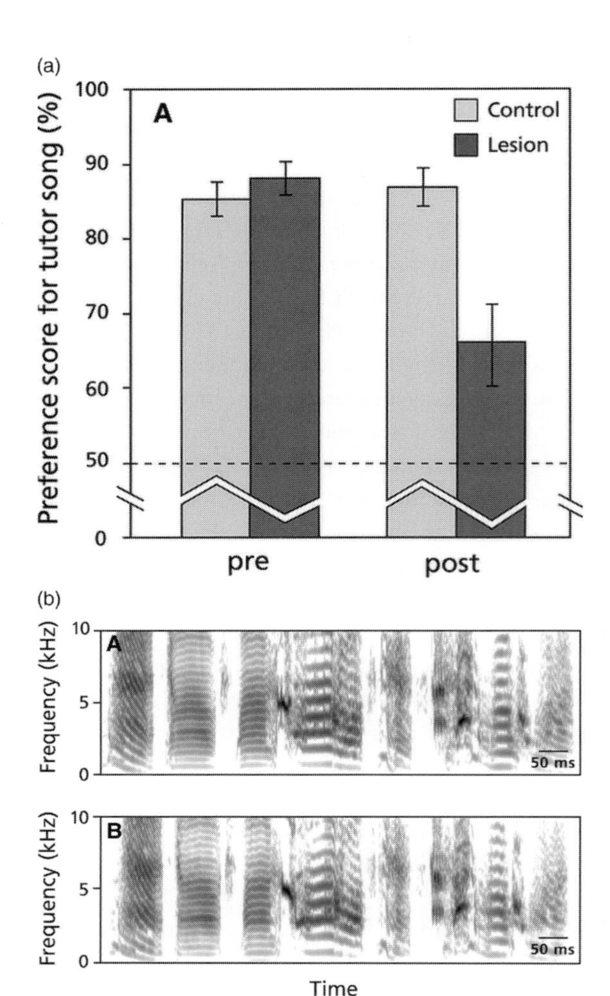

Figure 15.7 (a) Lesions of the NCM impaired recognition of the tutor song (compare post- to pre-lesion panels – dark grey bars). (b) Spectrograms of one motif of an experimental bird before (top) and 8 weeks after (bottom) NCM lesioning. There were no changes in song production after NCM lesions. S.M. Gobes and J.J. Bolhuis (2007), Birdsong memory: a neural dissociation between song recognition and production, *Curr Biol.* 17(9):789–93, with permission under Creative Commons Attribution License (http://creativecommons.org/licenses/by/4.0/).

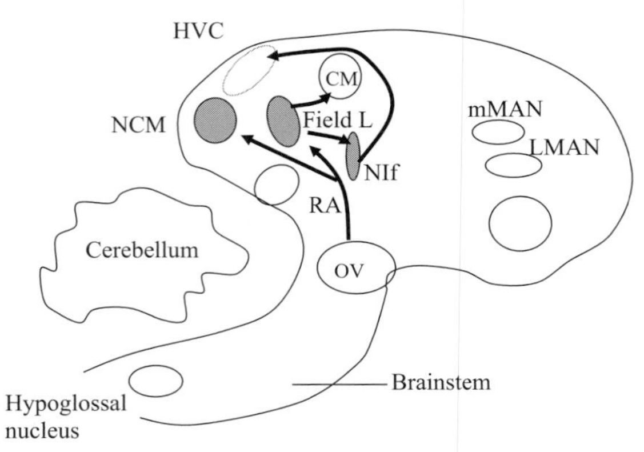

Figure 15.8 Main areas that are thought to be involved in audition (grey regions). Key: CM, caudal mesopallium; HVC, higher vocal centre; LMAN, lateral magnocellular nucleus of the anterior neostriatum; mMAN, medial magnocellular nucleus of the anterior neostriatum; NIf, nucleus interface; NCM, nidopallium caudal medial nucleus; OV, ovoidalis; RA, robust nucleus of the archistriatum.

the BOS. In an interesting study by Volman (1993), the author showed that in the white-crowned sparrow (where there is no overlap between the sensory phase and the sensorimotor phase, see Figure 15.2), neurons in the HVC start to respond strongly to the BOS during the sensorimotor phase, i.e. the phase when the birds start singing themselves. This idea of plasticity in the HVC has been confirmed more recently by Nick and Konishi (2005). These authors demonstrated that HVC neurons in the zebra finch are more responsive towards the tutor song 40 to 50 days after hatching, but this responsive pattern then switches, with a preference towards the BOS 70 days after hatching.

Another structure that exhibits a strong responsiveness to BOS is the nucleus interface (NIf; see Figure 15.8). As well as providing a primary input into the HVC, the neural activity of the NIf is time-locked to the production of the song. Thus, the NIf is in a strong position to act as both an auditory area as well as a premotor region. However, recent evidence suggests that the NIf is necessary for auditory processing and not for song production. This has been demonstrated by Cardin et al. (2005), who showed that lesions to the NIf reduce the auditory activity in the HVC but, more importantly, that the sound production was not affected. As Field L does not project directly to the HVC, auditory information from Field L must travel to the HVC via the NIf (see Figure 15.8). More research is required to determine the exact role of the NIf.

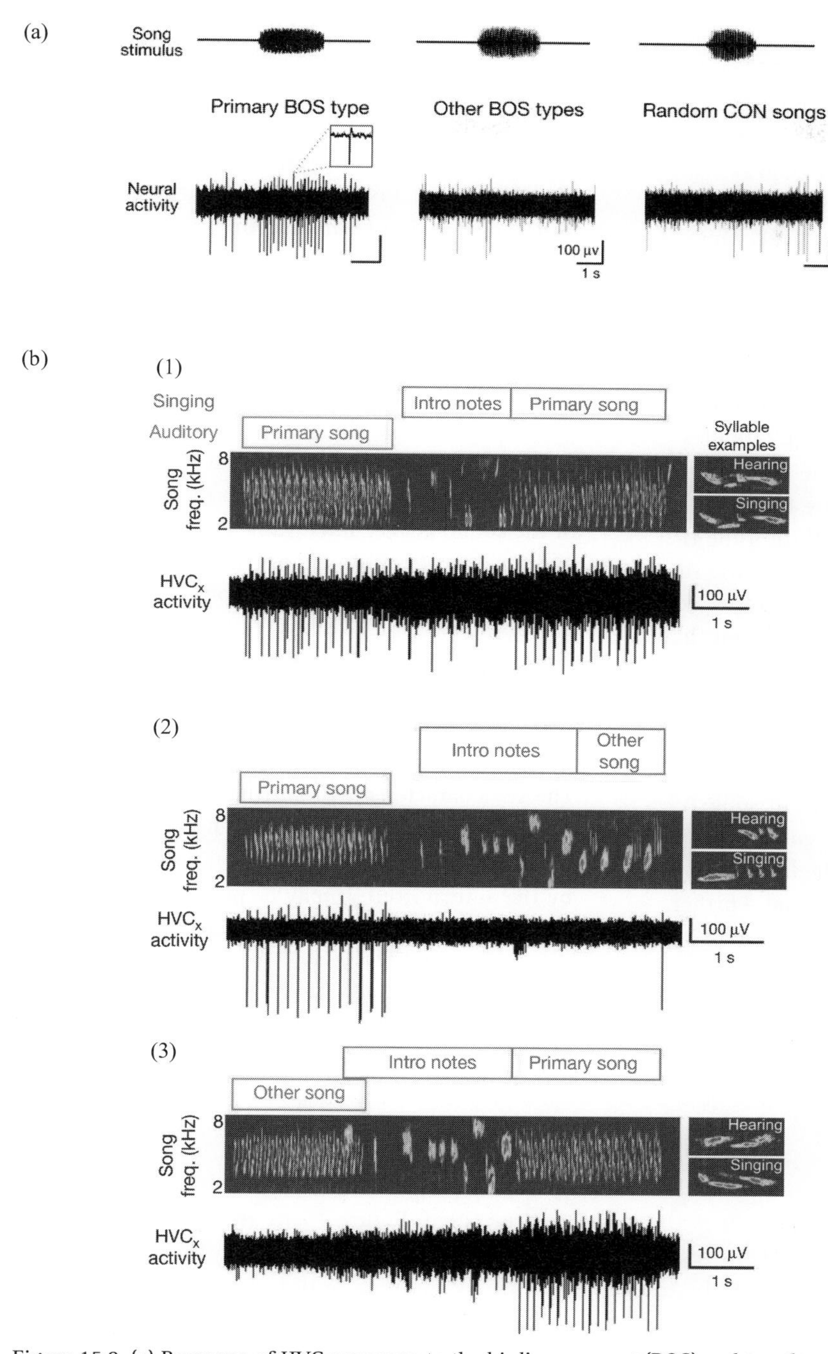

Figure 15.9 (a) Response of HVCx neurons to the bird's own song (BOS) and to other song types. CON = conspecific. (b) Response of HVCx neurons to both listening and singing (top), listening only (middle) and singing only (bottom). Top panel of each represents the sonograph of the song with the neural activity displayed in the lower panel. From J.F. Prather et al. (2008), Precise auditory-vocal mirroring in neurons for learned vocal communication, *Nature*. 451(7176):305–10. Copyright (2008). Reprinted with permission from Macmillan Publishers Ltd. For colour version, please refer to the plate section.

Auditory–Vocal Integration

If we think about how we learn to talk, we realise that there needs to be a strong correlation between what we hear and what we speak. The goal is to try and match the words/sounds we hear to the words/sounds that we speak. This is particularly important in songbirds, as the male bird must sing stable songs to attract a mate. As such, the birds need to rely on auditory feedback to maintain this stability – requiring a strong correspondence between the auditory input and vocal output. Prather et al. (2008) recently provided evidence that the HVC may be an excellent candidate structure for where such integration may take place. Within the HVC there are two distinct sets of neurons, those that project to area X (HVCx) and those that project to the RA (HVCra). Prather et al. (2008) found that many HVCx neurons (and not HVCra neurons) respond to a specific song from a selection of the bird's own song repertoire when these songs are replayed to the birds on loudspeakers (sparrows tend to sing 3 to 5 different songs); these neurons did not respond to conspecific songs or indeed other songs of the bird's own repertoire (Primary BOS, Figure 15.9a). Further, Prather et al. (2008) showed that some individual HVCx neurons responded to both listening to and singing the same song (Figure 15.9b [1]). Other neurons were active when the bird heard (Figure 15.9b [2]) or sang (Figure 15.9b [3]) the BOS, but were silent when the bird either heard or sang other types of songs. The firing rate of the HVC neurons is slightly different when the bird is responding to an auditory input from when it is singing the primary song; auditory-related activity tends to produce single action potentials whereas singing-related activity tended to produce a burst of firing (compare (2) to (3) in Figure 15.9b). However, both types of neurons fire at the identical syllables of a song.

The firing of the same HVC neuron to the auditory input of the song and to the actual singing of the sound is very interesting and may help in auditory-vocal coordination. Furthermore, such neurons resemble a special type of neurons called *mirror neurons*, first discovered in area F5 of the monkey by Rizzolatti and colleagues (Rizzolatti and Craighero, 2004; see also Roche and Commins, 2009). Area F5 is in the prefrontal cortex and is thought to be analogous to Broca's area in humans. The key characteristic of mirror neurons is that they fire both when the animal sees a particular action being done and also when the action is performed by the animal itself. It may be possible that neurons in the bird's HVC perform a similar function and therefore are critical in the song learning process.

Summary

Birdsong learning is very complex and although many elements of the system have been discovered, there are many gaps that remain to be filled. Figure 15.10 attempts to summarise some of the key features that have been discussed in this chapter. The higher vocal centre (HVC) seems to be the key structure involved in both learning and song production. The structure also contains very specialised neurons that respond to both song production and auditory input. Furthermore, the HVC gives rise to two critical projections, one to the RA that produces the song and the other projection to structure X, which via the LMAN is involved in learning and modifying the song. The question arises: How do neurons within the HVC interact? How do all the elements come together? One idea is that HVC_{ra} neurons send a copy of the song to HVCx neurons (1), where the motor copy and the auditory version of the song (from NIf) can be compared and modified (2), if necessary. Further, recent evidence posits that inhibitory interneurons within the HVC may play an important role (white neurons, 3). In this process these may serve to gate or switch neurons from an auditory role to a motor role; they may also serve to produce sparse bursts on the RA projection, refining elements of the song; they may also help select the specific BOS from the bird's full repertoire; or they may ensure the stability of the BOS through auditory feedback.

The details regarding each of these elements are still to be worked out. Other issues, such as how the bird compares its BOS to the stored memory of the tutor song, suggested

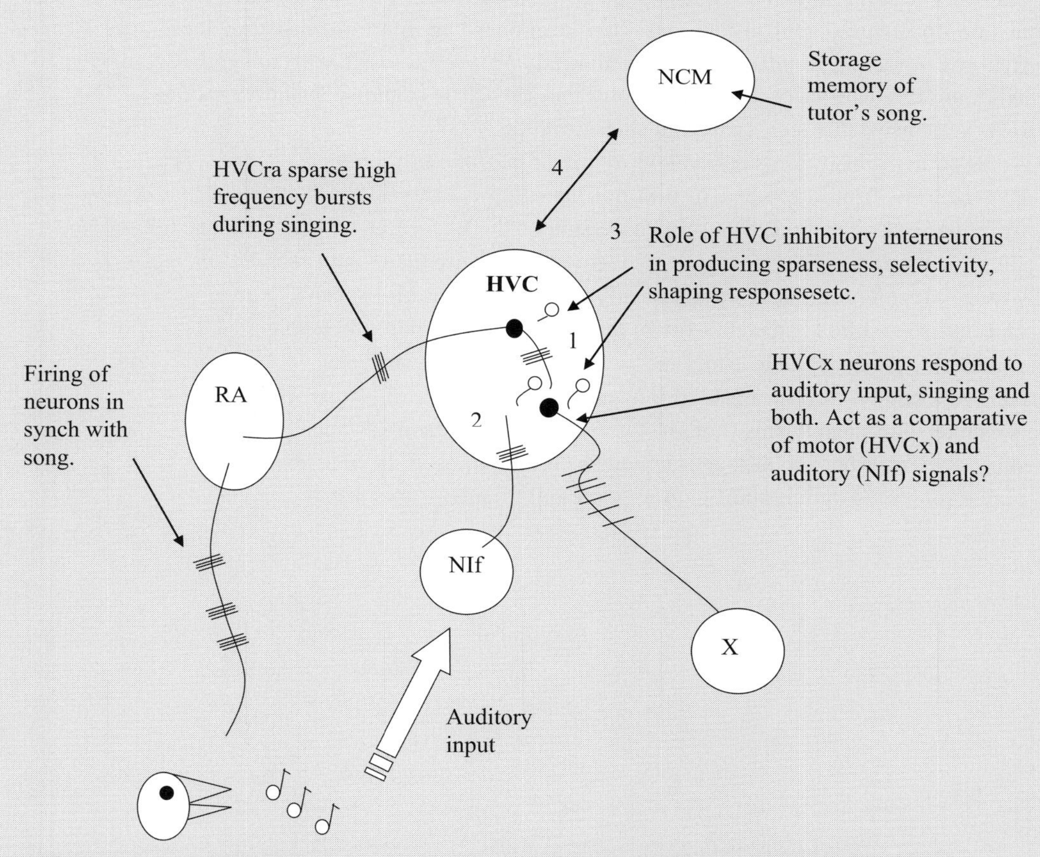

Figure 15.10 Summary diagram of the various structures and how they might interact during songbird learning (see text for details). Key: NCM, nidopallium caudal medial nucleus; NIf, nucleus interface; RA, robust nucleus of the archistriatum; X, area X.

to be in the NCM, also require further elucidation (4). The LMAN may play a role in this (Mooney, 2009). Other areas such as CM (caudal mesopallium; see Figure 15.8) have recently been shown to play an important role in driving auditory activity in NIf and HVC, as neurons in this region are active during singing and during the presentation of other birds' songs. This area, among others, is currently under investigation (Bauer et al., 2008).

Questions and Topics Under Current Investigation

- Known language genes in humans e.g. *FOXP2* are also expressed in finches and canaries, especially during periods of vocal plasticity (Scharff and Haesler, 2005). What other genes are involved, and what role do they play in song learning?
- Birds tend to sing at first light of day and during springtime. What role does the birds' circadian rhythm play in song learning? For example, it is known that different levels of melatonin (Bentley et al., 1999) and other hormones can interfere with song learning.
- Neurons in Field L respond stronger to natural conspecific songs than to synthetic ones. What are the features of the natural song that cause the response in these neurons?
- How are conspecific songs distinguished from the BOS?
- Are HVC neurons in birds equivalent to mirror neurons in primates?
- HVC neurons fire in response to BOS but the firing is sparse. Why is it sparse? Is the HVC firing pattern an enhanced or refined version of the BOS received from NIf, or are they simplified representations, so that easy comparisons are made with relevant memories?
- Given that songs can develop differently across different bird species and that some birds (e.g. zebra finch) are more sociable than others (e.g. canaries), how does the learning mechanism differ across species?
- NMDA is known to play a role in long-term memories of birdsongs. What are the molecular mechanisms thought to underpin song learning? There is also some evidence that the mitogen-activated protein kinase/ERK pathway might play a role (see London and Clayton, 2008 and Chapter 7 and 10).

References

Bauer, E.E., Coleman, M.J., Roberts, T.F., Roy, A., Prather, J.F., and Mooney, R. (2008). A synaptic basis for auditory-vocal integration in the songbird. *Journal of Neuroscience*, 28(6), 1509–1522.

Bentley, G.E., Van't Hof, T.J., and Ball, G.F. (1999). Seasonal neuroplasticity in the songbird telencephalon: A role for melatonin. *Proceedings of the National Academy of Sciences of the USA*, 96(8), 4674–4679.

Bolhuis, J.J. and Gahr M. (2006). Neural mechanisms of birdsong memory. Nature Review Neurosciences, 5, 347–357.

Bolhuis, J.J., Zijlstra, G.G., den Boer-Visser, A.M., and Van Der Zee, E.A. (2000). Localized neuronal activation in the zebra finch brain is related to the strength of song learning. *Proceedings of the National Academy of Sciences of the USA*, 97(5), 2282–2285.

Bottjer, S.W., Miesner, E.A., and Arnold, A.P. (1984). Forebrain lesions disrupt development but not maintenance of song in passerine birds. *Science*. 224(4651), 901–933.

Brainard, M.S., and Doupe, A.J. (2002). What songbirds teach us about learning. *Nature*, 417(6886):351–358.

Cardin, J.A., Raksin, J.N., and Schmidt, M.F. (2005). Sensorimotor nucleus NIf is necessary for auditory processing but not vocal motor output in the avian song system. *Journal of Neurophysiology*, 93(4), 2157–2166.

Gobes, S.M., and Bolhuis, J.J. (2007). Birdsong memory: a neural dissociation between song recognition and production. *Current Biology*, 17(9), 789–793.

Hahnloser, R.H., Kozhevnikov, A.A., and Fee, M.S. (2002). An ultra-sparse code underlies the generation of neural sequences in a songbird. *Nature*, 419(6902), 65–70.

Kao, M.H., and Brainard, M.S. (2006). Lesions of an avian basal ganglia circuit prevent context-dependent changes to song variability. *Journal of Neurophysiology*, 96(3), 1441–1455.

London, S.E., and Clayton, D.F. (2008). Functional identification of sensory mechanisms required for developmental song learning. *Nature Neuroscience*, 11, 579–586.

Marler, P. (1955). Characteristics of some animal calls. *Nature*, 176 (4470), 6–8.

Mello, C.V., Vicario, D.S., and Clayton, D.F. (1992). Song presentation induces gene expression in the songbird forebrain. *Proceedings of the National Academy of Sciences of the USA*, 89(15), 6818–6822.

Mooney, R. (2009). Neurobiology of song learning. *Current Opinion in Neurobiology*, 19, 654–660.

Newton, M. (2002). *Savage boys and wild girls: A history of feral children*. Faber and Faber, London.

Nick, T.A., and Konishi, M. (2005). Neural auditory selectivity develops in parallel with song. *Journal of Neurobiology*, 62(4), 469–481.

Nottebohm, F. (2005). The neural basis of birdsong. *PLoS Biology*, 3(5), e164.

Nottebohm, F., and Arnold, A.P. (1976). Sexual dimorphism in vocal control areas of the songbird brain. *Science*, 194 (4261), 211–213.

Olveczky, B.P., Andalman, A.S., and Fee, M.S. (2005). Vocal experimentation in the juvenile songbird requires a basal ganglia circuit. *PLoS Biology*, 3(5), e153.

Prather, J.F., Peters, S., Nowicki, S., and Mooney, R. (2008). Precise auditory-vocal mirroring in neurons for learned vocal communication. *Nature*, 451(7176), 305–310.

Rizzolatti, G., and Craighero, L. (2004). The mirror neuron system. *Annual Review of Neuroscience*, 27, 169–192.

Roche, R.A.P., and Commins, S. (2009). *Pioneering studies in cognitive neuroscience*. McGraw Hill.

Scharff, C., and Haesler, S. (2005). An evolutionary perspective on FoxP2: strictly for the birds? *Current Opinion in Neurobiology*, 15(6), 694–703.

Solis, M.M., Brainard, M.S., Hessler, N.A., and Doupe, A.J. (2000). Song selectivity and sensorimotor signals in vocal learning and production. *Proceedings of the National Academy of Sciences of the USA*, 97(22), 11836–11842.

Terpstra, N.J., Bolhuis, J.J., and den Boer-Visser, A.M. (2004). An analysis of the neural representation of birdsong memory. *Journal of Neuroscience*, 24(21), 4971–4977.

Thorpe, W. (1954). The process of song-learning in the chaffinch as studied by means of the sound spectrograph. *Nature*, 173(4402), 465–469.

Volman, S.F. (1993). Development of neural selectivity for birdsong during vocal learning. *Journal of Neuroscience*, 13(11), 4737–4747.

Vu, E.T., Mazurek, M.E., and Kuo, Y.C. (1994). Identification of a forebrain motor programming network for the learned song of zebra finches. *Journal of Neuroscience*, 14(11 Pt 2), 6924–6934.

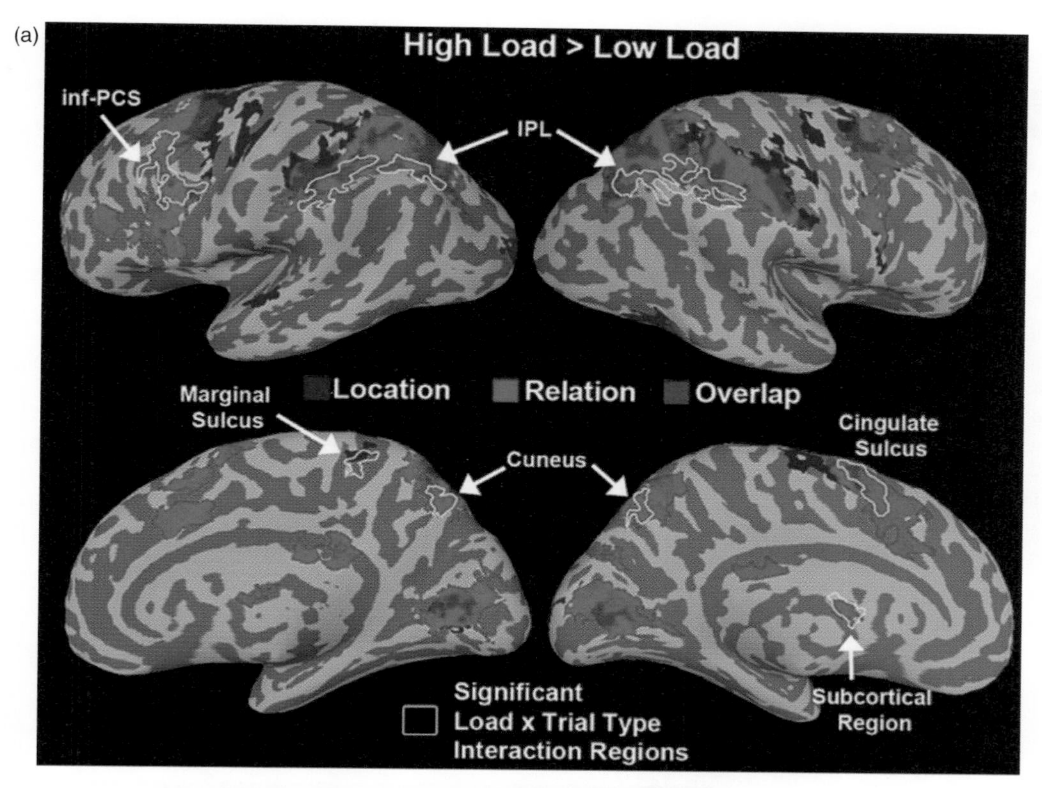

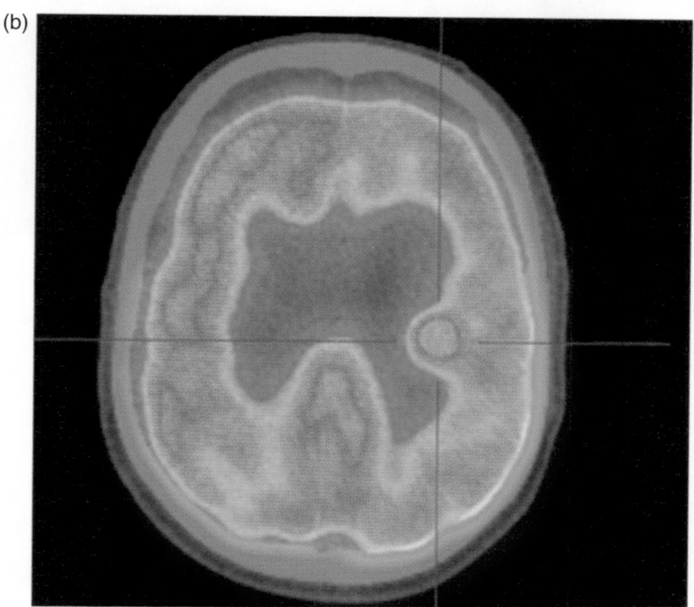

Figure 3.2 (a) An fMRI scan showing regions that were significantly more active for high vs. low load during the delay period of a working memory task. Image taken from K.J. Blacker and S.M. Courtney (2016), Distinct neural substrates for maintaining locations and spatial relations in working memory. *Front. Hum. Neurosci.* 10:594 (http://creativecommons.org/licenses/by/4.0/). (b) PET scan of a patient with cerebral lymphoma (cross hairs). Image taken from Lau et al. (2010), Comparative PET study using F-18 FET and F-18 FDG for the evaluation of patients with suspected brain tumour. *J Clin Neurosci.* 17(1):43–49 with permission from Elsevier.

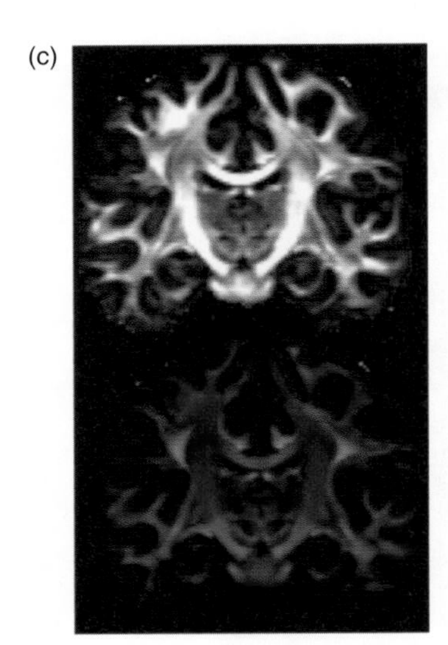

Figure 3.1 (c) A DTI scan showing how water diffuses in the brain. Diffusion directions are RGB colour encoded. Red, left–right; green, anterior–posterior; blue, inferior–superior. Image courtesy of the WU-Minn HCP consortium. (http://humanconnectome.org).

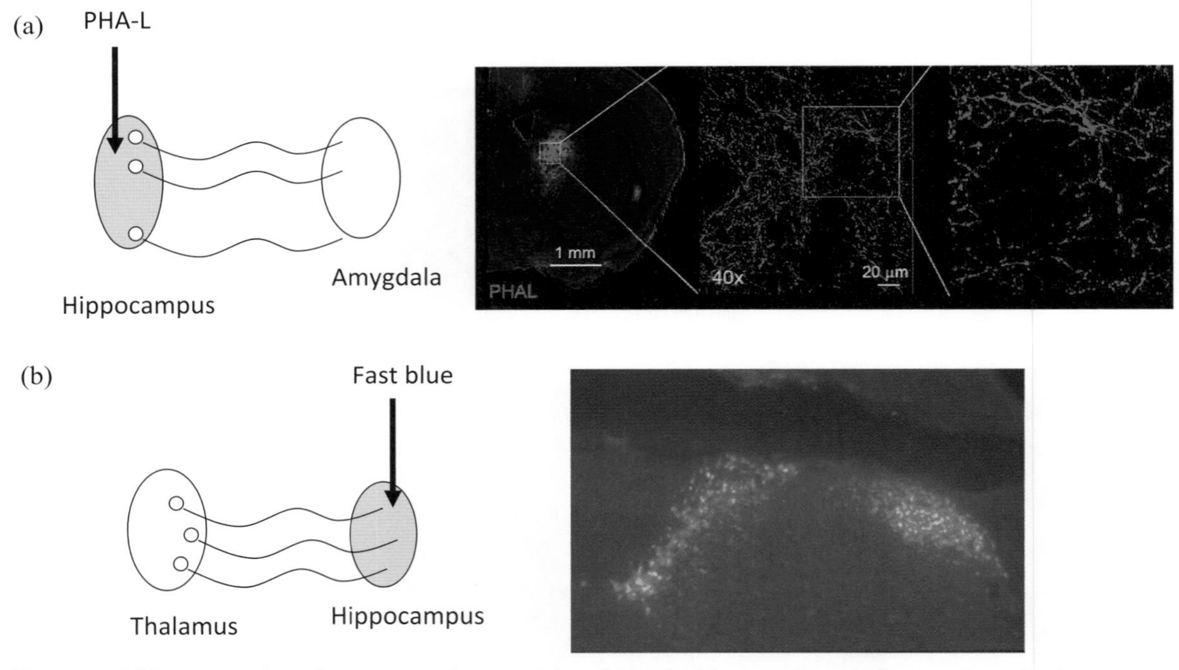

Figure 3.6 (a) Representation of an anterograde tracer injected into the hippocampus, which traces its projections to the amygdala (left). Increasing high- resolution confocal imaging at 40x magnification showing PHA-L-labelled axons (right). Image reprinted by permission from Macmillan Publishers Ltd: Nature Neuroscience, from Hintiryan, H. et al. (2016), The mouse cortico-striatal projectome. *Nat Neurosci.* 19(8):1100–1114), copyright 2016. (b) Representation of a retrograde tracer injected into the hippocampus, which traces its projections back to the thalamus (left). Example of cells in the thalamus labelled with fast blue following injection into hippocampal region (right). Image thanks to Drs Maria Cabellero-Bleda and Miroljub Popović, University of Murcia, Spain.

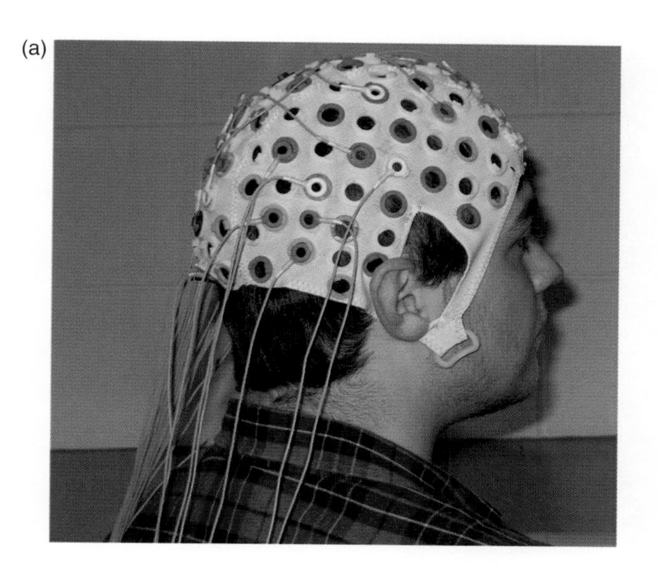

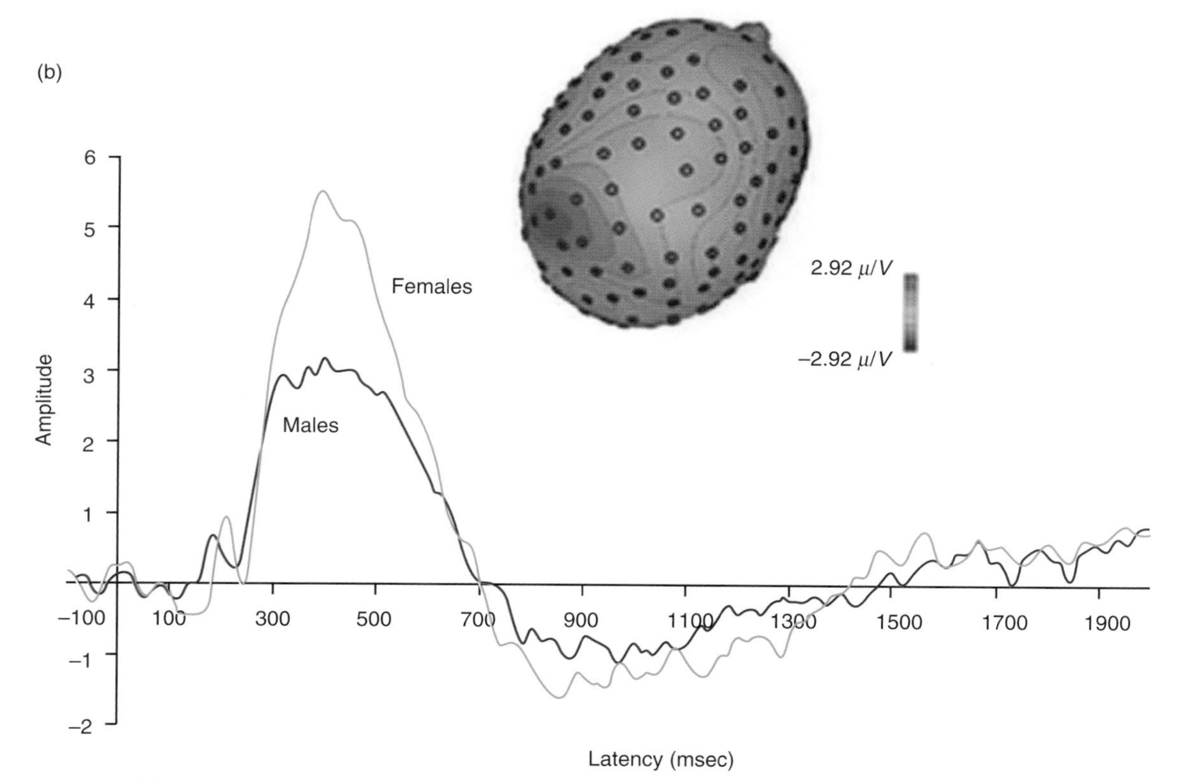

Figure 3.7 (a) Example of high-density EEG array. Image thanks to Mr. Jamie Cummins. (b) Averaged ERP waveforms over occipital regions for male and female participants following a spatial memory task. Reprinted from Murphy et al. (2009), High-resolution ERP mapping of cortical activation related to implicit object-location memory, *Biol. Psychol.* 82(3):234–245, with permission from Elsevier.

Figure 5.1 (a) *Aplysia californica* emitting an ink cloud. Image taken from G. Anderson, Marine Science, Santa Barbara City College, http://creativecommons.org/licenses/by/4.0.

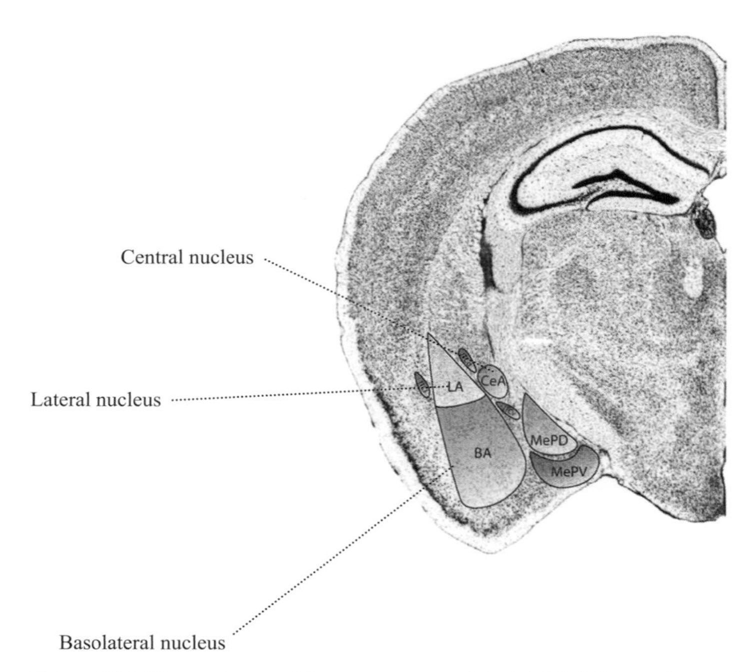

Figure 10.4 The anatomy of the mouse amygdala, highlighting various subregions, including basolateral (BA), lateral (LA) and central (CeA) nuclei. Image from Rob Hurt 2016, reproduced under Creative Commons Attribution License (CCA).

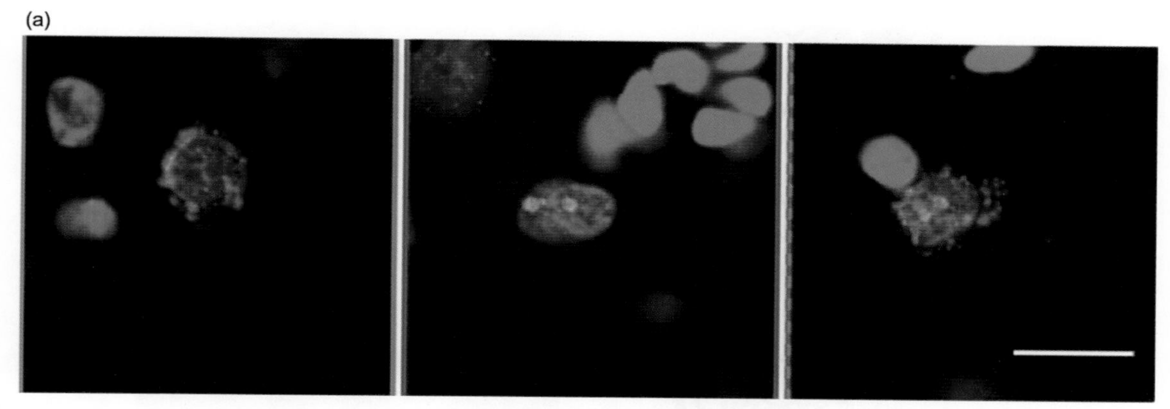

Figure 11.7 (a) Saccharin (CS) produces *Arc* in the cytoplasm (red, left) of amygdalar neurons; LiCl (US) induces *Arc* in the nucleus (middle); CS + US induce *Arc* in both the cytoplasm and nucleus (right).

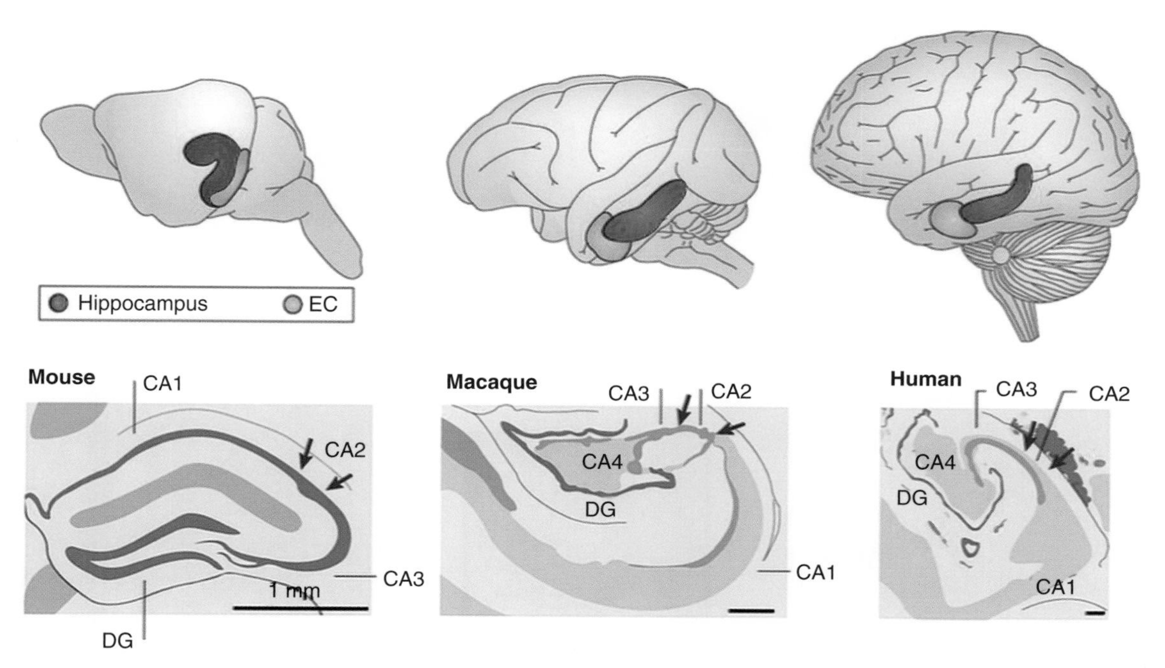

Figure 14.2 (top) Location of hippocampus and entorhinal cortex (EC) in mouse, macaque and human brains. (bottom) Section through the hippocampal region of each species, showing the various subregions. DG, dentate gyrus; CA1–4, cornu ammonis 1–4. From Strange et al. (2014), Functional organization of the hippocampal longitudinal axis, *Nature Reviews Neuroscience*, 15, 655–669. Reprinted by permission from Macmillan Publishers Ltd.

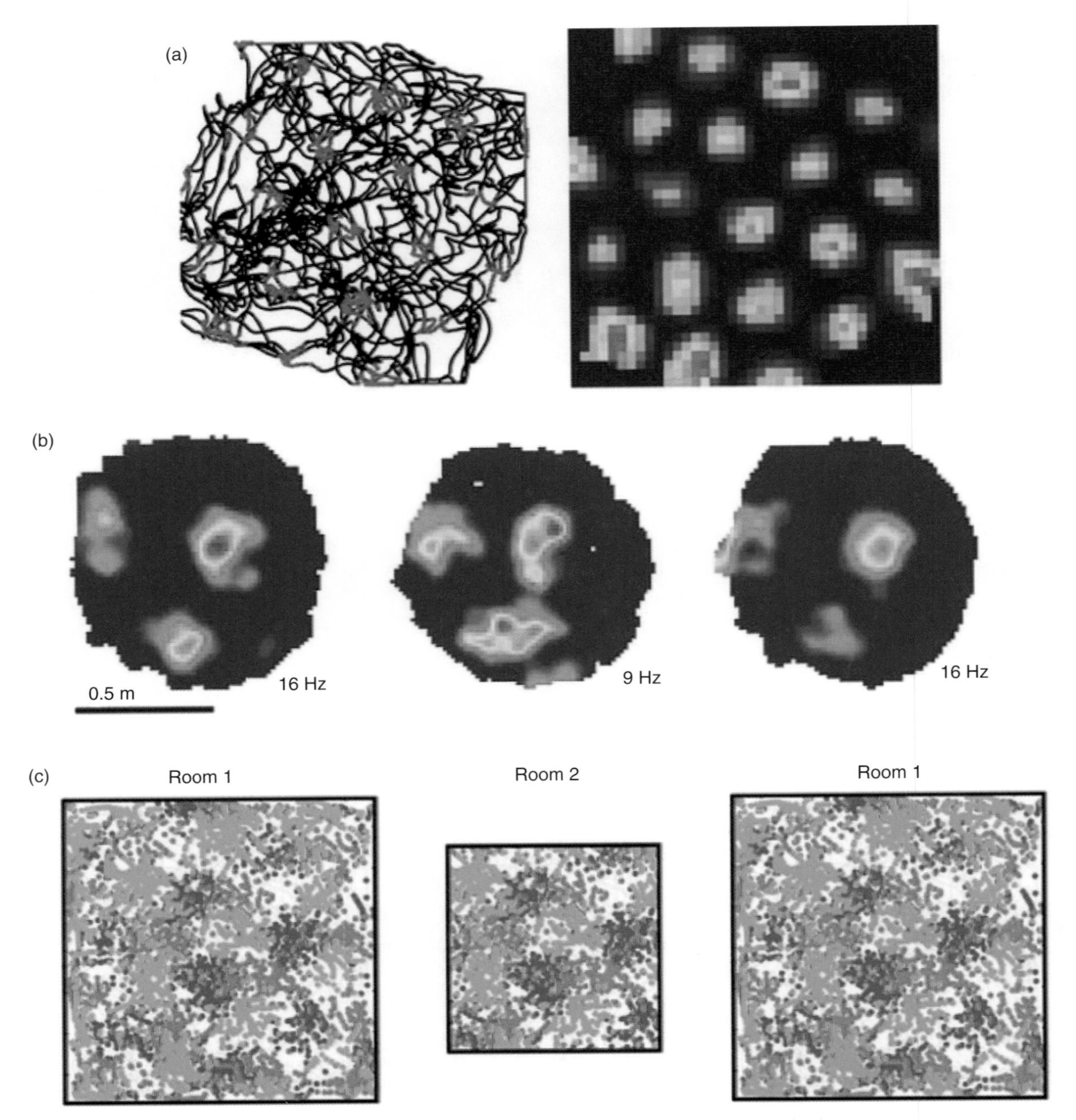

Figure 14.7 (a) Example of a grid cell. A neuron in the medial entorhinal cortex fires in multiple areas in a gridlike pattern (as represented by the red dots, left, and the red/yellow colours, right) as the animal traverses a square arena. Image taken from R.M. Grieves and J.K. Jeffery (2017), The representation of space in the brain, *Behav Processes* 135:113–131, with permission from Elsevier. (b) Grid cells maintain their firing pattern in the dark. Light condition (left), Dark condition (middle), Light condition (right). (c) Grid cells maintain their firing pattern across different environments. Images (b) and (c) from Professor Edvard Moser's Nobel lecture. Copyright © The Nobel Foundation (2014); source: http://Nobelprize.org.

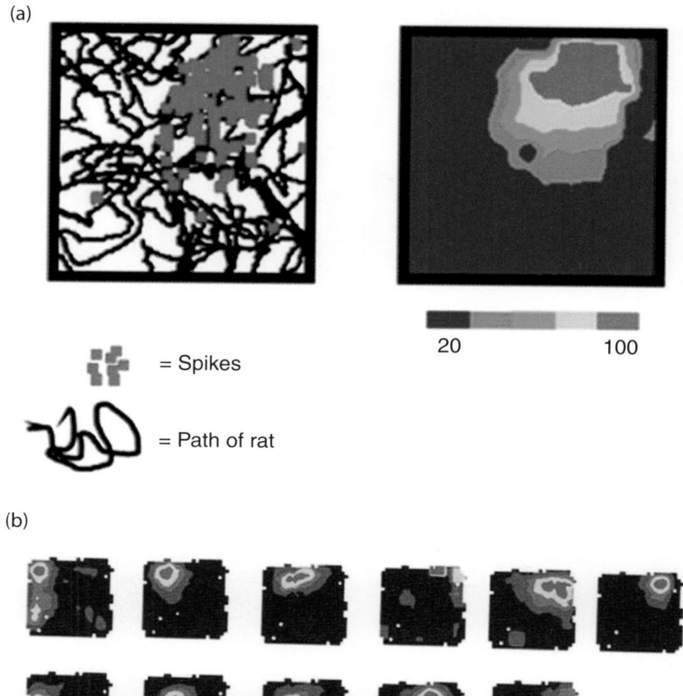

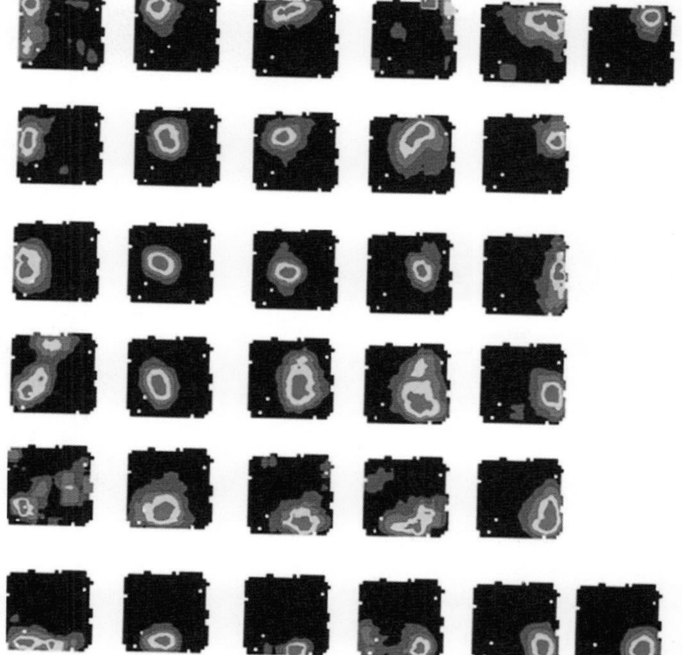

Figure 14.4 (a) Example of a place cell. A single neuron in the rat hippocampus fires (red dots, left image) when an animal is at a specific location in the square arena. Image taken from R.M. Grieves and K.J. Jeffery (2017), The representation of space in the brain, *Behav Processes*, 135:113–131, with permission from Elsevier. (b) Recordings of multiple place cells show that the entire environment is encoded. Image from Professor John O'Keefe Nobel lecture. Copyright © The Nobel Foundation (2014); Source: http://Nobelprize.org

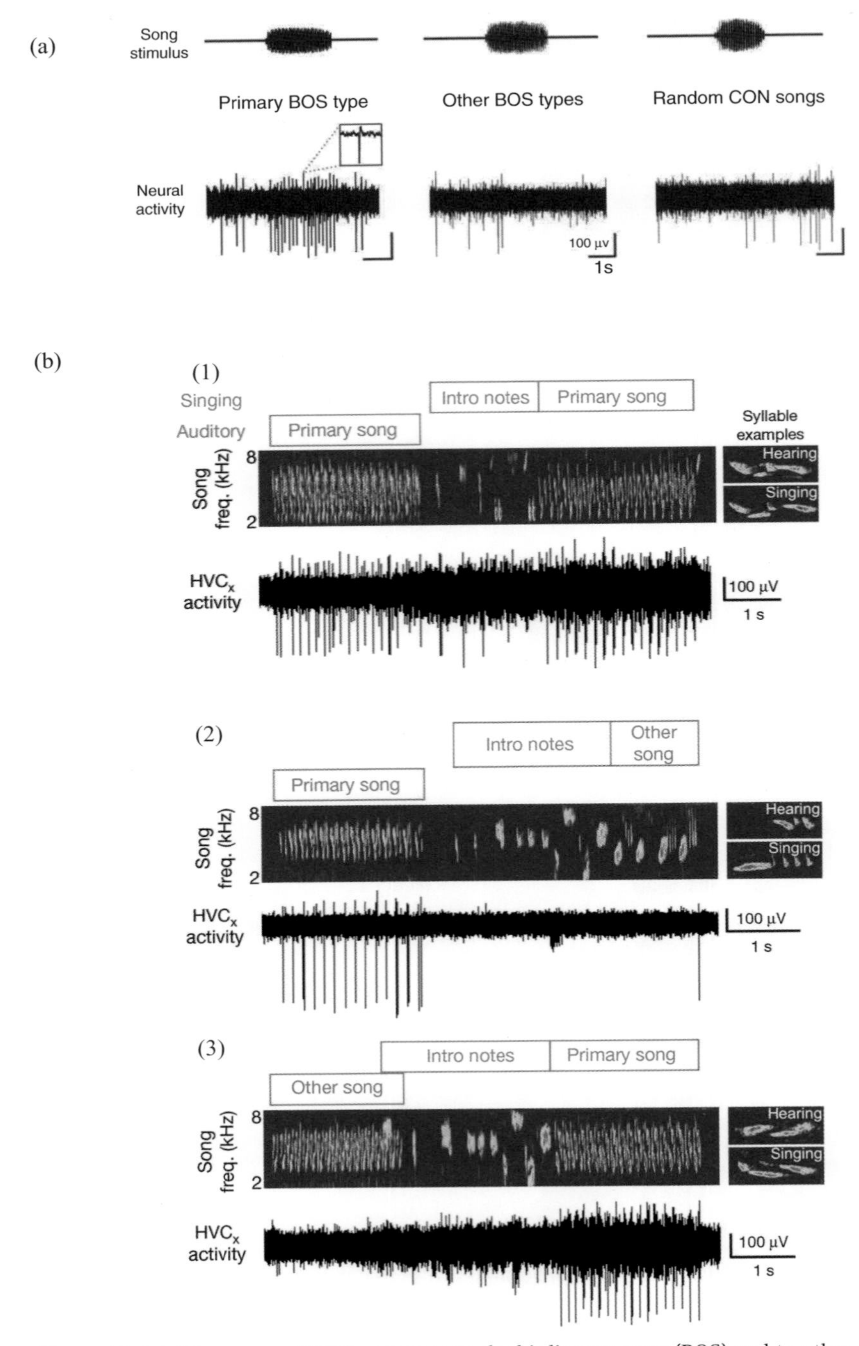

Figure 15.9 (a) Response of HVCx neurons to the bird's own song (BOS) and to other song types. CON = conspecific. (b) Response of HVCx neurons to both listening and singing (top), listening only (middle) and singing only (bottom). Top panel of each represents the sonograph of the song with the neural activity displayed in the lower panel. From J.F. Prather et al. (2008), Precise auditory-vocal mirroring in neurons for learned vocal communication, *Nature*. 451(7176):305–10. Copyright (2008). Reprinted with permission from Macmillan Publishers Ltd.

16 Circadian Rhythms

Introduction

Nature has a rhythm; night follows day, which follows night; tides ebb and flow; seasons change on an annual basis. Such natural cycles can directly affect our physiology and behaviour – birds migrate, animals hibernate, humans sleep and wake at certain times. Disruption of such cycles can have a serious effect on health and can be associated with a number of disorders including depression, schizophrenia, anxiety and even some neurodegenerative diseases (Wulff et al., 2010).

The brain too has an internal clock that keeps time. Such a clock can display a rhythm even in the absence of external environmental cycles. One of the first to discover an internal clock was the French scientist Jean-Jacques d'Ortous de Mairan (1678–1771), who in 1729 observed the daily opening and closing of the leaves of the mimosa plant (Figure 16.1a). Interested in exploring this further, de Mairan placed the plant in constant darkness in a cupboard and found that the leaves continued to open and close, even in the absence of sunlight. Later, in 1759, Henri-Louis Duhamel du Monceau (1700–1782) continued this exploration and demonstrated that leaf movements were also independent of temperature fluctuations.

Such internally driven cycles are not solely the preserve of plants. In a famous experiment conducted in 1965, Jurgen Aschoff (1913–1998) had an underground bunker built where he and some students had to live "as normal" for up to 3 or 4 weeks without access to the outside world or watches. Within the bunker there was a kitchen, a small bed-sitting area and a shower. Over the course of the experiment, each participant had to give a urine sample and take other measurements, including general levels of activity, time estimation and body temperature. Figure 16.1b shows the results from the experiment, and it is clear that even in the absence of external factors, including daylight, each measure taken displayed a rhythm, a regular rising and falling across a 24-hour period. Such rhythms that last approximately 24 hours are said to be *circadian*, derived from the Latin *circa diem* meaning 'around a day' (Aschoff, 1965).

Note the word 'approximately'. As such cycles are biological and not mechanical in nature, they are not exactly 24 hours. This slight deviation

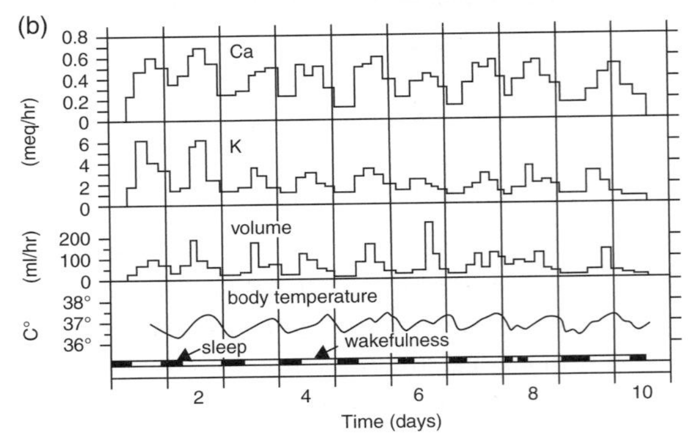

Figure 16.1 (a) Photograph of *Mimosa pudica* by Emőke Dénes, reproduced under Creative Commons Attribution License. (b) Circadian rhythm of urine excretion (calcium, potassium, and water), body temperature, and sleep–wake cycles in a human participant kept in isolation in a bunker. Image taken from Aschoff (1965), Circadian rhythms in Man, *Science*, 148, 3676, 1427–1432. Reprinted with permission from AAAS.

has consequences; if the cycle was only off slightly, the rhythm would start to shift and diverge. For example, if the cycle was 24.1 or 23.9 hours rather than 24 hours (a shift of 0.1 of an hour or 6 minutes), after just three months, nocturnal animals would be active during the day and vice versa! Thus there is a need to have a mechanism that keeps the internal clock entrained on a regular basis. As mentioned above, there are many external stimuli that can fulfil this role: food, tides, even social contact – but the most important of these is light, particularly daylight.

Circadian Behaviour

One of the easiest ways to study circadian rhythms in the laboratory is to place an animal (typically a mouse, hamster or rat) in a cage that also contains a running wheel. The running wheel can be hooked up to a computer that records the number of rotations. Animals, especially mice, love to run and hop readily into the wheel. As mice are nocturnal creatures, they will typically run during the night and sleep during the day (black bars, Figure 16.2a). If, like the mimosa plant, the mouse and running wheel are placed in an apparatus that is in constant darkness, the animal will tend to run over a 12-hour period and rest for the other 12 hours

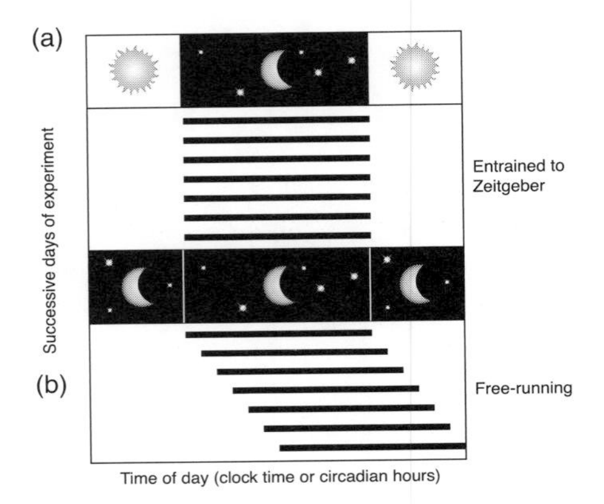

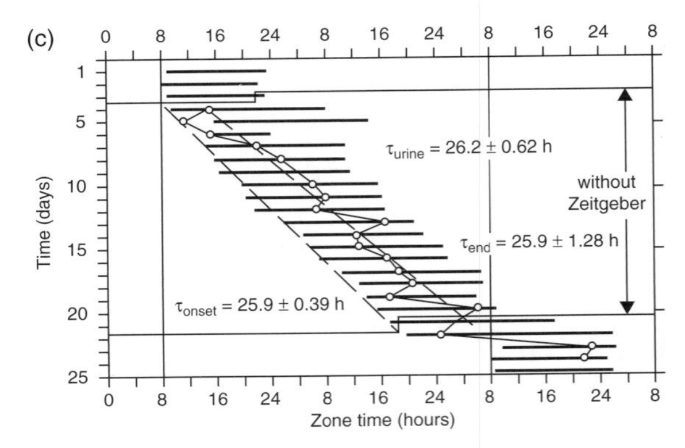

Figure 16.2 (a) Small rodents, such as mice, are nocturnal and typically are very active during the night and sleep during the day. (b) If animals are placed in complete darkness over an extended period, the animals will maintain their circadian cycle but will begin to shift with time. Image taken from D.A. Golombek and R.E. Rosenstein (2010), Physiology of circadian entrainment, *Physiol Rev.* 90(3):1063–1102 with permission. (c) The shift of the circadian cycle is also observed in humans. Image taken from J. Aschoff (1965), Circadian rhythms in man, *Science*, 148, 1427–1432. Reprinted with permission from AAAS.

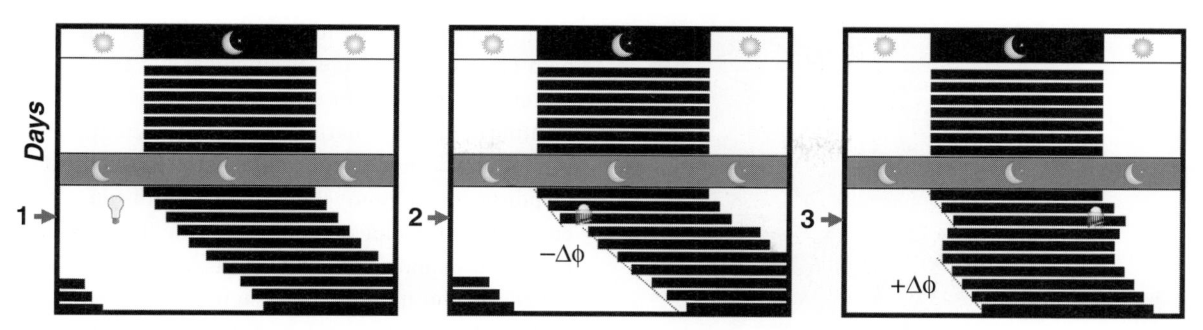

Figure 16.3 A light pulse given during the subjective day (left panel) has no effect on the circadian rhythm. Light pulses given early (middle panel) or late (right panel) cause a phase delay (represented by $-\Delta\Phi$) and a phase advance (represented by $+\Delta\Phi$), respectively. Image taken from D.A. Golombek and R.E. Rosenstein (2010), Physiology of circadian entrainment, *Physiol Rev.* 90(3):1063–1102 with permission.

(see Figure 16.2b). This suggests that their internal clock 'knows' when to be active and when to rest.

However, if the animal is kept in this constant darkness over an extended period of time, the activity phase begins to shift. This shift was also observed with the human participants of the Aschoff study (Figure 16.2c). If a light is switched on for a 12-hour period and turned off for another 12 hours, mimicking the day and night cycle, the animal returns to having a constant period of activity and rest (Figure 16.2a). The light *entrains* the behaviour; it acts as a pacemaker to the internal clock, controlling its rhythm. Any external stimulus that can entrain the clock, such as light, is referred to as a *Zeitgeber*, a word derived from the German meaning 'time giver'.

Using such an experimental setup allows scientists to further explore and manipulate the various properties of circadian rhythms. For example, a short burst of light (e.g. 30 minutes) presented at various stages during a constant darkness experiment can have very different consequences to the circadian rhythm. Light presented during the inactive period (subjective day) has no effect on the period of activity (Figure 16.3, left). Figure 16.3, middle, demonstrates that if the short pulse of light is presented *early* during the mouse's active period (subjective night), this causes an extra shift in the rhythm, delaying the period of activity slightly. This is known as a *phase delay*. If, however, the short pulse of light is presented *late* in the active period (Figure 16.3, right) this causes the opposite effect and is known as a *phase advance*.

Such shifts in the circadian rhythm may have profound implications for people on shift work or suffering from jet lag. In shift work, for example, people tend to work irregular hours, including at night, and often show higher rates of multiple illnesses including gastric problems, cardiovascular disorders and cancers (Foster and Kreitzman, 2014) compared to those that work in accordance with the natural cycle. Indeed, it has been also demonstrated that our alertness and concentration are at their lowest in the early hours of the morning, likewise there is a higher rate of car accidents at 3 a.m. than at other times of the day (Foster and Kreitzman, 2004).

Neural Basis of Circadian Rhythms

One of the major advances in circadian research was the discovery that a small nucleus called the superchiasmatic nucleus (SCN), located beside the hypothalamus, acts as our internal clock (Figure 16.4a). It is the SCN that keeps time and allows the cycle to continue, even in the absence of external cues. Destruction of this structure destroys the 24-hour cycle, and individuals that don't have it become arrhythmic. One of the first questions asked regarding the SCN was whether

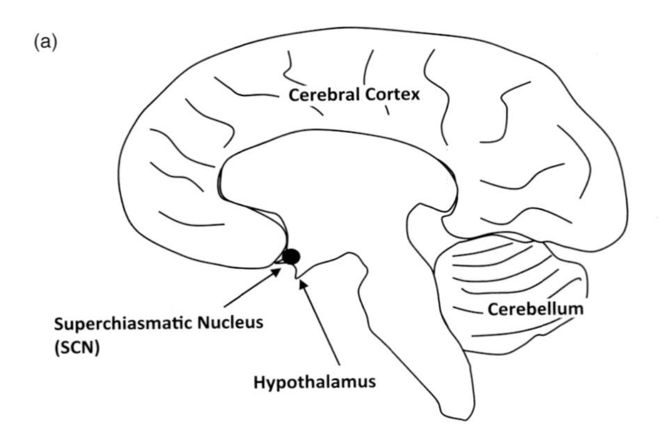

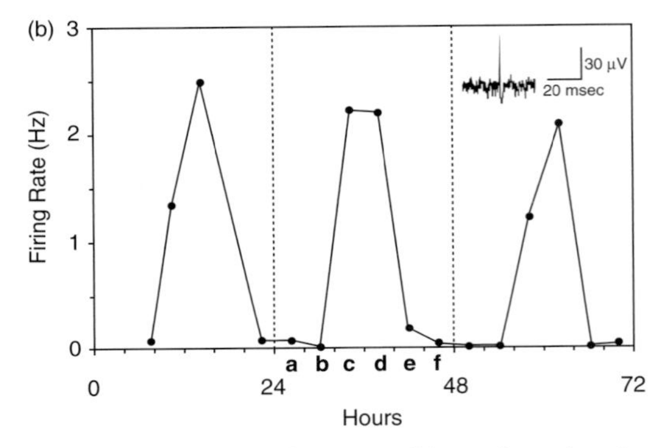

Figure 16.4 (a) Location of the superchiasmatic nucleus in the human brain. (b) Circadian firing pattern of an individual SCN neuron cultured for 1 week and recorded over 3 days. Reprinted from D.K. Welsh et al. (1995), Individual neurons dissociated from rat suprachiasmatic nucleus express independently phased circadian firing rhythms, *Neuron* 14(4):697–706, with permission from Elsevier.

the rhythm was produced from individual neurons within the structure or through a network of cells and connections. In the 1990s, Steven Reppert and his team cultured individual neurons from the SCN and recorded their spontaneous firing rate (action potentials) over a number of weeks (Welsh et al., 1995). The researchers found that each individual neuron shows an oscillatory pattern, whereby the neuron starts to fire, and then stops before firing again, continuing this pattern over a 24-hour period (Figure 16.4b). Such rhythmic behaviour can continue for many days and weeks. As the SCN contains approximately 20,000 neurons, each individual neuron can act as a clock.

However, it is now known that most, if not all, structures throughout our body (stomach, liver, kidneys etc.) contain cells that have their own clocks. Having localised clocks aid the systematic production of urine, the regular feeling of hunger, the fluctuations in temperature, the production of hormones and other physiological responses throughout the day. The brain too contains many clocks (apart from the SCN), which may help explain how our moods, emotions and concentration may fluctuate across a 24-hour period. Despite the various clocks, it is the SCN that is considered the master clock that helps to coordinate all other local clocks. The SCN acts like a conductor of an orchestra, or like a pacemaker. If the conductor is removed, all the various sections would start playing slightly out of tune.

How neurons within the SCN continue to fire in a rhythmic fashion is quite complex and involves multiple steps. Figure 16.5 describes the process in a simplified fashion (see Foster and Kreitzman; 2014 and Hastings et al., 2014 for a more detailed description). Although up to 20 genes and their proteins have been discovered to be involved in the process, there are a number of key ones including *Per* (*Period*) and *Cry* (*Cryptochrome*). Note that the *Period* gene was first cloned and sequenced in the 1980s by Jeffrey Hall and Michael Rosbash, who along with Michael Young (who discovered another circadian gene, *Timeless*), won the 2016 Nobel Prize for Physiology or Medicine (Reddy et al., 1984; Myers et al., 1995).

(1) Within the nucleus of an SCN neuron, clock genes such as *Per* and *Cry* are transcribed as a result of the binding of the CLOCK:Bmal protein complex to the two gene promoters.
(2) *Per* and *Cry* are then translated into Per and Cry proteins in the cell cytoplasm.
(3) These protein products are involved in the generation of the cell's action potentials. Once their work has been done, the proteins are degraded by casein kinase 1 through phosphorylation (addition of a phosphate group).
(4) However, some Per and Cry proteins interact with each other and form a new protein complex.
(5) This Per/Cry complex moves into the nucleus, where it inhibits the CLOCK:Bmal protein complex. This in turn inhibits the

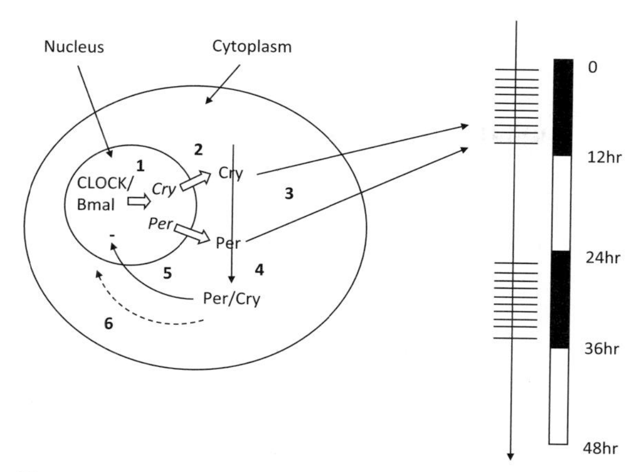

Figure 16.5 Sequence of events describing how SCN neurons can continue to fire in a rhythmic fashion (see text for details).

production of the clock genes, their protein products and the firing of action potentials.

(6) Over time the Per/Cry complex degrades, allowing the CLOCK:BMal to transcribe the clock genes again. The process of activation and inhibition and reactivation occurs and recurs over each 24-hour period, giving the cell its rhythm (represented by the firing of the neuron only during the dark phase in Figure 16.5, right).

Entrainment of the Clock

In order to stop the cycle from drifting across time, the internal clock requires an external pacemaker. The clock needs to be entrained. As mentioned, there are many stimuli that can fulfil this role, but the most important zeitgeber or pacemaker is light. Light enters our eyes and passes through a number of structures, including the pupil and lens, before reaching the retina, located at the back of the eye (see Figure 16.6a). The retina itself is composed of a number of layers, each containing different cell types. These layers include the receptor layer, layers containing horizontal cells, bipolar cells, amacrine cells and a final layer containing retinal ganglion cells (see Figure 16.6a). Note that the horizontal cell layer (located between the receptor layer and bipolar layer) and the amacrine layer (located between the bipolar layer and retinal ganglion layer) are not included in Figure 16.6a but play an important role in communication across the various cells. The receptor layer of the retina contains two kinds of specialised light-sensitive receptors, called the rods and cones, named for their respective shapes. The rods are more numerous than cones, approximately 100 million compared to less than 10 million cones. In addition, rods are more sensitive and are typically used under low light conditions. They are not, however, very sensitive to colour. In contrast, cones are sensitive to colour and are very useful for visual acuity, particularly under high light conditions. Typically, light reaches the receptor cell layer by passing through the other layers first before activating the rods and cones. Signals from the rods and cones pass information through the bipolar cells before reaching the ganglion cells. From here, signals are sent to the lateral geniculate nucleus, located in the centre of the brain, and onto the visual cortex via the optic nerve.

Within the ganglion cell layer of the retina, non-image-forming cells that are sensitive to light have been only recently discovered, and these are called *photosensitive retinal ganglion cells* (pRGC; Berson et al., 2002; Hattar et al., 2002). It is thought that in every 100 ganglion cells that form the optic nerve about 1 or 2 are pRGCs (Foster and Kreitzman, 2014). These cells express melanopsin, a pigment sensitive to visible light; importantly, *these cells send light information directly to the SCN* via the retinohypothalamic tract (RHT) and are thought to entrain the clock. If the melanopsin gene is deleted, ganglion cells that were photosensitive are now no longer sensitive to light; furthermore some properties of circadian rhythm, e.g. light-induced phase delays, are also impaired (Panda et al., 2002). However, animals that lack rod, cone and melanopsin photoreception show a complete lack of entrainment (see Figure 16.6b).

Glutamate is the main neurotransmitter released from the RHT and, together with the neuropeptide pituitary adenylate cyclase-activating protein (PACAP), activates several signalling pathways in the SCN. One possible signalling pathway

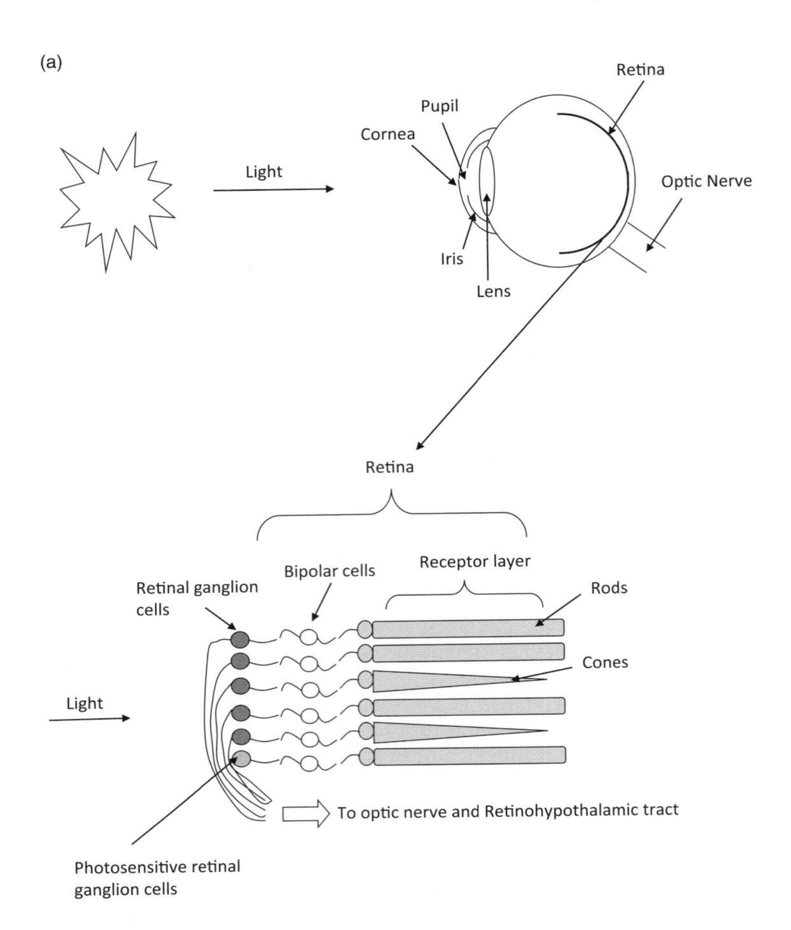

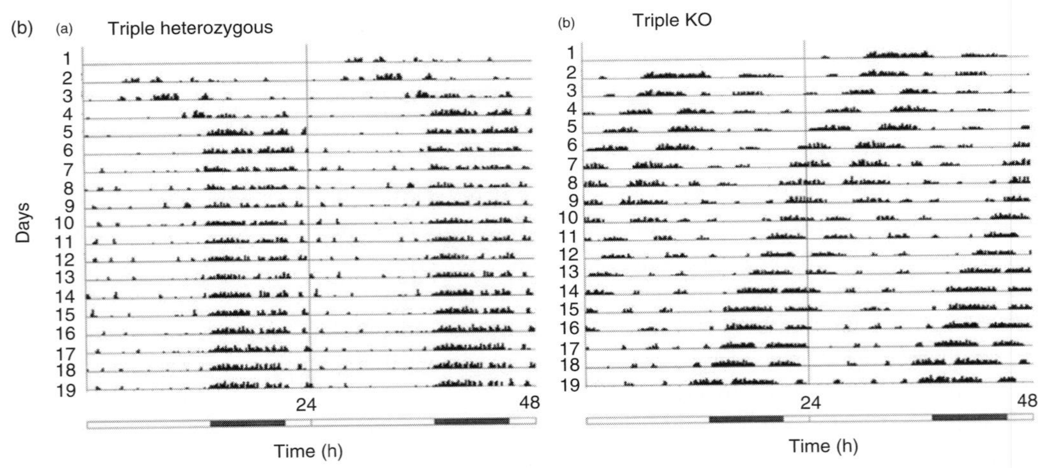

Figure 16.6 (a) Anatomy of the eye and the various cells contained within the retina in mice. (b) Triple heterozygote (Opn4+/-Gnat1+/-Cnga3+/-) similar to wild type (left) show normal circadian patterns. Triple-knockout (Opn4-/-Gnat1-/-Cnga3-/-) mice, i.e. animals that lack rod, cone and melanopsin photoreception show a lack of photo-entrainment (right). S. Hattar et al. (2003), Melanopsin and rod–cone photoreceptive systems account for all major accessory visual functions in mice, *Nature* 424(6944):76–81. Reprinted by permission from Macmillan Publishers Ltd. Nature, copyright (2003).

is the extracellular-signal-regulated kinase (ERK) pathway. Activation of ERK leads to the phosphorylation of CREB (cAMP response element-binding protein). This in turn binds to CRE (cAMP response elements) located in targeted genes. Interestingly, both *Per1* and *Per2* genes contain a CRE site and therefore these genes may be activated independently of the CLOCK:Bmal protein complex discussed earlier (Dibner et al., 2010) and offer one possible mechanism by which the SCN clock genes are activated during entrainment.

Note, although the RHT is the main direct pathway into the SCN, the SCN also receives indirect light information via the geniculohypothalamic tract (GHT). Fibres originating in the retina pass through the intergeniculate nucleus before projecting to the SCN. This pathway may also help with the entrainment process through its many neurotransmitters, including GABA and neuropeptide Y. A third projection to the SCN, also indirect, arises in the raphe nuclei and seems to play a role in nonphotic signalling through the neurotransmitter serotonin. How these indirect pathways interact with each other and with the direct RHT pathway to influence the circadian process is still being elucidated.

Other Brain Clocks

Identification of various clock genes in the SCN such as *Clock*, *Per*, *Cry*, and *Bmal* amongst others has led researchers to look for such genes in other parts of the brain. To date, daily oscillations of clock gene expression have been found in many regions including the hypothalamus, thalamus, amygdala, hippocampus, olfactory bulb, cerebellum and many more. Structures that seem to show strong rhythmic expression include the olfactory bulb, the paraventricular nucleus of the hypothalamus, habenula and pituitary gland (see Dibner et al., 2010). Although many of these regions depend on the 'master' SCN clock for their rhythm, some, including the olfactory bulb, can maintain their daily oscillation even in the absence of the SCN (Granados-Fuentes et al., 2004). The olfactory bulb may play an important role in the entrainment of the clock to nonphotic stimuli such as food. For example, being able to smell coffee in the morning or food in the evening may serve as just as powerful a stimulus as light, under certain circumstances.

Other structures many play very different roles. For example, the habenula is a structure that is considered part of the reward system. It projects directly to the ventral tegmental area and receives input from the nucleus accumbens – structures of the reward system that use dopamine as the main neurotransmitter. Food, alcohol, drugs and sex all impact directly on the reward system. Addictive behaviours are thought to arise from the repetitive action towards one or many of these stimuli; interestingly, such repetitive behaviours often exhibit a 24-hour pattern. It is thought that modulation of dopamine in the reward system via the expression of clock genes in the habenula may contribute to the rhythmic nature of some addictive behaviours. The expression of clock genes in another structure, the hippocampus, is thought to contribute to the ability of some people to learn more effectively in the morning, and of others to learn better later in the day (see Winocur and Hasher, 2004). Furthermore, in a recent study, a correlation between hippocampal activity, circadian rhythm consistency and accurate memory recall has been demonstrated (Sherman et al., 2015). Indeed, disruption of the internal clock, whereby the normal sleep pattern may become disturbed, is the hallmark of many disorders, including Alzheimer's disease, a disease characterised by memory deficits.

Passing SCN Information to the Body

In addition to locations in the brain, clock genes have also been found in many structures throughout the body, including the liver, lungs, stomach, heart, spleen and kidneys (see Figure 16.7). Urine production, lipid metabolism, carbohydrate metabolism and

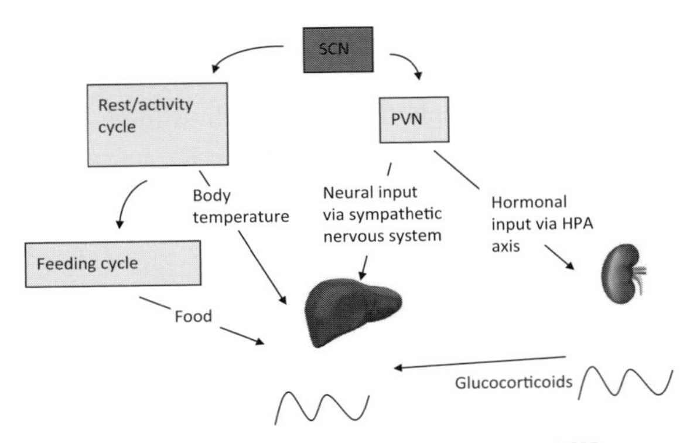

Figure 16.7 As the master circadian clock, the SCN influences other organs in the body. Figure adapted from C. Dibner et al. (2010), The mammalian circadian timing system: organization and coordination of central and peripheral clocks, *Annual Rev Physiol.* 72:517–549, with permission of *Annual Review of Physiology*, Volume 72; © by Annual Reviews, http://www.annualreviews.org

digestion all show rhythmic behaviour. Although most of these peripheral clocks operate in the similar fashion to that described earlier for the SCN, they are dependent on the SCN for synchronisation. For example, cultured cells from these organs show robust oscillations, but they tend to dampen after a number of cycles and do not last. Furthermore, the rhythmic behaviour in peripheral tissue quickly becomes desynchronised in SCN-lesioned animals.

Although how the SCN synchronises peripheral clocks is largely unknown, there are a number of mechanisms by which this could happen. The SCN may influence some peripheral organs through direct hormonal and neural projections. For example, one of the target projections of the SCN is the paraventricular nucleus (PVN) of the thalamus. From this the SCN can influence the release of glucocorticoids from the adrenal glands, located above the kidneys, via the traditional hypothalamus–pituitary–adrenal (HPA) hormonal axis. Glucocorticoids can themselves directly affect the clock in other peripheral structures including the liver, kidney and heart (Balsalobre et al., 2000). The SCN can also influence the autonomic nervous system via the PVN. Here, the SCN innervates the PVN, which in turn directly projects to the liver, resulting in the daily production of glucose (Kalsbeek et al. 2004).

The SCN may also entrain peripheral clocks more indirectly. For example, body temperature, the rest–active cycle and the feeding–fasting cycle may all act to entrain peripheral tissue, leading to the cyclic metabolism of amino acids, cholesterol and fatty acids and the release of glucose and melatonin, amongst others. The liver clock is particular influenced by feeding. Haptic clock gene and protein expression follow the timings of meals and may become entrained by the secretion of hormones released upon feeding, such as glucose, cholesterol, leptin and ghrelin (Dibner et al., 2010).

Clocks and Sleep

Sleep is an extremely important behaviour and a topic that has fascinated scientists, artists and philosophers throughout the ages. All mammals sleep, although the amount of sleep varies from one mammal to another. A giant sloth, for example, sleeps on average 20 out of 24 hours, while at the other extreme, horses and giraffes sleep for less than an hour. Average humans spend approximately 8 out of 24 hours in the sleeping state, which roughly translates as one-third of our lives! This length of sleeping time is shared with other animals, including rabbits and pigs. The sleep–wake behaviour is rhythmic. Nocturnal animals sleep during the day and are active at night; humans are diurnal and sleep at night and are active during the day. Although we may try to stay awake, eventually sleep will overcome us. Indeed, prolonged sleep deprivation can lead to many serious illnesses (see next section) and can even cause death in some small animals.

Sleep itself is rhythmic, cycling between rapid eye movement (REM) and non-REM states. It is predominantly during the REM state that we tend to dream, and non-REM is what is generally considered to be deep sleep.

Sleep must serve an important function; otherwise, we would have lost it through evolution. Indeed, many mammals that live in very dangerous environmental conditions, where you might imagine that it would be advantageous to remain awake and alert, have adapted their sleeping patterns. For example, several species of dolphin that live in very turbulent waters, e.g. the Indus River dolphin, have the ability to switch off one brain hemisphere, while keeping the other one awake, and then reverse this pattern over a 24-hour period. Figure 16.8 shows the activity in the right hemisphere of the brain of the bottlenose dolphin, while the left remains relatively quiet. Later, there is a lot of activity in the left hemisphere, while the right remains quiet. Sleep is thought to fulfil many functions, described in a recent review paper (Krueger et al., 2015) discussing at least 6 different functions for sleep. These functions include the role of sleep in immune functions, caloric conservation, the replenishment of brain energy stores, the removal of toxic byproducts that may have accumulated during the waking state, restoration of performance and particularly, its role in the facilitation of neural plasticity during childhood development and also in relation to the consolidation of memories (Frank, 2006).

Although sleep is a very complex process that requires the interaction between multiple brain regions and different neurotransmitters, it is thought that sleep is governed by two mechanisms; homeostatic drive and the circadian system (Foster and Kreitzman, 2014). *Homeostatic drive* refers to the increased pressure on an individual to sleep, the more he or she remains awake. The circadian system provides the timing for the wake–sleep cycle. It is this regular sleep–wake pattern that dictates our lives. We tend to go to sleep at the same time

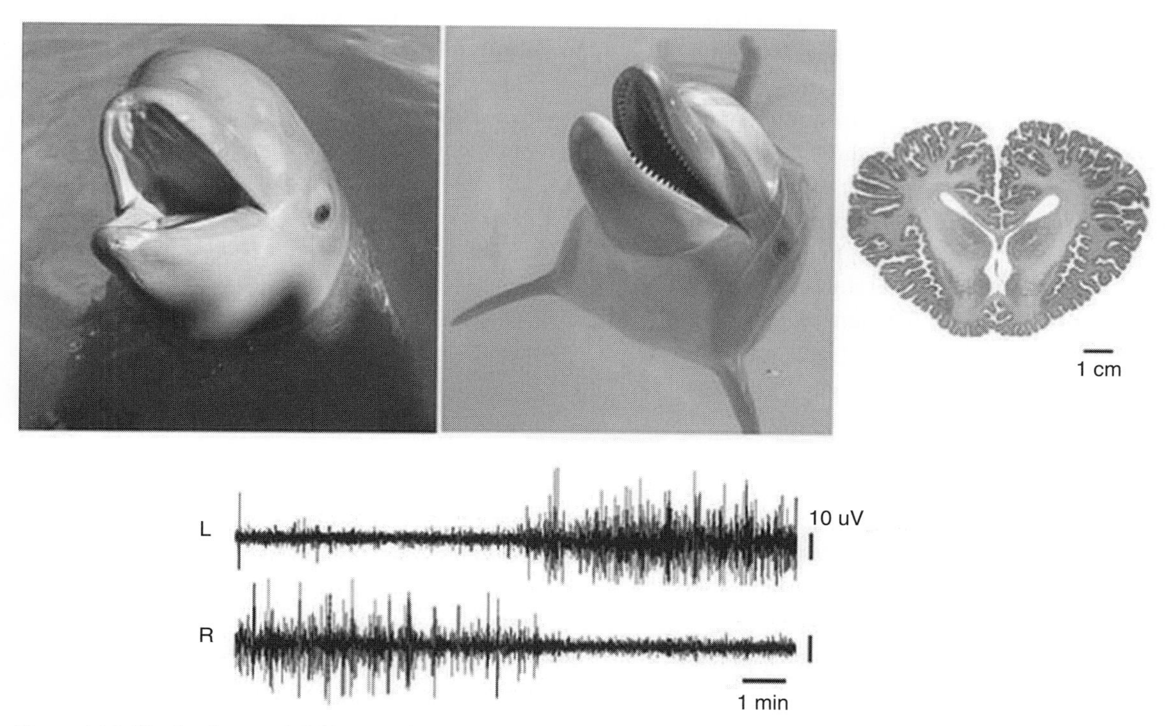

Figure 16.8 The bottlenose dolphin can sleep with one brain hemisphere at a time. From J.M. Siegel, Clues to the functions of mammalian sleep, *Nature,* 437(7063):1264–1271, copyright (2005). Reprinted by permission from Macmillan Publishers Ltd.

every day; we also tend to wake at the same time each morning, even in the absence of an alarm. Perhaps more frustrating, we even tend to wake at the same time even when we do not need to, when on holidays or on the weekend! However, usually there is a difference between our preferred time to get up and the time we are required to get up, and this difference is termed *social jet lag* (see Roenneberg et al., 2013).

Furthermore, although humans are diurnal there is variation from one person to another in terms of their preferred waking and sleeping times. Morning people, or *larks*, tend to wake early and go to bed early, while evening people, or *owls*, tend to wake late and stay up later. Such difference in preference is referred to as *chronotype*, and some reports suggest up to an 18-hour difference between the extreme ends of the two chronotypes (Gamble and Young, 2015). Body temperature, urine production and other biochemical metabolites are also known to differ between chronotypes. In addition, it has been reported that evening types are more creative, less conscientious, less agreeable and less anxious compared to morning types, who tend to achieve more academically and to be more satisfied with their lives (Tsaousis, 2010; Gamaldo et al., 2014). How the different chronotypes, their innate biological timing and their preferred period of activity match the daily schedule of school and work is a current major topic of research, as any mismatch can have serious consequences on performance, behaviour and health.

Clocks and Health

Disruption to sleep of various forms including sleep deprivation, poor sleep quality, and circadian misalignment can cause many serious illnesses. Too much and too little sleep can have negative effects. For example, Gamaldo et al. (2014) reported on a number of studies that have shown that with less than 6 hours of sleep there is a 15% higher risk of cardiovascular disease, a 1.4-fold higher risk of hypertension and 1.9-fold higher risk of stroke. Having less than 6 hours of sleep combined with poor quality sleep increases the risk of cardiovascular disease to 63%. On the other hand, people having more than 9 hours of sleep have a 1.3-fold higher risk of a diagnosis of hypertension.

Shift work too can have health consequences. Shift work can include rotating shifts, permanent night work, early morning work or indeed random work hours. Such working arrangements may have the consequence of disrupting the circadian patterns of workers, leading many workers to have poor sleep quality, a lack of sleep generally and problems readjusting to the various times. Many occupational accidents, such as the nuclear accident at Chernobyl and the Exxon Valdez oil spill, are thought to have occurred as a result of sleep deprivation related to shift work. Aside from accidents and mistakes, many studies have associated shift work with long-term health concerns. In a recent meta-analysis examining shift work and cardiovascular disease, Vyas et al. (2012) found that shift workers had a 1.23-fold increase of risk of infarction and a 1.24-fold risk of having a coronary event compared to non-shift workers. Other studies have reported that shift workers were more at a risk of having hypertension, diabetes and obesity (see Niedhammer et al., 1996). The link between mental disorders, including schizophrenia and depression, and disturbed sleep is increasingly being recognised. There are a number of treatments that may help with disorders associated with circadian rhythm and sleep disruption. For example, a patient may undergo chronotherapy, including bright-light therapy, or cognitive behavioural therapy that focuses on trying to shift sleep time so that it suits the preferred schedule of the patient. Alternatively, melatonin may be used, as this medication is known to have phase-shifting effects. Such behavioural modification and pharmacological interventions are gaining acceptance and go some way to relieve various symptoms.

Summary

Life has a rhythm. In this chapter we explored the internal mechanism of timekeeping and how cells in one particular brain structure, the suprachiasmatic nucleus (SCN), can fire on a regular basis across a 24-hour period. This regular firing helps synchronise the activity of cells in other brain regions, as well as the metabolic processes in the lungs, kidneys, liver and other peripheral organs. We also examined the importance of light and how this and other external stimuli can entrain or pace the clock to prevent the internal timing mechanism from going out of synch. This is of critical importance, as disruption of the circadian cycle, including the regular sleep–wake cycle, can result in many disorders and cause long-term poor health.

Questions and Topics Under Current Investigation

- What role is played by the various brain clocks?
- How does the suprachiasmatic nucleus (SCN) interact with other clocks in the brain?
- How does the SCN interact with the many peripheral clocks?
- How do the different chronotypes affect cognition and behaviour? Do work schedules, school schedules, and other schedules need to accommodate various chronotypes?
- How does disruption of the circadian lead to many diseases and disorders? How can these be treated?

References

Aschoff, J. (1965). Circadian rhythms in man. *Science*, 148, 3676, 1427–1432.

Balsalobre, A., Brown, S.A., Marcacci, L., Tronche, F., Kellendonk, C., Reichardt, H.M., Schütz, G., and Schibler, U. (2000). Resetting of circadian time in peripheral tissues by glucocorticoid signalling. *Science*, 289(5488), 2344–7.

Berson, D.M., Dunn, F.A., and Takao, M. (2002). Phototransduction by retinal ganglion cells that set the circadian clock. *Science*, 295(5557), 1070–1073.

Dibner, C., Schibler, U., and Albrecht, U. (2010). The mammalian circadian timing system: organization and coordination of central and peripheral clocks. *Annual Review in Physiology*, 72, 517–549.

Foster, R.G., and Kreitzman, L. (2004). *Rhythms of life: the biological clock that control the daily lives of every living thing*. Profile, London.

Foster, R.G., and Kreitzman, L. (2014). The rhythms of life: what your body clock means to you! *Experimental Physiology*, 99(4), 599–606.

Frank, M.G. (2006). The mystery of sleep function: current perspectives and future directions. *Reviews in Neuroscience*, 17(4), 375–392.

Gamaldo, C.E., Chung, Y., Kang, Y.M., and Salas, R.M. (2014). Tick-tock-tick-tock: the impact of circadian rhythm disorders on cardiovascular health and wellness. *Journal of the American Society of Hypertension*, 8(12), 921–929.

Gamble, K.L., and Young, M.E. (2015). Circadian biology: the early bird catches the morning shift. *Current Biology*, 25(7), R269–71.

Golombek, D.A., and Rosenstein, R.E. (2010). Physiology of circadian entrainment. *Physiol Rev.* 90(3):1063–1102.

Granados-Fuentes, D., Prolo, L.M., Abraham, U., and Herzog, E.D. (2004). The suprachiasmatic nucleus entrains, but does not sustain, circadian rhythmicity in the olfactory bulb. *Journal of Neuroscience,* 24(3), 615–9.

Hastings, M.H., Brancaccio, M., and Maywood, E.S. (2014). Circadian pacemaking in cells and circuits of the suprachiasmatic nucleus. *Journal of Neuroendocrinology*, 26(1), 2–10.

Hattar, S., Liao, H.W., Takao, M., Berson, D.M., and Yau, K.W. (2002). Melanopsin-containing retinal ganglion cells: architecture, projections, and intrinsic photosensitivity. *Science*, 295(5557), 1065–1070.

Hattar, S., Lucas, R.J., Mrosovsky, N., Thompson, S., Douglas, R.H., Hankins, M.W., Lem, J., Biel, M., Hofmann, F., Foster, R.G., and Yau, K-W. (2003). Melanopsin and rod–cone photoreceptive systems account for all major accessory visual functions in mice. *Nature,* 424, 75–81.

Kalsbeek, A., La Fleur, S., Van Heijningen, C., and Buijs, R.M. (2004). Suprachiasmatic GABAergic inputs to the paraventricular nucleus control plasma glucose concentrations in the rat via sympathetic innervation of the liver. *Journal of Neuroscience*, 24, 7604–7613.

Krueger, J.M., Frank, M.G., Wisor, J.P., and Roy, S. (2015). Sleep function: Toward elucidating an enigma. *Sleep Medicine Reviews*, 28, 42–50.

Myers, M.P., Wager-Smith, K., Wesley, C.S., Young, M.W., and Sehgal, A. (1995). Positional cloning and sequence analysis of the Drosophila clock gene, timeless. Science 270, 805–808.

Niedhammer, I., Lert, F., and Marne, M.J. (1996). Prevalence of overweight and weight gain in relation to night work in a nurses' cohort. *International Journal of Obesity and Related Metabolic Disorders*, 20(7), 625–633.

Panda, S., Sato, T.K., Castrucci, A.M., Rollag, M.D., DeGrip, W.J., Hogenesch, J.B., Provencio, I., and Kay, S.A. (2002) Melanopsin (Opn4) requirement for normal light-induced circadian phase shifting. *Science*, 298(5601), 2213–2216.

Reddy, P., Zehring, W.A., Wheeler, D.A., Pirrotta, V., Hadfield, C., Hall, J.C., and Rosbash, M. (1984). Molecular analysis of the period locus in Drosophila melanogaster and identification of a transcript involved in biological rhythms. Cell, 38, 701–710.

Roenneberg, T., Kantermann, T., Juda, M., Vetter, C., and Allebrandt, K.V. (2013). Light and the human circadian clock. *Handbook of Experimental Pharmacology*, 217, 311–331.

Sherman, S.M., Mumford, J.A., and Schnyer, D.M. (2015). Hippocampal activity mediates the relationship between circadian activity rhythms and memory in older adults. *Neuropsychologia*, 75, 617–625.

Siegel, J.M. (2005). Clues to the functions of mammalian sleep. *Nature*, 437, 1264–1271.

Tsaousis, I. (2010). Circadian preferences and personality traits: A meta-analysis. *European Journal of Personality*, 24, 356–373.

Vyas, M.V., Garg, A.X., Iansavichus, A.V., Costella, J., Donner, A., Laugsand, L.E., Janszky, I., Mrkobrada, M., Parraga, G., and Hackam, D.G. (2012). Shift work and vascular events: systematic review and meta-analysis. *British Medical Journal*, 345, e4800.

Welsh, D.K., Logothetis, D.E., Meister, M., and Reppert, S.M. (1995). Individual neurons dissociated from rat suprachiasmatic nucleus express independently phased circadian firing rhythms. *Neuron*, 14(4), 697–706.

Winocur, G., and Hasher, L. (2004). Age and time-of-day effects on learning and memory in a non-matching-to-sample test. *Neurobiology of Aging*, 25(8), 1107–1115.

Wulff, K., Gatti, S., Wettstein, J.G., and Foster, R.G. (2010). Sleep and circadian rhythm disruption in psychiatric and neurodegenerative disease. *Nature Review Neuroscience*, 11(8), 589–599.

INDEX